Delphi High Performance

Master the art of concurrency, parallel programming, and memory management to build fast Delphi apps

Primož Gabrijelčič

<packt>

BIRMINGHAM—MUMBAI

Delphi High Performance

Group Product Manager: Kunal Sawant

Publishing Product Manager: Teny Thomas

Senior Editor: Nithya Sadanandan

Technical Editor: Jubit Pincy

Copy Editor: Safis Editing

Project Coordinator: Manisha Singh

Proofreader: Safis Editing

Indexer: Rekha Nair

Production Designer: Vijay Kamble

Senior Developer Relations Marketing Executive: Rayyan Khan

Developer Relations Marketing Executive: Sonia Chauhan

Business Development Executive: Samriddhi Murarka

First published: February 2018

Second edition: June 2023

Production reference: 1160623

Published by Packt Publishing Ltd.

Livery Place

35 Livery Street

Birmingham

B3 2PB, UK.

ISBN 978-1-80512-587-7

www.packtpub.com

Contributors

About the author

Primož Gabrijelčič started coding in Pascal on 8-bit micros in the 1980s and he never looked back. In the last 25 years, he was mostly programming high-availability server applications used in the broadcasting industry. A result of this focus was the open source parallel programming library for Delphi—OmniThreadLibrary. He's also an avid writer and has written several hundred articles, and he is a frequent speaker at Delphi conferences, where he likes to talk about complicated topics ranging from memory management to creating custom compilers.

About the reviewers

Bruce McGee has been an Embarcadero MVP since 2013 and an enthusiastic, prolific, and (sometimes) opinionated user of Delphi since it was released in 1995. Much of that time has been spent as the owner and operator of Glooscap Software – building, fixing, and rehabilitating software for any number of companies.

Stefan Glienke was first introduced to programming with Turbo Pascal in upper school, after which it was clear to him that he wanted to make his career in this field. Apart from Turbo Pascal and later, Delphi, he also learned C/C++ and other languages. For more than 20 years, he has been working as a Delphi developer and has contributed to several open source projects, such as Spring4D, and started his own.

Because developer productivity is very important to him, this eventually led him to write TestInsight. During his work on Spring4D and other runtime library code, he learned how to write efficient code and how to analyze and optimize existing code. As an award-winning MVP, he has been a speaker at numerous conferences and is an active member of the online community.

Table of Contents

4

Don't Reinvent, Reuse

73

5

Fine-Tuning the Code

125

6

7

8

Working with Parallel Tools   265

9

Exploring Parallel Practices   299

10

More Parallel Patterns   345

11

Using External Libraries   387

12

Best Practices   407

Index   421

Other Books You May Enjoy   430

Preface

Performance matters!

I started programming on 8-bit micros, and boy, was that an interesting time! Memory was typically not a problem, as we didn't write big programs, but they certainly weren't running fast, especially if you run them with a built-in BASIC interpreter. It is not surprising that I quickly learned assembler and spent lots of early years shifting bits and registers around. So did almost everybody else who wanted to release a commercial application written for one of those computers. There were, more or less, no games and applications written in BASIC simply because they would run too slowly and nobody would use them.

Times have changed; computers are now fast — incredibly fast! If you don't believe me, check the code examples for this book. A lot of times, I had to write loops that spin over many million iterations so that the result of changing the code would be noticed at all. The raw speed of processors has also changed the software development world. Low-level languages such as assembler and C mostly gave way to more abstract approaches — C#, C++, Delphi, F#, Java, Python, Ruby, JavaScript, Go, and so on. The choice is yours. Almost anything you write in these languages runs fast or at least fast enough.

Computers are so fast that we sometimes forget the basic rule — performance matters.

Customers like programs that operate so fast that they don't have to think about it. If they have to wait 10 seconds for a form to appear after clicking on a button, they won't be very happy. They'll probably still use the software, though, provided that it works for them and doesn't crash. On the other hand, if you write a data processing application that takes 26 hours to do a job that needs to execute daily, you'll certainly lose them.

I'm not saying that you should switch to assembler. Low-level languages are fast, but coding in them is too slow for modern times, and the probability of introducing bugs is just too high. High-level languages are just fine, but you have to know how to use them. You have to know what is fast and what is not, and ideally, you should take this into account when designing code.

This book will walk you through the different approaches that will help you write better code. Writing fast code is not the same as optimizing a few lines of your program to the max. Most of the time, that is, in fact, the completely wrong approach! However, I'm getting ahead of myself. Let the book speak for itself.

Who this book is for

This book was written for all Delphi programmers out there. You will find something interesting inside, whether you are new to programming or a seasoned old soul. I'm talking about basic stuff, such as strings, arrays, lists, and objects, but I'm also discussing parallel programming, memory manager internals, and object linking. There are also plenty of dictionaries, pointers, algorithmic complexities, code inlining, parameter passing, and whatnot.

So, whoever you are, dear reader, I'm pretty sure you'll find something new in this book. Enjoy!

What this book covers

Chapter 1, About Performance, discusses performance. We'll dissect the term itself and try to find out what users actually mean when they say that a program performs (or doesn't perform) well. Then, we will move into the area of algorithm complexity. We'll skip all the boring mathematics and just mention the parts relevant to programming.

Chapter 2, Profiling the Code, gives you a basic idea about measuring program performance. We will look at different ways of finding the slow (non-performant) parts of the program, from pure guesswork to measuring tools of a different sophistication, homemade and commercial.

Chapter 3, Fixing the Algorithm, examines a few practical examples where changing an algorithm can speed up a program dramatically. In the first part, we'll look at graphical user interfaces and what we can do when a simple update to TListBox takes too long. The second part of the chapter explores the idea of caching and presents a reusable caching class with very fast implementation.

Chapter 4, Don't Reinvent, Reuse, admits that sometimes other people will write better code than we may do and that using a good external library is preferred to any other solution. This chapter looks at the Spring4D collections library and explains when and why you should use lists, dictionaries, sets, hashmaps, or other collection types.

Chapter 5, Fine-Tuning the Code, deals with lots of small things. Sometimes, performance lies in many small details, and this chapter shows how to use them to your advantage. We'll check the Delphi compiler settings and see which ones affect the code speed. We'll look at the implementation details for built-in data types and method calls. Using the correct type in the right way can mean a lot. Of course, we won't forget about the practical side. This chapter will give examples of different optimization techniques, such as extracting common expressions, using pointers to manipulate data, and implementing parts of the solution in assembler.

Chapter 6, Memory Management, is all about memory. It starts with a discussion on strings, arrays, and how their memory is managed. After that, we will move to the memory functions exposed by Delphi. We'll see how we can use them to manage memory. Next, we'll cover records — how to allocate them, how to initialize them, and how to create useful dynamically allocated generic records. We'll then move into the murky waters of memory manager implementation. I'll sketch a very rough overview of FastMM, the default memory manager in Delphi. First, I'll explain why FastMM is excellent, and then I'll show when and why it may slow you down. We'll see how to analyze memory performance problems and how to switch the memory manager to a different one. In the last part, we'll revisit the `SlowCode` program and reduce the number of memory allocations it makes.

Chapter 7, Getting Started with the Parallel World, explores the topic of parallel programming. In the introduction, I'll talk about processes, threads, multithreading, and multitasking to establish some common ground for discussion. After that, you'll start learning what not to do when writing parallel code. I'll explain how a user interface must be handled from background threads and what problems are caused by sharing data between threads. Then, I'll start fixing those problems by implementing various kinds of synchronization mechanisms and interlocked operations. We'll also deal with the biggest problem synchronization brings to code — deadlocking. As synchronization inevitably slows a program down, I'll explain how to achieve the highest possible speed using data duplication, aggregation, and communication. Finally, I'll introduce two third-party libraries that contain helpful parallel functions and data structures.

Chapter 8, Working with Parallel Tools, focuses on a single topic, Delphi's `TThread` class. In the introduction, I'll explain why I believe that `TThread` is still important even in this modern age. I will explore different ways in which `Tthread`-based threads can be managed in your code. After that, I'll go through the most important `TThread` methods and properties and explain what they're good for. In the second part of the chapter, I'll extend `TThread` into something more modern and easier to use. First, I'll add a communication channel so that are able to send messages to the thread. After that, I'll implement a derived class designed to handle one specific usage pattern and show how this approach simplifies writing parallel code to the extreme.

Chapter 9, Exploring Parallel Practices, moves the multithreaded programming to more abstract terms. In this chapter, we'll discuss modern multithreading concepts – tasks and patterns. We'll look into Delphi's own implementation, the Parallel Programming Library, and demonstrate the use of `TTask`/`ITask`. We'll look at topics such as task management, exception handling, and thread pooling. After that, we'll move on to patterns and talk about all Parallel Programming Library patterns – Join, Future, and Parallel For. We will also introduce three custom patterns — Async/Await, Join/Await, and Pipeline.

Chapter 10, More Parallel Patterns, looks into another external programming library, OmniThreadLibrary, and explores its parallel programming patterns. In this chapter, we will revisit Async/Await, Join/Await, Future, and Pipeline from the OmniThreadLibrary perspective and introduce four new patterns – Parallel Task, Background Worker, Parallel Map, and Timed Task.

Chapter 11, Using External Libraries, admits that sometimes Delphi is not enough. Sometimes, a problem is too complicated to be efficiently solved by a human. Sometimes, Pascal is just lacking speed. In such cases, we can try finding an existing library that solves our problem. In most cases, it will not support Delphi directly but will provide some kind of C or C++ interface. This chapter looks into linking with C object files and describes typical problems that you'll encounter on the way. In the second half, we'll present a complete example of linking to a C++ library, from writing a proxy DLL to using it in Delphi.

Chapter 12, Best Practices, wraps everything up and revisits all the important topics we explored in previous chapters. At the same time, we'll drop in some additional tips, tricks, and techniques.

To get the most out of this book

You will need basic proficiency with the Delphi development tool. All programs were tested with Delphi 11.3 Alexandria. Most of the examples, however, should also work with older versions, going back to Delphi XE.

Software/hardware covered in the book	Operating system requirements
Embarcadero RAD Studio or Embarcadero Delphi (11.3 or newer preferred)	Windows
Microsoft Visual Studio 2019 or 2022	

Microsoft Visual Studio is only required to compile some of the examples in *Chapter 11, Using External Libraries*.

If you are using the digital version of this book, we advise you to type the code yourself or access the code from the book's GitHub repository (a link is available in the next section). Doing so will help you avoid any potential errors related to the copying and pasting of code.

Download the example code files

You can download the example code files for this book from GitHub at `https://github.com/PacktPublishing/Delphi-High-Performance---Second-Edition`. If there's an update to the code, it will be updated in the GitHub repository.

We also have other code bundles from our rich catalog of books and videos available at `https://github.com/PacktPublishing/`. Check them out!

Download the color images

We also provide a PDF file that has color images of the screenshots and diagrams used in this book. You can download it here: `https://packt.link/GLK5S`.

Conventions used

There are a number of text conventions used throughout this book.

`Code in text`: Indicates code words in text, database table names, folder names, filenames, file extensions, pathnames, dummy URLs, user input, and Twitter handles. Here is an example: "This default value can be changed in the code by inserting `{$INLINE ON}`, `{$INLINE OFF}`, or `{$INLINE AUTO}` into the source."

A block of code is set as follows:

```
{$IFOPT O+}{$DEFINE OPTIMIZATION}{$ELSE}{$UNDEF OPTIMIZATION}{$ENDIF}
{$OPTIMIZATION ON}
```

When we wish to draw your attention to a particular part of a code block, the relevant lines or items are set in bold:

```
{$IFOPT O+}{$DEFINE OPTIMIZATION}{$ELSE}{$UNDEF OPTIMIZATION}{$ENDIF}
{$OPTIMIZATION ON}
```

Any command-line input or output is written as follows:

```
3
3,6
```

Bold: Indicates a new term, an important word, or words that you see on screen. For instance, words in menus or dialog boxes appear in **bold**. Here is an example: "They all always work in an **unsorted** mode, so adding an element takes $O(1)$, while finding and removing an element takes $O(n)$ steps."

> **Tips or important notes**
> Appear like this.

Get in touch

Feedback from our readers is always welcome.

General feedback: If you have questions about any aspect of this book, email us at `customercare@packtpub.com` and mention the book title in the subject of your message.

Errata: Although we have taken every care to ensure the accuracy of our content, mistakes do happen. If you have found a mistake in this book, we would be grateful if you would report this to us. Please visit `www.packtpub.com/support/errata` and fill in the form.

Piracy: If you come across any illegal copies of our works in any form on the internet, we would be grateful if you would provide us with the location address or website name. Please contact us at copyright@packtpub.com with a link to the material.

If you are interested in becoming an author: If there is a topic that you have expertise in and you are interested in either writing or contributing to a book, please visit authors.packtpub.com.

Share Your Thoughts

Once you've read *Delphi High Performance*, we'd love to hear your thoughts! Scan the QR code below to go straight to the Amazon review page for this book and share your feedback.

https://packt.link/r/1805125877

Your review is important to us and the tech community and will help us make sure we're delivering excellent quality content.

Download a free PDF copy of this book

Thanks for purchasing this book!

Do you like to read on the go but are unable to carry your print books everywhere?

Is your eBook purchase not compatible with the device of your choice?

Don't worry, now with every Packt book you get a DRM-free PDF version of that book at no cost.

Read anywhere, any place, on any device. Search, copy, and paste code from your favorite technical books directly into your application.

The perks don't stop there, you can get exclusive access to discounts, newsletters, and great free content in your inbox daily

Follow these simple steps to get the benefits:

1. Scan the QR code or visit the link below

https://packt.link/free-ebook/9781805125877

2. Submit your proof of purchase
3. That's it! We'll send your free PDF and other benefits to your email directly

1

About Performance

"My program is not fast enough. Users are saying that it is not performing well.
What can I do?"

These are the words I hear a lot when consulting on different programming projects. Sometimes the answer is simple, sometimes hard, but almost always, the critical part of the answer lies in the question. More specifically, in one word: *performing*.

What do we mean when we say that a program is performing well? Actually, nobody cares. What we have to know is what users mean when they say that the program is *not* performing well. And users, you'll probably admit, look at the world in a very different way than us programmers.

Before starting to measure and improve the performance of a program, we have to find out what users really mean by the word *performance*. Only then can we do something productive about it.

We will cover the following topics in this chapter:

- What is performance?
- What do we mean when we say that a program performs well?
- What can we tell about the code speed by looking at the algorithm?
- How does knowledge of compiler internals help us write fast programs?

Technical requirements

All code in this chapter was written with Delphi 11.3 Alexandria. It does not use the latest additions to the language, so most of the code could still be executed on Delphi XE and newer. You can find all the examples on GitHub at `https://github.com/PacktPublishing/Delphi-High-Performance---Second-Edition/tree/main/ch1`.

What is performance?

To better understand what we mean when we say that a program is *performing* well, let's take a look at a user story. In this book, we will use a fictitious person, namely Mr. Smith, chief of the Antarctica Department of Forestry. Mr. Smith is stationed in McMurdo Base, Antarctica, and he doesn't have much real work to do. He has already mapped all the forests in the vicinity of the station, and half of the year, it is too dark to be walking around and counting trees, anyway. That's why he spends most of his time behind a computer. And that's also why he is very grumpy when his programs are not performing well.

Some days, he writes long documents analyzing the state of forests in Antarctica. When he is doing that, he wants the document editor to *perform* well. By that, he actually means that the editor should work *fast enough* so that he doesn't feel any delay (or *lag*, as we call the delay when dealing with user input) while typing, preparing graphs, formatting tables, and so on.

In this scenario, performance simply means *working fast enough* and nothing else. If we speed up the operation of the document editor by a factor of 2, or even by a factor of 10, that would make no noticeable improvement for our Mr. Smith. The document editor would simply stay *fast enough* as far as he is concerned.

The situation completely changes when he is querying a large database of all of the forests on Earth and comparing the situation across the world to the local specifics of Antarctica. He doesn't like to wait, and he wants each database query to complete in as short a time as possible. In this case, performance translates to *speed*. We will make Mr. Smith a happier person if we find a way to speed up his database searches by a factor of 10, or even a factor of 5, or 2. He will be happy with any speedup and he'd praise us up to the heavens.

After all this hard work, Mr. Smith likes to play a game. While the computer is thinking about the next move, a video call comes in. Mr. Smith knows he's in for a long chat and he starts resizing the game window so that it will share the screen with a video call application. But the game is thinking hard and is not processing user input and poor Mr. Smith is unable to resize it, which makes him unhappy.

In this example, Mr. Smith simply expects that the application's user interface will *respond* to his commands. He doesn't care whether the application takes some time to find the next move, as long as he can do with the application what he wants to. In other words, he wants a user interface that doesn't *block*.

Different types of speed

It is obvious from the previous example that we don't always mean the same thing when we talk about a program's speed. There is a **real speed**, as in the database example, and there is a **perceived speed**, hinted at in the document editor and game scenario. Sometimes, we don't need to improve the program speed at all. We just have to make it not stutter while working (by making the user interface responsive at all times) and users will be happy.

We will deal with two types of performance in this book:

- Programs that react quickly to user input
- Programs that perform computations quickly

As you'll see, the techniques to achieve the former and the latter are somewhat different. To make a program react quickly, we can sometimes just put a long operation (as was the calculation of the next move in the fictitious game) into a background thread. The code will still run as long as in the original version but the user interface won't be blocked and everybody will be happy.

To speed up a program (which can also help with a slowly reacting user interface), we can use different techniques, from changing the algorithm to changing the code so that it will use more than one CPU at once, to using a hand-optimized version, either written by us or imported from an external library.

To do anything, we have to know which part of the code is causing a problem. If we are dealing with a big legacy program, the problematic part may be hard to find. In the rest of this chapter, we will look at different ways to locate such code. We'll start by taking an educated *guess* and then we'll improve that by *measuring* the code speed, first by using home-grown tools and then with a few different open source and commercial programs.

Algorithm complexity

Before we start with the dirty (and fun) job of improving program speed, I'd like to present a bit of computer science theory, namely, **Big O** notation.

You don't have to worry, I will not use pages of mathematical formulas and talk about *infinitesimal asymptotes*. Instead, I will just present the essence of Big O notation, the parts that are important to every programmer.

In the literature and, of course, on the web, you will see expressions such as *O(n)*, *O(n^2)*, *O(1)*, and similar. This fancy-looking notation hides a really simple story. It tells us how much slower the algorithm will become if we increase the data size by a factor of *n*.

> **Information**
> The *n^2* notation means "*n* to the power of two," or n^2. This notation is frequently used on the internet because it can be written with standard ASCII characters. This book uses the more readable variant, $O(n^2)$.

Let's say we have an algorithm with a complexity of *O(n)*, which on average takes *T* seconds to process input data of size *N*. If we increase the size of the data by a factor of 10 (to *10*N*), then the algorithm will (on average) also take 10 times more time (that is, *10*T*) to process the data. If we process 1,000 times more data, the program will also run 1,000 times slower.

If the algorithm complexity is $O(n^2)$, increasing the size of the data by a factor of 10 will cause the algorithm to run 10^2 or 100 times longer. If we want to process 1,000 times more data, then the algorithm will take $1,000^2$ or a million times longer, which is quite a hit. Such algorithms are typically not very useful if we have to process large amounts of data.

Note

Most of the time, we use Big O notation to describe how the computation time relates to the input data size. When this is the case, we call Big O notation **time complexity**. On the other hand, sometimes the same notation is used to describe how much **storage** (memory) the algorithm is using. In that case, we are talking about **space complexity**.

You may have noticed that I was using the word *average* a lot in the last few paragraphs. When talking about the algorithm complexity, we are mostly interested in the average behavior, but sometimes we will also need to know about the worst behavior. We rarely talk about the best behavior because users don't really care much if the program is sometimes faster than average.

Let's look at an example. The following function checks whether a `string` parameter `value` is present in a string list:

```
function IsPresentInList(strings: TStrings; const value: string):
Boolean;
var
   i: Integer;
begin
  Result := False;
  for i := 0 to strings.Count - 1 do
    if SameText(strings[i], value) then
       Exit(True);
end;
```

What can we tell about this function? The best case is really simple—it will find that `value` is equal to `strings[0]` and it will exit. Great! The best behavior for our function is *O(1)*. That, sadly, doesn't tell us much as that won't happen frequently in practice.

The worst behavior is also easy to find. If `value` is not present in the list, the code will have to scan all of the `strings` list before deciding that it should return `False`. In other words, the worst behavior is *O(n)*, if the *n* represents the number of elements in the list. It is also quite obvious (and simple to prove) that, on average, this algorithm finds an element in *n/2* steps. Is the Big O limit for this search, therefore, *O(n/2)*?

> **Note**
>
> Big O limits don't care about constant factors. If an algorithm would use *n/2* steps on average, or even just *0.0001 * n* steps, we would still write this down as *O(n)*. Of course, an *O(10 * n)* algorithm is slower than an *O(n)* algorithm, and that is absolutely important when we fine-tune the code, but no constant factor *C* will make *O(C * n)* faster than *O(log n)* if *n* gets sufficiently large.

There are better ways to check whether an element is present in some data than searching the list sequentially. We will explore one of them in the next section, *Big O and Delphi data structures*.

While the function of *n* inside the *O()* notation can be anything, there are some O functions that appear constantly in standard programming problems. The following table shows those Big O limits and the most common examples of problems that belong to each class:

Time complexity	Common examples of problems with that time complexity
O(1)	Accessing array elements
O(log n)	Search in an ordered list
O(n)	Linear search
O(n log n)	Quick sort (average behavior)
O(n²)	Quick sort (worst behavior), naive sort (bubble sort, insertion sort, selection sort)
O(cⁿ)	Recursive Fibonacci, traveling salesman problem using dynamic programming (c is some numeric constant)

Table 1.1 – Big O limits of the most commonly encountered programming problems

If we care about program performance, then *O(1)* algorithms are of special interest to us as they present algorithms that don't get slower (at least, not noticeably) when we increase the problem size. We'll see an example of such *O(1)* algorithms in the next section.

When we deal with algorithms that search in some datasets, we usually try to make them behave as *O(log n)*, not *O(n)*, as the former slows down much, much slower than the latter.

Another big class of problems deals with sorting the data. While the naive approaches sort in *O(n²)*, better algorithms (such as merge sort and quicksort) need on average just *O(n log n)* steps.

The following figure shows how the time complexity for these typical limits (we have used 2^n as an example of a more generic c^n) grows when we increase the problem size up to 20-fold:

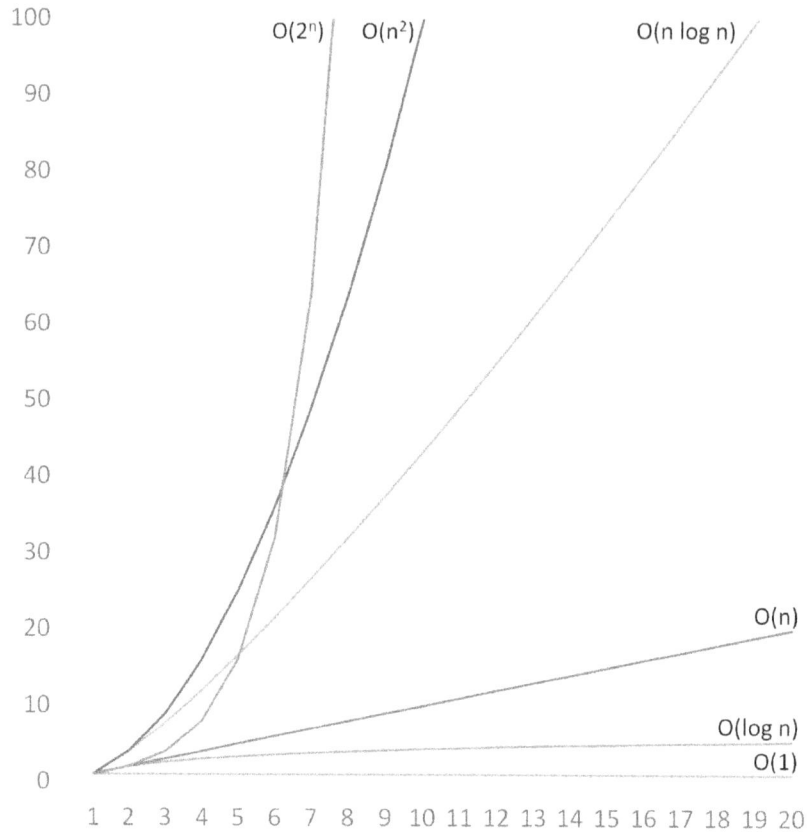

Figure 1.1 – Comparison of growth for the most commonly encountered Big-O limits

We can see that *O(1)* and *O(log n)* grow very slowly. While *O(n log n)* grows faster than *O(n)*, it also grows much slower than *O(n²)*, which we had to stop plotting when data was increased nine-fold.

O(2ⁿ) starts slowly and looks like a great solution for small data sizes (small *n*), but then it starts rising terribly fast, much faster than *O(n²)*.

The following table shows how fast *O(n log n)* and *O(n²)* are growing if we compare them with *O(n)*, and how quickly *O(2ⁿ)* explodes.

The **Data size** column shows the data size increase factor. The number **10** in this column, for example, represents input with 10 times more elements than in the original data:

Data size	O(1)	O(log n)	O(n)	O(n log n)	O(n²)	O(2ⁿ)
1	1	1	1	1	1	1
2	1	2	2	4	4	2
10	1	4	10	43	100	512
20	1	5	20	106	400	524,288
100	1	8	100	764	10,000	10^{29}
300	1	9	300	2,769	90,000	10^{90}

Table 1.2 – Examples of a growth factor for the most commonly encountered Big-O limits

We can see from this table that *O(log n)* algorithms present a big improvement over *O(n)* algorithms (8 versus 100 times increase in time when data increases 100-fold). We can also see that *O(2ⁿ)* quickly becomes completely unmanageable.

The last cell in this table is particularly interesting. There are different estimates for the number of elementary particles (electrons, protons, neutrons, and so on) in the visible universe, but they all lie somewhere around 10^{90}. Suppose we have a computer that can solve an *O(2ⁿ)* algorithm in a reasonable time. If we were to increase the input data by a factor of just 300, then we would need 10^{90} computers to solve the new problem in the same time. That is as much as the number of particles in the visible universe!

> Tip
>
> Don't use algorithms that have time complexity *O(2ⁿ)*. It won't end well.

Big O and Delphi data structures

Delphi's **Run-Time Library** (**RTL**) contains many **data structures** (classes that are specifically designed to store and retrieve data), mostly stored in `System.Classes` and `System.Generics.Collections` units that greatly simplify everyday work. We should, however, be aware of their good and bad sides.

Every data structure in the world is seeking a balance between four different types of data access: accessing data, inserting data, searching for data, and deleting data. Some data structures are good in some areas, others in different ones, but no data structure in this world can make all four operations independent of data size.

When designing a program, we should therefore know what our needs are. That will help us select the appropriate data structure for the job.

The most popular data structure in Delphi is undoubtedly `TStringList`. It can store a large number of strings and assign an object to each of them. It can—and this is important—work in two modes, **unsorted** and **sorted**. The former, which is a default, keeps strings in the same order as they were added, while the latter keeps them alphabetically ordered.

This directly affects the speed of some operations. While accessing any element in a string list can always be done in a constant time (*O(1)*), adding to a list can take *O(1)* when the list is not sorted (and the underlying storage does not need to be increased), and *O(log n)* when the list is sorted.

Why that big difference? When the list is unsorted, `Add` just adds a string at its end. If the list is, however, sorted, `Add` must first find a correct insertion place. It does this by executing a **bisection search**, which needs *O(log n)* steps to find the correct place.

The reverse holds true for searching in a string list. If it is not sorted, `IndexOf` needs to use a linear search – that is, compare each element one by one to find a match. In the worst case, this search will have to examine all elements in the list. In a sorted list, it can do it much faster (again, by using a bisection) in *O(log n)* steps.

We can see that `TStringList` offers us two options – either a fast addition of elements or a fast lookup, but not both. In a practical situation, we must look at our algorithm and think wisely about what we really need and what will behave better.

To sort a string list, you can call its `Sort` method, or you can set its `Sorted` property to `True`. There is, however, a subtle difference that you should be aware of. While calling `Sort` sorts the list, it doesn't set its internal *is sorted* flag, and all operations on the list will proceed as if the list is unsorted. Setting `Sorted := True`, on the other hand, does both – it sets the internal flag and calls the `Sort` method to sort the data.

To store any (non-string) data, we can use traditional `TList` and `TObjectList` classes or their more modern generic counterparts, `TList<T>` and `TObjectList<T>`. They all always work in an **unsorted** mode, and so adding an element takes *O(1)* while finding and removing an element takes *O(n)* steps.

All provide a `Sort` function that sorts the data with a quicksort algorithm (*O(n log n)* on average) but only generic versions have a `BinarySearch` method, which searches for an element with a bisection search taking *O(log n)* steps. Be aware that `BinarySearch` requires the list to be sorted but doesn't make any checks to assert that. It is your responsibility to sort the list before you use this function.

If you need a very quick element lookup, paired with fast addition and removal, then `TDictionary` is the solution. It has methods for adding (`Add`), removing (`Remove`), and finding a key (`ContainsKey` and `TryGetValue`) that, on average, function in a constant time, *O(1)*. Their worst behavior is actually quite bad, *O(n)*, but that will only occur on specially crafted sets of data that you will never see in practical applications.

I've told you before that there's no free lunch and so we can't expect that TDictionary is perfect. The big limitation is that we can't access the elements it is holding in a direct way. In other words, there is no TDictionary[i]. We can walk over all elements in a dictionary by using a for statement, but we can't access any of its elements directly. Another limitation of TDictionary is that it does not preserve the order in which elements were added.

Delphi also offers two simple data structures that mimic a standard queue (TQueue<T>) and stack (TStack<T>). Both have very fast *O(1)* methods for adding and removing the data and a simple way to access the elements. The List property enables direct access to the array containing the data. To insert data, we use Enqueue (queue) or Push (stack), and to remove data, we use Dequeue (queue) or Pop (stack).

To help you select the right tool for the job, I have put together a table showing the most important data structures and their most important methods, together with average and (when they differ from the average) worst-case time complexities:

Data structure	Operation	Average	Worst
TStringList	Direct access	O(1)	
	Add	O(1) / O(log n)	
	Insert	O(1)	
	Delete	O(1)	
	IndexOf	O(n) / O(log n)	
	Sort	O(n log n)	O(n^2)
TList, TObjectList	Direct access	O(1)	
	Add	O(1)	
	Insert	O(1)	
	Delete	O(1)	
	Remove	O(n)	
	IndexOf	O(n)	
	Sort	O(n log n)	O(n^2)

Data structure	Operation	Average	Worst
TList<T>, TObjectList<T>	Direct access	O(1)	
	Add	O(1)	
	Insert	O(1)	
	Delete	O(1)	
	Remove	O(n)	
	IndexOf	O(n)	
	BinarySearch	O(log n)	
	Sort	O(n log n)	O(n²)
TDictionary	Direct access	Not possible	
	Add	O(1)	O(n)
	Remove	O(1)	O(n)
	TryGetValue	O(1)	O(n)
	ContainsKey	O(1)	O(n)
	ContainsValue	O(n)	
TQueue<T>	Direct access	Not possible	
	Enqueue	O(1)	
	Dequeue	O(1)	
TStack<T>	Direct access	Not possible	
	Push	O(1)	
	Pop	O(1)	

Table 1.3 – Time complexity of the most important operations on built-in Delphi data structures

The table shows the time complexity of the most important operations on built-in Delphi data structures. Complexity for the worst case is only listed if it differs from the average complexity. Insertion into a sorted TStringList is not possible, hence the table lists only the complexity for inserting into an unsorted list.

Data structures in practice

Enough with the theory already! I know that you, like me, prefer to talk through the code. As one program explains more than a thousand words could, I have prepared a simple demo project: RandomWordSearch.

This program functions as a very convoluted random word generator. When started, it will load a list of 370,101 English words from a file. It will also prepare three internal data structures preloaded with these words:

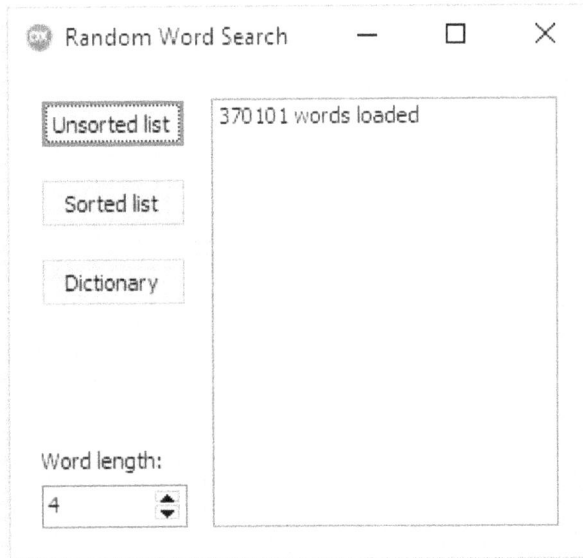

Figure 1.2 – Random word generator

The program shows three buttons to the user. All three run basically the same code. The only difference is the test function, which is passed to the centralized word generator as a parameter:

```
procedure TfrmRandomWordSearch.FindGoodWord(
  const wordTest: TWordCheckDelegate);
var
  word: string;
  isWordOK: boolean;
  time: TStopwatch;
begin
  time := TStopwatch.StartNew;
  repeat
    word := GenerateWord;
    isWordOK := wordTest(word);
```

```
    until isWordOK or (time.ElapsedMilliseconds > 10000);
    if isWordOK then
      lbWords.ItemIndex := lbWords.Items.Add(Format('%s (%d ms)',
        [word, time.ElapsedMilliseconds]))
    else
      lbWords.ItemIndex := lbWords.Items.Add('timeout'); end;
```

The core of the FindGoodWord method can be easily described:

1. Generate a random word by calling GenerateWord.

2. Call the test function, wordTest, on that word. If this function returns False, repeat *Step 1*. Otherwise, show the word.

The code is a bit more complicated because it also checks that the word generation part runs for at most 10 seconds, and reports a timeout if no valid word was found in that time.

The GenerateWord random word generator is incredibly simple. It just appends together lowercase English letters until the specified length (settable in the user interface) is reached:

```
function TfrmRandomWordSearch.GenerateWord: string;
var
  pos: integer;
begin
  Result := '';
  for pos := 1 to inpWordLength.Value do
    Result := Result +
              Chr(Ord('a') + Random(Ord('z') - Ord('a') + 1));
end;
```

Let's now check the data preparation phase. The not very interesting (and not shown here) OnCreate handler loads data from a file into TStringList and calls the LoadWords method:

```
procedure TfrmRandomWordSearch.LoadWords(wordList: TStringList);
var
  word: string;
begin
  FWordsUnsorted := TStringList.Create;
  FWordsUnsorted.Assign(wordList);
  FWordsSorted := TStringList.Create;
  FWordsSorted.Assign(wordList);
  FWordsSorted.Sorted := True;
  FWordsDictionary := TDictionary<string,boolean>.Create(
    wordList.Count);
```

```
   for word in wordList do
     FWordsDictionary.Add(word, True);
 end;
```

The first data structure is an unsorted TStringList, FWordsUnsorted. Data is just copied from the list of all words by calling the Assign method.

The second data structure is a sorted TStringList, FWordsSorted. Data is firstly copied from the list of all words. The list is then sorted by setting FWordsSorted.Sorted := True.

The last data structure is TDictionary, FWordsDictionary. TDictionary always stores *pairs* of keys and values. In our case, we only need the *keys* part as there is no data associated with any specific word, but Delphi doesn't allow us to ignore the *values* part and so the code defines the value as a Boolean and always sets it to True.

Although the dictionaries can grow, they work faster if we can initially set the number of elements that will be stored inside. In this case, that is simple—we can just use the length of wordList as a parameter to TDictionary.Create.

The only interesting part of the code left is the OnClick handlers for all three buttons. All three call the FindGoodWord method, but each passes in a different test function.

When you click on the **Unsorted list** button, the test function checks whether the word can be found in the FWordsUnsorted list by calling the IndexOf function. As we will mostly be checking non-English words (remember, they are just random strings of letters), this IndexOf function will typically have to compare all 370,101 words before returning -1:

```
procedure TfrmRandomWordSearch.btnUnsortedListClick(Sender: TObject);
begin
  FindGoodWord(
    function (const word: string): boolean
    begin
      Result := FWordsUnsorted.IndexOf(word) >= 0;
    end);
end;
```

When you click on the **Sorted list** button, the test function calls FWordsSorted.IndexOf. As this TStringList is sorted, IndexOf will use a binary search that will need at most *log(307101)* = *19* (rounded up) comparisons to find out that a word is not found in the list. As this is much less than 370,101, we can expect that finding words with this approach will be much faster:

```
procedure TfrmRandomWordSearch.btnSortedListClick(Sender: TObject);
begin
  FindGoodWord(
    function (const word: string): boolean
    begin
```

```
      Result := FWordsSorted.IndexOf(word) >= 0;
    end);
end;
```

A click on the last button, **Dictionary**, calls FWordsDictionary.ContainsKey to check whether the word can be found in the dictionary, and that can usually be done in just one step. Admittedly, this is a bit of a slower operation than comparing two strings, but still, the TDictionary approach should be faster than any of the TStringList methods:

```
procedure TfrmRandomWordSearch.btnDictionaryClick(Sender: TObject);
begin
  FindGoodWord(
    function (const word: string): boolean
    begin
      Result := FWordsDictionary.ContainsKey(word);
    end);
end;
```

If we use the terminology from the last section, we can say that the *O(n)* algorithm (unsorted list) will run much slower than the *O(log n)* algorithm (sorted list), and that the *O(1)* algorithm (dictionary) will be the fastest of them all. Let's check this in practice.

Start the program and click on the **Unsorted list** button a few times. You'll see that it typically needs from a few hundred milliseconds to a few seconds to generate a new word. As the process is random and dependent on CPU speed, your numbers may differ quite a lot from mine. If you are only getting *timeout* messages, you are running on a slow machine and you should decrease **Word length** to **3**.

If you increment **Word length** to **5** and click the button again, you'll notice that the average calculation time will grow up to a few seconds. You may even get an occasional timeout. Increase it to **6** and you'll mostly be getting timeouts. We are clearly hitting the limits of this approach.

Prepare now to be dazed! Click on the **Sorted list** button (while keeping **Word length** at **6**) and words will again be calculated blazingly fast. On my computer, the code only needs 10 to 100 milliseconds to find a new word:

Figure 1.3 – Testing with different word lengths and with the first two algorithms

To better see the difference between a sorted list and a dictionary, we have to crank up the word length again. Setting it to 7 worked well for me. The sorted list needed from a few hundred milliseconds to a few seconds to find a new word, while the dictionary approach mostly found a new word in under 100 milliseconds.

Increase **Word length** to **8** and the sorted list will start to time out while the dictionary will still work. Our *O(1)* approach is indeed faster than the *O(log n)* code:

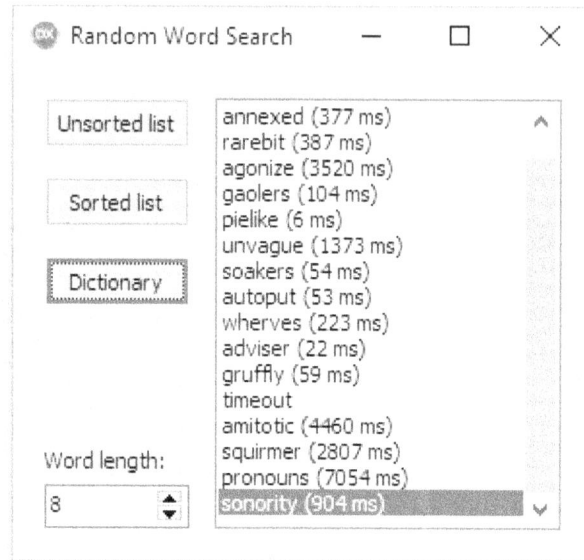

Figure 1.4 – Comparing the sorted list (first six words and the two after
"timeout") with the dictionary approach (next five and last two)

While knowledge of algorithms and their complexity is useful, it will not help on every occasion. Sometimes, you're already using the best possible algorithm but your program is still too slow. At such times, you'll want to find the slow parts of the program and speed them up with whatever trick possible. To do that, you have to find them first, which is the topic of the rest of this chapter.

Mr. Smith's first program

While I was talking about theory and data structures and best practices, our friend Mr. Smith read everything about polar forests on the internet and then got really bored. In a moment of sheer panic— when he was thinking about watching videos of cute kittens for the whole polar night—he turned to programming. He checked various internet sources and decided that Delphi is the way to go. After all, he could use it to put together a new game for iOS and Android and make lots of money while waiting for the sun to show up!

As he didn't have any programming literature on the Antarctic base, he learned programming from internet resources. Sadly, he found most of his knowledge on Experts Exchange and Yahoo Answers, and his programs sometimes reflect that.

As of this moment, he has written one working program (and by *working*, I mean that it compiles), but he is not really sure what the program actually does. He only knows that the program is a bit slow and because of that, he named it `SlowCode`.

His program is a console mode program, which, upon startup, calls the `Test` method:

```
procedure Test;
var
  data: Tarray<Integer>;
  highBound: Integer;
begin
  repeat
    Writeln('How many numbers (0 to exit)?');
    Write('> ');
    Readln(highBound);
    if highBound = 0 then
      Exit;
    data := SlowMethod(highBound);
    ShowElements(data);
  until false;
end;
```

That one, at least, is easy to grasp. It reads some number, passes it to something called `SlowMethod` (hmm, Mr. Smith really should work on naming techniques), and then passes the result (which is of the `TArray<Integer>` type) to a method called `ShowElements`. When the user types in 0, the program exits.

Let's check the `SlowMethod` function:

```
function SlowMethod(highBound: Integer): TArray<Integer>;
var
  i: Integer;
  temp: TList<Integer>;
begin
  temp := TList<Integer>.Create;
  try
    for i := 2 to highBound do
      if not ElementInDataDivides(temp, i) then
        temp.Add(i);
    Result := Filter(temp);
  Finally
    FreeAndNil(temp);
  end;
end;
```

The code creates a list of integers. Then it iterates from a value of 2 to the value that was entered by the user, and for each value, calls `ElementInDataDivides`, passing in the list and the current value. If that function returns `True`, the current value is entered into the list.

After that, SlowMethod calls the Filter method, which does something with the list and converts it into an array of integers that is then returned as a function result:

```
function ElementInDataDivides(data: TList<Integer>; value: Integer):
boolean;
var
  i: Integer;
begin
  Result := True;
  for i in data do
    if (value <> i) and ((value mod i) = 0) then
      Exit;
  Result := False;
end;
```

The ElementInDataDivides function iterates over all the numbers in the list and checks whether any element in the list divides the value (with the additional constraint that this element in the list must not be equal to the value).

Let's check the last part of the puzzle—the Filter function together with a helper function, Reverse:

```
function Reverse(s: string): string;
var
  ch: char;
begin
  Result := '';
  for ch in s do
    Result := ch + Result;
end;

function Filter(list: TList<Integer>): TArray<Integer>;
var
  i: Integer;
  reversed: Integer;
begin
  SetLength(Result, 0);
  for i in list do
  begin
    reversed := StrToInt(Reverse(IntToStr(i)));
    if not ElementInDataDivides(list, reversed) then
    begin
      SetLength(Result, Length(Result) + 1);
      Result[High(Result)] := i;
    end;
```

```
      end;
   end;
```

This once again iterates over the list, reverses the numbers in each element (changes `123` to `321`, `3341` to `1433`, and so on), and calls `ElementInDataDivides` on the new number. If it returns `True`, the element is added to the returned result in a fairly inefficient way.

I agree with Mr. Smith—it is hard to tell what the program does. Maybe it is easiest to run it and look at the output:

Figure 1.5 – Output of Mr. Smith's first program

It looks like the program is outputting prime numbers. Not all prime numbers, just some of them. (For example, 19 is missing from the list, and so is 23.) Let's leave it at that for the moment.

Looking at code through Big O eyes

We can tell more about the program, about its good and bad parts, if we look at it through the eyes of time complexity, in terms of the Big O notation.

We'll start where the code starts—in the `SlowMethod` method. It has a loop iterating from 2 to the user-specified upper bound, `for i := 2 to highBound do`. The size of our data, or n, is therefore equal to `highBound`, and this `for` loop has a time complexity of $O(n)$:

```
for i := 2 to highBound do
   if not ElementInDataDivides(temp, i) then
      temp.Add(i);
```

Inside this loop, the code calls `ElementInDataDivides` followed by an occasional `temp.Add`. The latter will execute in *O(1)*, but we can't say anything about `ElementInDataDivides` before we examine it.

This method also has a loop iterating over the `data` list. We can't guess how many elements are in this list, but in the short test that we just performed, we know that the program writes out 13 elements when processing values from 2 to `100`:

```
for i in data do
  if (value <> i) and ((value mod i) = 0) then
    Exit;
```

For the purpose of this very rough estimation, I'll just guess that the `for i in data do` loop also has a time complexity of *O(n)*.

In `SlowMethod`, we therefore have an *O(n)* loop executing another *O(n)* loop for each element, which gives us *O(n²)* performance.

`SlowMethod` then calls the `Filter` method, which also contains *O(n)* for loop calling `ElementInDataDivides`, which gives us *O(n²)* complexity for this part:

```
for i in list do
begin
  reversed := StrToInt(Reverse(IntToStr(i)));
  if not ElementInDataDivides(list, reversed) then
  begin
    SetLength(Result, Length(Result) + 1);
    Result[High(Result)] := i;
  end;
end;
```

There's also a conversion to a string, some operation on that string, and conversion back to the `StrToInt(Reverse(IntToStr(i)))` integer. It works on all elements of the list (*O(n)*), but in each iteration, it processes all characters in the string representation of a number. As the length of the number is proportional to *log n*, we can say that this part has a complexity of *O(n log n)*, which can be ignored as it is much less than the *O(n²)* complexity of the whole method.

There are also some operations hidden inside `SetLength`, but at this moment, we don't know yet what they are and how much they contribute to the whole program. We'll cover that area in *Chapter 6, Memory Management*.

`SlowMethod`, therefore, consists of two parts, both with complexity *O(n²)*. Added together, that would give us *2*n²*, but as we ignore constant factors (that is, *2*) in Big O notation, we can only say that the time complexity of `SlowMethod` is *O(n²)*.

So what can we say simply by looking at the code?

- The program probably runs in $O(n^2)$ time. It will take around 100 times longer to process 10,000 elements than 1,000 elements.

- There is a conversion from the integer to the string and back (`Filter`), which has a complexity of only $O(n \log n)$, but it would still be interesting to know how fast this code really is.

- There's a time complexity hidden behind the `SetLength` call, which we know nothing about.

- We can guess that, most probably, `ElementInDataDivides` is the most time-consuming part of the code and any improvements in this method would probably help.

- Fixing the terrible idea of appending elements to an array with `SetLength` could probably speed up a program, too.

As the code performance is not everything, I would also like to inform Mr. Smith about a few places where his code is less than satisfactory:

- The prompt *How many numbers* is misleading. A user would probably expect it to represent the number of numbers output, while the program actually wants to know how many numbers to test.

- Appending to `TArray<T>` in that way is not a good idea. Use `TList<T>` for temporary storage and call its `ToArray` method at the end. If you need `TArray<T>`, that is. You can also do all processing using `TList<T>`.

- `SlowMethod` has two distinctive parts – data generation, which is coded as a part of `SlowMethod`, and data filtering, which is extracted in its own method, `Filter`. It would be better if the first part is extracted into its own method, too.

- Console program? Really? Console programs are good for simple tests, but that is it. Learn VCL or FireMonkey, Mr. Smith!

We can now try and optimize parts of the code (`ElementInDataDivides` seems to be a good target for that) or, better, we can do some measuring to confirm our suspicions with hard numbers.

In a more complicated program (what we call *real life*), it would usually be much simpler to measure the program's performance than to do such an analysis. This approach, however, proves to be a powerful tool if you are using it while designing code. Once you hear a little voice nagging about the time complexities all the time while you're writing code, you'll be on the way to becoming an excellent programmer.

Summary

This chapter provided a broad overview of the topics we'll be dealing with in this book. We took a look at the very definition of performance. Next, we spent some time describing Big O notation for describing time and space complexity and we used it in a simple example. We learned how to analyze existing code and how to estimate its execution time. This enables us to locate potential problems without even executing the code.

In the next chapter, I will look into the topic of profiling. We will see how we can get more detailed information about the program speed, which is essential before we can start optimizing the code.

2
Profiling the Code

Now that we know what performance is and how to measure it, we can start speeding up our programs. This is, however, not an easy process. For starters, we must know which parts of the program are slowing us down, and that can sometimes be hard to determine.

We can, of course, start by *guessing*. This will sometimes work, but it will often just result in wasted time and frustration. It turns out that it is usually hard to tell what will be a slow part of a complex program. Even if we start by guessing, it is good if we can then *confirm* our assumption before we start rewriting the code.

We can do that by adding measurement code, which works great if a program is small or if we know which specific part we want to improve, but most of the time, it is better to use a specialized tool: a *profiler*.

In this chapter, we'll look into both approaches, manual profiling and using an automated tool.

We will cover the following topics:

- Why is it better to measure than to guess?
- How to manually determine which part of the program is the slowest
- What tools can we use to find the slow parts of a program?

Technical requirements

All code in this chapter was written with Delphi 11.3 Alexandria. It does not use the latest additions to the language, so most of the code can still be executed on Delphi XE and newer versions. You can find all the examples on GitHub at `https://github.com/PacktPublishing/Delphi-High-Performance---Second-Edition/tree/main/ch2`.

Don't guess, measure!

There is only one way to get a good picture of the fast and slow parts of a program—by measuring it. We can do it manually by inserting time-measuring calls in the code, or we can use specialized tools. We have a name for measuring—**profiling**—and we call specialized tools for measuring **profilers**.

In the rest of this chapter, we'll look at different techniques for measuring the execution speed. First, we will measure the now familiar program, SlowCode, with a simple software stopwatch, and then we'll look at a few open source and commercial profilers.

Before we start, I'd like to point out a few basic rules that apply to all profiling techniques:

- Always profile *without* the debugger. The debugger will slow the execution in unexpected places, and that will skew the results. If you are starting your program from the Delphi **integrated development environment** (IDE), just press *Ctrl + Shift + F9* instead of *F9*.

- Try not to do anything else on the computer while profiling. Other programs will take the CPU away from the measured program, which will make it run slower.

- Take care that the program doesn't wait for user action (data entry, button click) while profiling. This will completely skew the report.

- Repeat the tests a few times. Execution times will differ because Windows (and any other **operating system** (OS) that Delphi supports) will always execute other tasks besides running your program.

- All the preceding points especially hold true for multithreaded programs, which is an area explored in *Chapters 7 to 10*.

Profiling with TStopwatch

Delphi includes a helpful unit called System.Diagnostics, which implements a TStopwatch record. It allows us to measure time events with better precision than 1 millisecond and has a pretty exhaustive public interface, as shown in the following code fragment:

```
type
  TStopwatch = record
  public
    class function Create: TStopwatch; static;
    class function GetTimeStamp: Int64; static;
    procedure Reset;
    procedure Start;
    class function StartNew: TStopwatch; static;
    procedure Stop;
    property Elapsed: TTimeSpan read GetElapsed;
    property ElapsedMilliseconds: Int64 read
```

```
    GetElapsedMilliseconds;
  property ElapsedTicks: Int64 read GetElapsedTicks;
  class property Frequency: Int64 read FFrequency;
  class property IsHighResolution: Boolean read
    FIsHighResolution;
  property IsRunning: Boolean read FRunning;
end;
```

To use a stopwatch, you first have to create it. You can call `TStopwatch.Create` to create a new *stopped* stopwatch or `TStopwatch.StartNew` to create a newly *started* stopwatch. As `TStopwatch` is implemented as `record`, there's no need to destroy a `stopwatch` object.

When a stopwatch is started, it measures time. To start a stopwatch, call the `Start` method, and to stop it, call the `Stop` method. The `IsRunning` property will tell you whether the stopwatch is currently started. Call the `Reset` method to reset the stopwatch to zero.

`TStopwatch` contains a few functions that return the currently measured time. The most precise of them is `ElapsedTicks`, but as there is no built-in (public) function to convert this into standard time units, this function is hard to use. My recommendation is to just use the `ElapsedMilliseconds` property, which will give you elapsed (measured) time in milliseconds.

For a simple demo, this code will return 1,000 or a bit more:

```
function Measure1sec: int64;
var
  sw: TStopwatch;
begin
  sw := TStopwatch.StartNew;
  Sleep(1000);
  Result := sw.ElapsedMilliseconds;
end;
```

Let's now use this function to measure the `SlowMethod` method.

First, you have to add the `System.Diagnostics` unit to the uses list:

```
uses
  System.SysUtils,
  System.Generics.Collections,
  System.Classes,
  System.Diagnostics;
```

Next, you have to create this stopwatch inside `SlowMethod`, stop it at the end, and write out the elapsed time:

```
function SlowMethod(highBound: Integer): TArray<Integer>;
var
 // existing variables
 sw: TStopwatch;
begin
  sw := TStopwatch.StartNew;

  // existing code

  sw.Stop;
  Writeln('SlowMethod: ', sw.ElapsedMilliseconds, ' ms');
end;
```

We can use this code to verify the theory that `SlowCode` has time complexity $O(n^2)$. To do this, we have to measure the execution times for different counts of processed numbers (different values entered at the *How many numbers* prompt).

I did some testing for selected values from 10,000 to 1,000,000 and got the following numbers:

Highest number	Execution time [ms]
10,000	15
25,000	79
50,000	250
75,000	506
100,000	837
250,000	4,515
500,000	15,564
750,000	30,806
1,000,000	54,219

Table 2.1 – Measured speed of the SlowCode program

If you repeat the tests, you will, of course, measure different values, but the growth rate should be the same.

For quick confirmation, I have plotted a **scatter chart** of this data in Excel and the result looks like a square function. To be more sure, I have added a **power trendline**, which created a function in the form of n^c, where c was a constant that Excel calculated from the data. In the case of my specific measurements, this *fitting* function was $y = 10^{-6} * x^{1.7751}$, which is not that far from x^2:

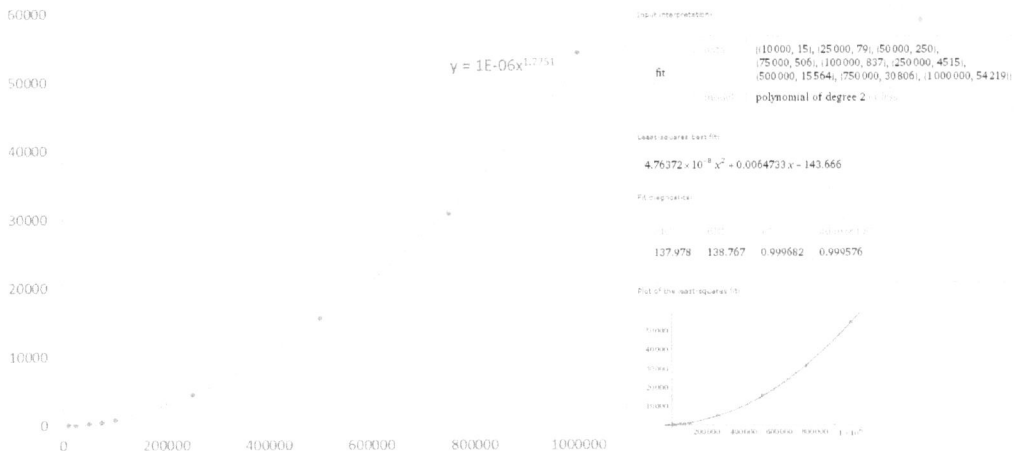

Figure 2.1 – Curve fitting in Excel (left) and Wolfram Alpha (right)

Next, I repeated this *curve-fitting* process on the wonderful Wolfram Alpha (www.wolframalpha.com), where you can find a *regression calculator widget*, a tool designed specifically for this task. I entered measurements into the widget, and it calculated a fitting function of $4.76372 \times 10^{-8} * x2 + 0.0064733 * x - 143.666$. If we ignore all unimportant factors and constants, the only part left is $x2$. Another confirmation for our analysis!

Now we know how the program behaves in global terms, but we still have no idea of which part of the code is the slowest. To find that out, we also have to measure the execution times of the ElementInDataDivides and Filter methods.

These two methods are called multiple times during the execution of the program, so we can't just create and destroy a stopwatch each time Filter (for example) is executed. We have to create a global stopwatch, which is a bit inconvenient in this program because we have to introduce a global variable.

If you check the SlowCode_Stopwatch program, you'll see that it actually creates three global stopwatches, one for each of the functions that we want to measure:

```
var
  Timing_ElementInData: TStopwatch;
  Timing_Filter: TStopwatch;
  Timing_SlowMethod: TStopwatch;
```

All three stopwatches are created (but not started!) when the program starts:

```
Timing_ElementInData := TStopwatch.Create;
Timing_Filter := TStopwatch.Create;
Timing_SlowMethod := TStopwatch.Create;
```

When the program ends, the code logs the elapsed time for all three stopwatches:

```
Writeln('Total time spent in SlowMethod: ',
  Timing_SlowMethod.ElapsedMilliseconds, ' ms');
Writeln('Total time spent in ElementInDataDivides: ',
  Timing_ElementInData.ElapsedMilliseconds, ' ms');
Writeln('Total time spent in Filter: ',
  Timing_Filter.ElapsedMilliseconds, ' ms');
```

In each of the three methods, we only have to *start* the stopwatch at the beginning and *stop* it at the end:

```
function SlowMethod(highBound: Integer): TArray<Integer>;
var
 // existing variables
 sw: TStopwatch;
begin
  sw := TStopwatch.StartNew;

  // existing code

  sw.Stop;
  Writeln('SlowMethod: ', sw.ElapsedMilliseconds, ' ms');
end;
```

The only tricky part is the ElementInDataDivides function, which calls Exit as soon as one element divides the value parameter. The simplest way to fix that is to wrap the existing code in a try .. finally handler and to stop the stopwatch in the finally part:

```
function ElementInDataDivides(data: TList<Integer>; value: Integer):
boolean;
var
  i: Integer;
begin
  Timing_ElementInData.Start;
  try
    Result := True;
    for i in data do
      if (value <> i) and ((value mod i) = 0) then
```

```
      Exit;
    Result := False;
  finally
    Timing_ElementInData.Stop;
  end;
end;
```

If you run the program and play with it for a while and then exit, you'll get a performance report. In my case, I got the following result:

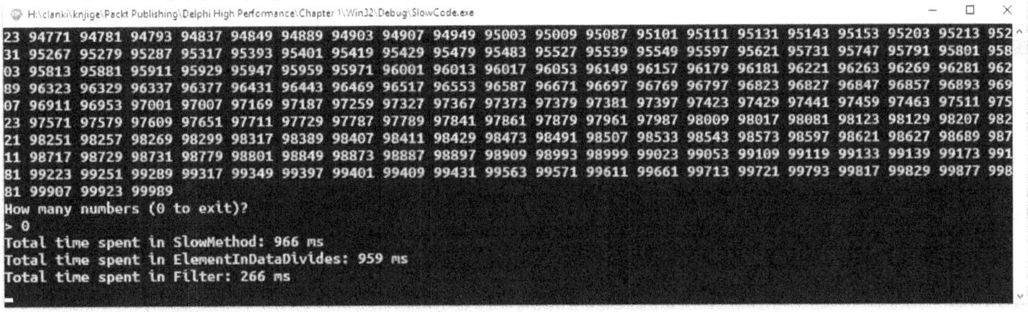

Figure 2.2 – Time spent in various parts of the SlowCode program

We now know that most of the time is spent in `ElementInDataDivides`, but we don't know how many calls to it were made directly from `SlowMethod` and how many from the `Filter` method. To find that out, we have to add two new global variables and some more code:

```
var
  Generate_ElementInData_ms: int64;
  Filter_ElementInData_ms: int64;

function SlowMethod(highBound: Integer): TArray<Integer>;
var
  i: Integer;
  temp: TList<Integer>;
begin
  Timing_SlowMethod.Start;
  temp := TList<Integer>.Create;
  try

    Timing_ElementInData.Reset;
    for i := 2 to highBound do
      if not ElementInDataDivides(temp, i) then
        temp.Add(i);
```

```
        Generate_ElementInData_ms := Generate_ElementInData_ms +
          Timing_ElementInData.ElapsedMilliseconds;

      Timing_ElementInData.Reset;
      Result := Filter(temp);
      Filter_ElementInData_ms := Filter_ElementInData_ms +
        Timing_ElementInData.ElapsedMilliseconds;

    finally
      FreeAndNil(temp);
    end;
    Timing_SlowMethod.Stop;
  end;
```

The code (which can be found in the SlowCode_Stopwatch2 program) now resets the Timing_ElementInData stopwatch before the data generation phase and adds the value of the stopwatch to Generate_ElementInData_ms afterward. Then it resets the stopwatch again for the Filter phase and adds the value of the stopwatch to Filter_ElementInData_ms afterward.

In the end, that will give us the cumulative execution time for ElementInDataDivides called directly from SlowMethod in Generate_ElementInData_ms and the cumulative execution time for ElementInDataDivides called from Filter in Filter_ElementInData_ms.

A test run with an upper bound of 100,000 produced the following output:

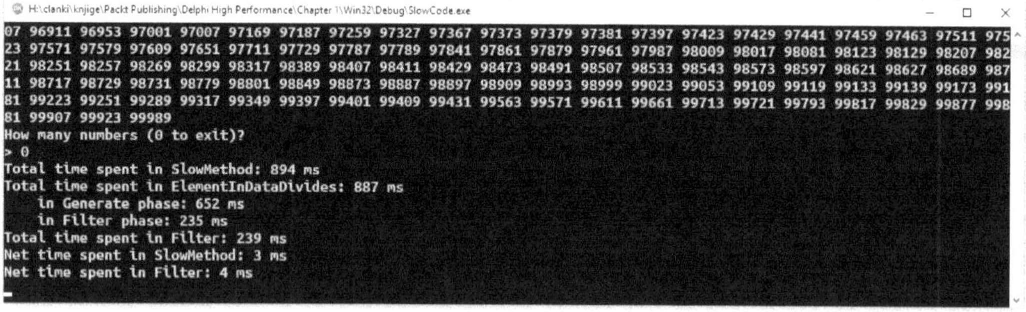

Figure 2.3 – Detailed measurements of the SlowCode execution time

Now we can be really sure that almost all the time is spent in ElementInDataDivides. We also know that approximately 75% of the time, it is called directly from SlowMethod and 25% of the time from the Filter method.

We are now ready to optimize the program. We can either improve the implementation to make it faster, replace it with a faster algorithm, or both.

Profilers

Measuring the speed of a program by inserting special code into the program by hand is perfect if you want to measure a very specific part of the code, but it becomes cumbersome if you don't know exactly which part of the program you should focus on. In such cases, it is best to use specialized software: profilers.

Profilers can measure all kinds of parameters. Sure, they are mostly used for measuring execution speed, at a method or even on a line level, but they can do much more. They can display the *call tree*—a graph showing how methods call one another. They can show memory usage in a program so you can quickly find the one method constantly eating the memory. They can show you the *coverage* of your tests so you can see which code was tested and which was not. And much more.

They do that magic in two ways: *sampling* or *instrumentation*.

The sampling profiler looks at the state of your program at regular intervals (for example, 100 times per second) and, each time, checks which line of the code is currently being executed. Statistically, it will predominantly *see* lines of code that are executed most of the time.

Sampling profiling will give us only a rough overview of behavior inside the program, but it will do that without affecting the program speed, and because of that, it is excellent for taking a first look at some code.

Instrumenting profilers do their magic by changing—*instrumenting*—the code. They, in fact, do almost exactly the same kind of changing the code as we did by inserting the stopwatch calls.

There are two ways to perform instrumentation. A profiler can change the source code, or it can change the program binary. Source instrumenting profilers are rare because they are less safe to use (there's always the possibility that a profiler will mess up the source) and because you have to recompile the code after it is instrumented (modified by the profiler).

Most of the instrumenting profilers on the market modify the binary code, which is a bit trickier to implement but doesn't require recompilation and cannot destroy the source code.

The advantage of instrumentation over sampling is that the former can collect everything that is going on inside the program and not just a few samples here and there. The disadvantage is that instrumentation reduces the speed of the program. Although instrumenting profilers take extra care to optimize the code that is inserted into your program, executing this code in many instrumented methods can still take a long time.

The other problem is the amount of collected data. A short and fast method called 100,000 times in 1 second will generate 100,000 data samples in an instrumenting profiler and only 100 samples in a sampling profiler (provided that it samples the code 100 times a second).

Because of all that, instrumenting profilers are best used once we already know which part(s) of the program we want to focus on.

I'll end this chapter with an overview of four profilers, two are open source and free (AsmProfiler and Sampling Profiler) and two are commercial (AQTime and Nexus Quality Suite (NQS)), to make it easier for you to choose the best fit for your situation. These four profilers are, of course, not your only options. If you need very precise profiling of short methods, check out ProDelphi (`https://www.prodelphi.de`). And if you need a high-end tool that will not only profile your application but help optimize it to its fullest potential, take a look at Intel's VTune Amplifier (`https://www.intel.com/content/www/us/en/developer/tools/oneapi/vtune-profiler.html`).

AsmProfiler

AsmProfiler is a 32-bit instrumenting and sampling profiler written by André Mussche. Its source, along with the Windows exe, can be found at `https://github.com/andremussche/asmprofiler`. Although the latest version was released in 2019, it works well with the newest Delphi at the time of writing this book, 11.3 Alexandria.

The sampling and instrumenting profilers are used in different ways, so I'll cover them separately.

To use AsmProfiler, you first have to unpack the release ZIP into a folder on your disk. That will create two subfolders—`Sampling` and `Instrumenting`. To start the sampling profiler, start `AsmProfiling_Sampling` from the `Sampling` folder and then click **Start profiling**.

AsmProfiler has two ways of starting a profiling session. You can start the program manually by clicking **Select process** in AsmProfiler and selecting your program from the list of running processes. Alternatively, you can click **Select exe**, browse to the compiled EXE of your program, and then click **Start process now**, which will start the program, or click **Start sampling**, which will start the program and also start sampling the code.

The reasoning behind the two different ways of starting the program is that you mostly want to profile a specific part of the program, so you would load it into the profiler, navigate to the specific part, click **Start sampling**, do the required steps in your program, click **Stop sampling**, and analyze the results. Sometimes, however, you would want to profile the very startup of the program, so you want sampling to start immediately after your program is launched. The following screenshot shows the initial AsmProfiler screen:

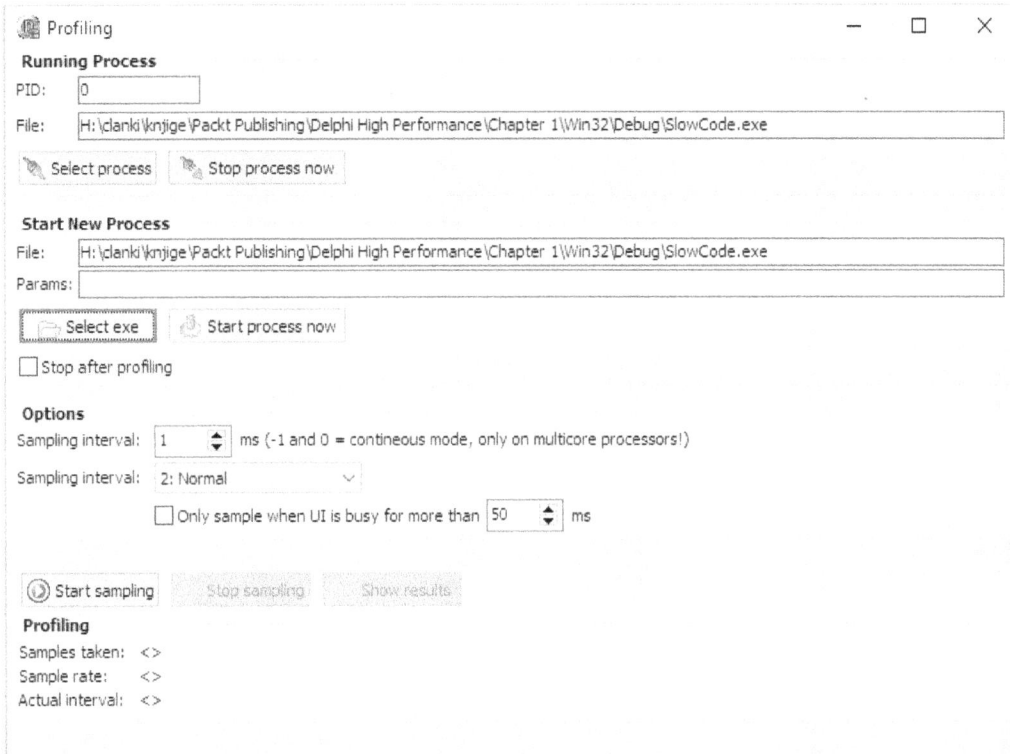

Figure 2.4 – Starting the profiling session in AsmProfiler

Sampling options can be configured in the **Options** group. You can set the **Sampling interval** value—how many milliseconds will pass between two samples. Setting this value to **4**, for example, will generate 250 (1000/4) samples per second.

Setting the sampling interval to **0** enables a continuous sampling mode, which still leaves some time for threads running on the same CPU. (The sampling code calls `Sleep(0)` between taking two samples.) Setting it to **-1** causes the samples to be taken as fast as possible. In effect, AsmProfiler will use one CPU core for itself.

You can also set the priority of sample-taking threads from the default, **Normal**. I will discuss threads and priorities in *Chapter 7, Getting Started with the Parallel World*.

After you have taken the samples and clicked **Stop sampling**, click the **Show results** button to analyze the results.

If you want to compare your results with mine, here are the steps to produce the results (as shown in the following screenshot):

1. Start AsmProfiler.

2. Click **Select exe** and select **SlowCode.exe**.

3. Click **Start process now**.

4. Click **Start sampling**.

5. Enter 100,000 into the program.

6. Click **Stop sampling**.

7. Enter 0 into the program to exit.

8. Click **Show results**.

9. Click **Results** on the **Sampling Results** form:

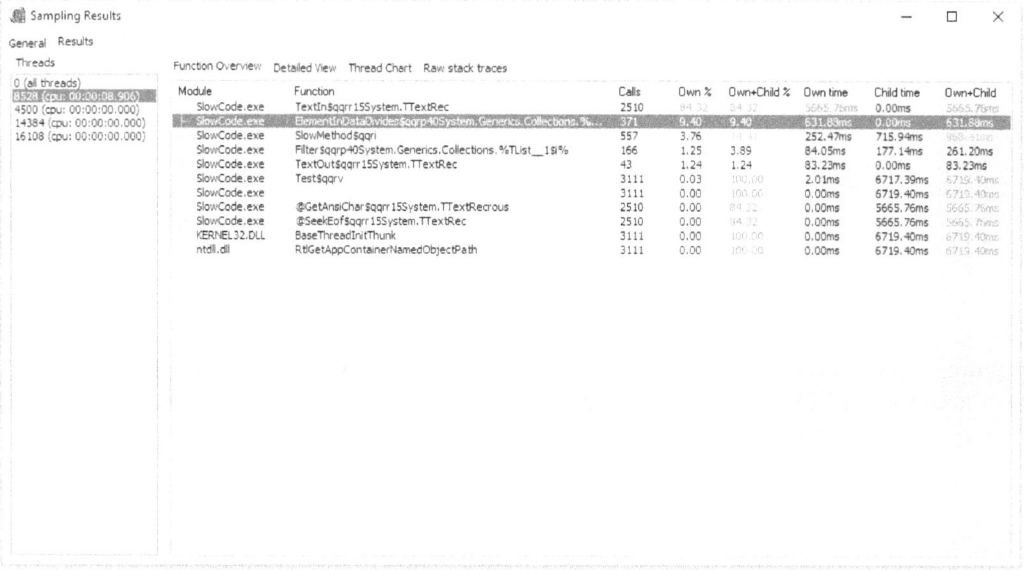

Figure 2.5 – The result of a test run profiled with AsmProfiler's Sampling profiler

AsmProfiler displays results organized by threads, but it automatically highlights the main thread, so you don't have to care about that detail if your program doesn't use multithreading.

In the result grid, it shows the module (main EXE or DLL), the name of the function, how many times the code was found to be in this function, and how many times (absolute and percentage) the code spent in that function (**Own time**), in all functions called from it (**Child time**), and both in that function and in all functions called from it (**Own+Child**).

If we sort results by the time spent only in the function (**Own time**), we'll see that the `TextIn$qqrr15System.TTextRec` function comes to the top. This a function that reads console input (in the `Readln` statement) and we can safely ignore it.

The next one on the list, `ElementInDataDivides$qqrp40...`, is the one that interests us.

We can see that it was sampled 371 times (*calls*) and that it needed 0.6 seconds to execute. If you switch to the **Detailed View** tab and select this function, you'll see in the **Parent calls (called by ...)** panel that it was called **258** times from `SlowMethod` and **113** times from the `Filter` method. In reality, of course, it was called many more times, but most of them were not *seen* by the sampling profiler.

In this view, we can also see how many times each line of a method was hit, which will give us a good idea about where the program spends most of the time. Unfortunately, we cannot sort the data on different criteria:

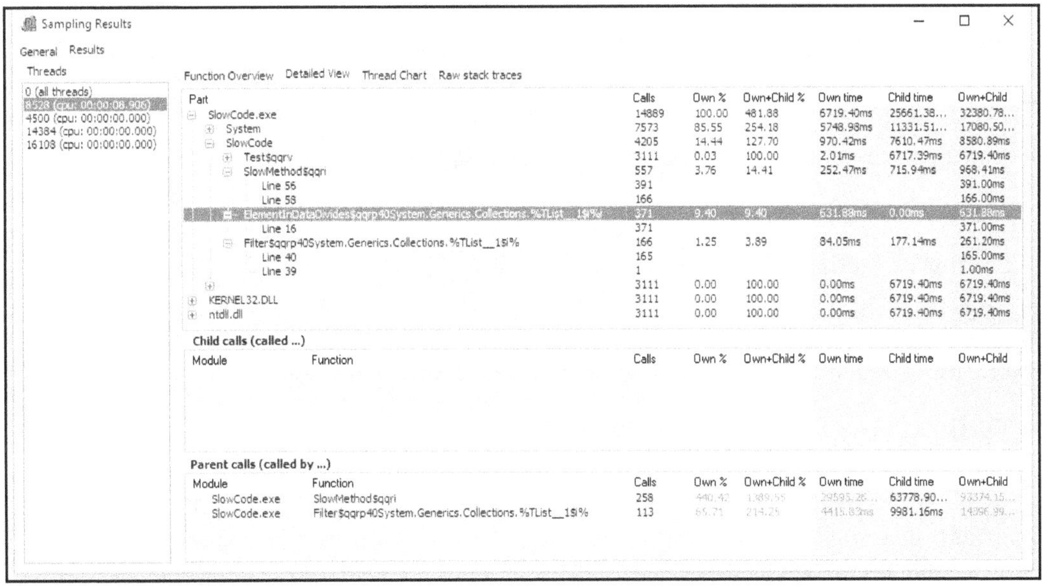

Figure 2.6 – A detailed view showing information for each line of the program

The names of the methods in these outputs are very weird, but that is how the Delphi compiler calls these methods internally. The part after the $ character encodes the parameter types. This process is called **name mangling** (and the part after $ is sometimes referred to as a **decoration**), and it enables us to use overloaded methods with the same name and different parameters—internally, they all have different names.

For example, the `SlowMethod(highBound: Integer)` function is internally known as `SlowMethod$qqri`. The `qqr` part specifies the *fastcall* calling convention (it describes how parameters are passed to the function in registers and on the stack), and `i` identifies one `integer` parameter.

AsmProfiler's instrumenting profiler requires a bit more work. First, you have to copy `AsmProfiler.dll` from the `Instrumenting` subfolder into a folder on the Windows environment path or into the exe folder. Second, you have to copy `Instrumenting\API\_uAsmProfDllLoader.pas` into Delphi's library path, into your project's folder, or add this folder to your project's search path.

Third, you have to add the `_uAsmProfDllLoader` unit to your project and call the following code. This will show the profiler's main form on the screen:

```
if _uAsmProfDllLoader.LoadProfilerDll then
  _uAsmProfDllLoader.ShowProfileForm;
```

To start profiling, call `_uAsmProfDllLoader.StartProfiler(False)`. To stop collecting data, call `_uAsmProfDllLoader.StopProfiler`. These two calls are not strictly necessary. You can also start and stop profiling from the profiler's user interface. Modifying the code will, however, give you more control over the profiling process.

Before your program exits, unload the profiler DLL with `_uAsmProfDllLoader.UnLoadProfilerDll`.

Make sure that your program has the following compiler options correctly set:

- **Linker, Map file = detailed**
- **Compiler, Optimization = off**
- **Compiler, Stack frames = on**

If you are using Delphi 11, you will also have to disable following linker options:

- Data Execution Prevention compatible = off
- Support address space layout randomization (ASLR) = off

The instrumenting profiler requires your program to process messages, making it mostly useless when used in Mr. Smith's program, as it spends most of the time inside a `Readln` call (and is not processing messages). As I still wanted to show you how this profiler works, I have converted `SlowCode` into a more modern VCL version, `SlowCode_VCL`.

At first, I wanted to start/stop profiling right in `SlowMethod`:

```
function TfrmSlowCode.SlowMethod(highBound: Integer): TArray<Integer>;
var
  i: Integer;
  temp: TList<Integer>;
begin
  _uAsmProfDllLoader.StartProfiler(False);
```

```
  // existing code

  _uAsmProfDllLoader.StopProfiler;
end;
```

However, that attempt misfired, as AsmProfiler didn't want to show profiling results for SlowCode. It turned out to be better to move the start/stop calls out of this method and into the method that calls SlowCode:

```
procedure TfrmSlowCode.btnTestClick(Sender: TObject);
var
  data: TArray<Integer>;
begin
  outResults.Text := '';
  outResults.Update;

  _uAsmProfDllLoader.StartProfiler(False);
  data := SlowMethod(inpHowMany.Value);
  _uAsmProfDllLoader.StopProfiler;

  ShowElements(data);
end;
```

A version of the program, ready for profiling with AsmProfiler, is stored in the SlowCode_VCL_ Instrumented project. You will still have to download AsmProfiler and store AsmProfiler. dll and _uAsmProfDllLoader.pas in appropriate places.

When you start the program, a small form will appear alongside the program's main form. From here, you can start and stop profiling, select items (methods) that should be profiled (**Select items**), and open the results of the profiling session (**Show results**):

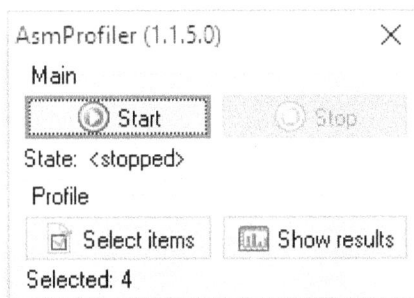

Figure 2.7 – AsmProfiler's instrumenting profiler

We are interested only in three methods, so click the **Select items** button, select the `ElementInDataDivides`, `Filter`, and `SlowMethod` methods, and click **OK**:

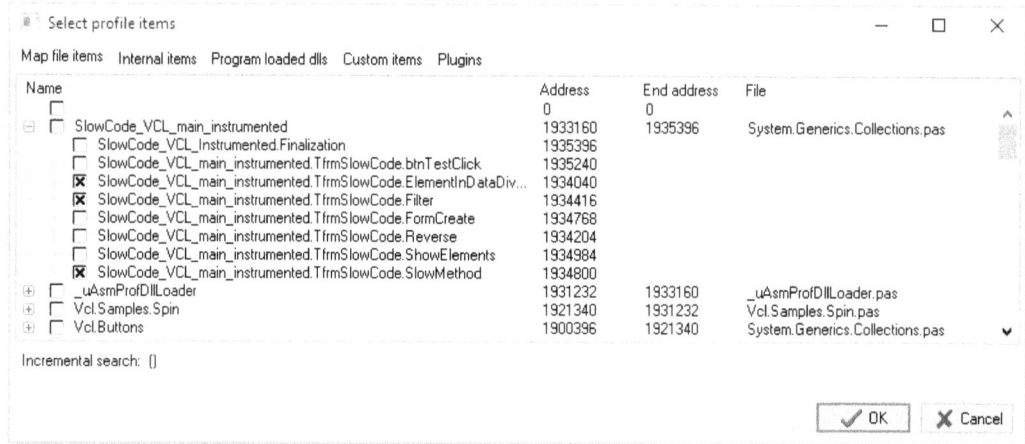

Figure 2.8 – Selecting the methods to be profiled

Next, enter `100000` into the **How many numbers** field and click the **Test** button. You don't have to start and stop the profiler, as the program will do that. When the values are calculated and displayed on the screen, click the **Show results** button. Don't close the profiled program, as that would close the profiler form, too.

The results form of the instrumenting profiler is very similar to the equivalent form of the sampling profiler. The most interesting feature is the **Unit overview** tab, which combines detailed timing information and a call tree:

Figure 2.9 – The Unit overview display

We can see that ElementInDataDivides is, in fact, called **99,999** times directly from SlowMethod and only **9592** times from the Filter method, not **258** and **113** times, as shown by the sampling profiler.

AsmProfiler gives a good combination of a global overview and detailed analysis, although it is rough around the edges and requires more effort on your part than more polished commercial profilers.

Sampling Profiler

Sampling Profiler is, as its name suggests, a sampling profiler for Delphi, written by Eric Grange. You can find it at https://www.delphitools.info. After years of inactivity, it was updated in 2023 and now officially supports all modern Delphis. It supports both 32- and 64-bit applications.

The strongest part of Sampling Profiler is its ability to be configured for multithreaded sampling. You can specify which CPUs will execute the profiler and which CPUs will run the profiled application. You can also focus on a specific thread by issuing a OutputDebugString('SAMPLING THREAD threadID') command from your code (replace threadID with the real ID of the thread you want to profile). It is also very simple to turn profiling on or off by calling OutputDebugString('SAMPLING ON') and OutputDebugString('SAMPLING OFF').

An interesting feature of Sampling Profiler, which other profilers don't provide, is the ability to enable a web server in the profiler. After that, we can use a browser to connect to the profiler (if firewalls allow us, of course), and we get an instant insight into the currently most executed lines of our program:

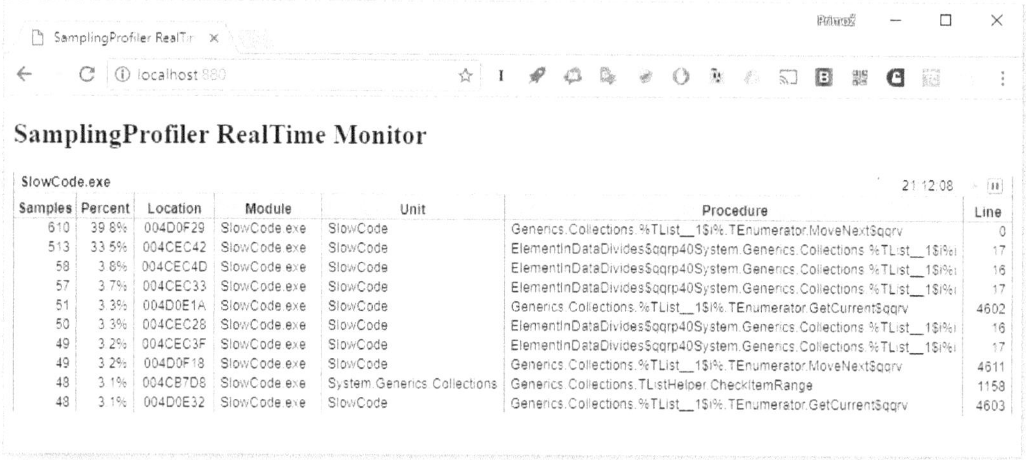

Samples	Percent	Location	Module	Unit	Procedure	Line
610	39.8%	004D0F29	SlowCode.exe	SlowCode	Generics.Collections.%TList__1$i%.TEnumerator.MoveNext$qqrv	0
513	33.5%	004CEC42	SlowCode.exe	SlowCode	ElementInDataDivides$qqrp40System.Generics.Collections.%TList__1$i%i	17
58	3.8%	004CEC4D	SlowCode.exe	SlowCode	ElementInDataDivides$qqrp40System.Generics.Collections.%TList__1$i%i	16
57	3.7%	004CEC33	SlowCode.exe	SlowCode	ElementInDataDivides$qqrp40System.Generics.Collections.%TList__1$i%i	17
51	3.3%	004D0E1A	SlowCode.exe	SlowCode	Generics.Collections.%TList__1$i%.TEnumerator.GetCurrent$qqrv	4602
50	3.3%	004CEC28	SlowCode.exe	SlowCode	ElementInDataDivides$qqrp40System.Generics.Collections.%TList__1$i%i	16
49	3.2%	004CEC3F	SlowCode.exe	SlowCode	ElementInDataDivides$qqrp40System.Generics.Collections.%TList__1$i%i	17
49	3.2%	004D0F18	SlowCode.exe	SlowCode	Generics.Collections.%TList__1$i%.TEnumerator.MoveNext$qqrv	4611
48	3.1%	004CB7D8	SlowCode.exe	System.Generics.Collections	Generics.Collections.TListHelper.CheckItemRange	1158
48	3.1%	004D0E32	SlowCode.exe	SlowCode	Generics.Collections.%TList__1$i%.TEnumerator.GetCurrent$qqrv	4603

Figure 2.10 – A live status view from a remote location

The weakest point of Sampling Profiler is its complete inability to select methods that are of interest to us. As we can see in the following screenshot, we get some methods from `System.Generics.Collections` mixed between methods from `SlowCode`. This only distracts us from our task—trying to find the slow parts of `SlowCode`.

Saying all that, I must admit that the display of profiling results is really neatly implemented. The results view is simple, clean, and easy to use:

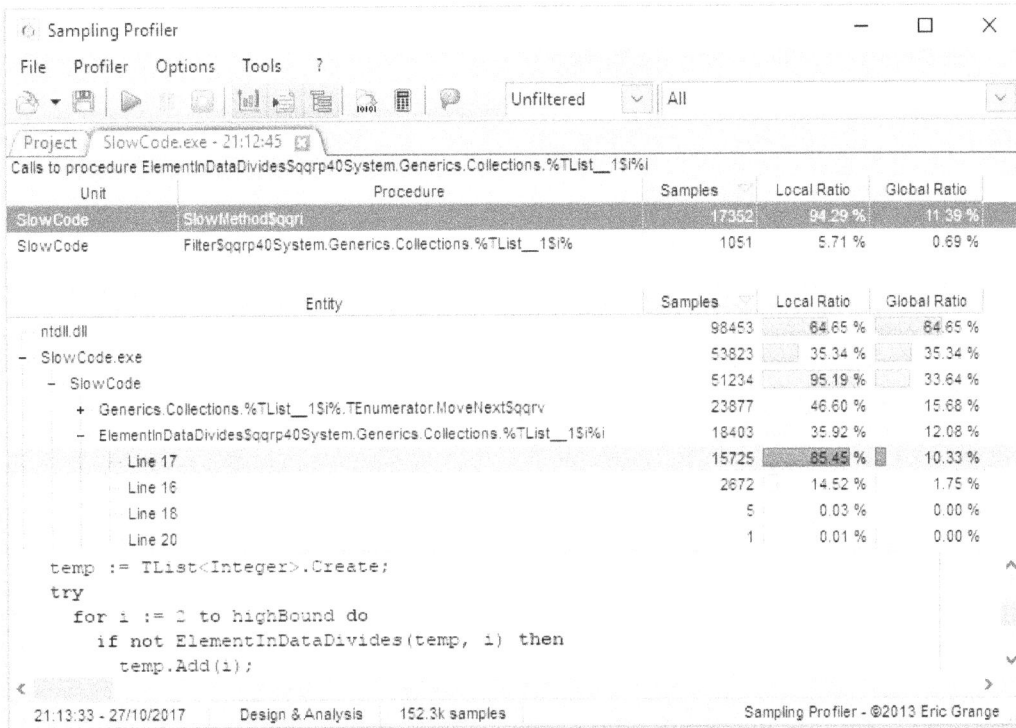

Figure 2.11 – The simple and effective result view

Sampling Profiler would be a perfect solution for occasional profiling if it would only allow us to select topics of interest.

AQTime

AQTime is a performance and memory profiler for C/C++, Delphi, .NET, Java, and Silverlight, produced by SmartBear Software. It supports 32- and 64-bit applications and can be found at https://www. smartbear.com.

Previously, a special Standard version of AQTime was included with RAD Studio, C++Builder, and Delphi. This offer was only available for releases XE to XE8 and the licensing was not renewed after that. If you want to use AQTime with any other Delphi release, you have to buy AQTime Professional.

For testing purposes, you can install a trial version of AQTime Professional, which will only run for six days. Dedicate some time to testing and use it wisely!

AQTime Professional supports all Delphi versions from 2006 to 10.4 Sydney, and you can even use it in Visual Studio, which is a great plus for multiplatform developers. It contains a variety of profilers—from the *performance profiler* (binary instrumenting profiler) and the *sampling profiler* to the *coverage profiler* (to see which parts of the program were executed) and more specific tools such as *BDE SQL profiler, static analysis* (a code analysis tool, which is not really a profiler), and more.

It integrates nicely into the Delphi IDE, but you can also use it as a standalone application. That gives you more flexibility during the profiling and result analysis, and that's why I also used the standalone AQTime Professional for the examples. Version 8 (current at the time of writing) does not integrate with the Delphi 11 IDE, but you can still profile Delphi 11 programs with the standalone profiler.

To prepare your program for profiling, make sure that the following compiler options are set:

- **Compiler, Stack frames = on**
- **Compiler, Debug information = Debug information**
- **Compiler, Local symbols = true**
- **Linker, Debug information = true**

In order for AQTime to be able to find the source file for your project, you have to specify a search path. Go to **Options | Options**, then select **General | Search directory**, and add all the folders with your source files.

Next, you can choose to profile all units, but unless you are using a sampling profiler, this will slow down the execution in a typical program a lot. It is better to select just a few units or, as in our example, just a few methods.

The easiest way to do that is to create a new *profiling area*, and the easiest way to do that is to select one or more methods in the left tree (use *Shift* + click and *Ctrl* + click), then right-click and select **Add selected to | New profiling area**. After that, you can add additional methods to that profiling area by right-clicking and selecting **Add selected to | Existing profiling area** or simply with drag-and-drop.

When creating a new profiling area, you also have to choose whether to profile on a method or on a line level by checking or unchecking the **Collect info about lines** checkbox:

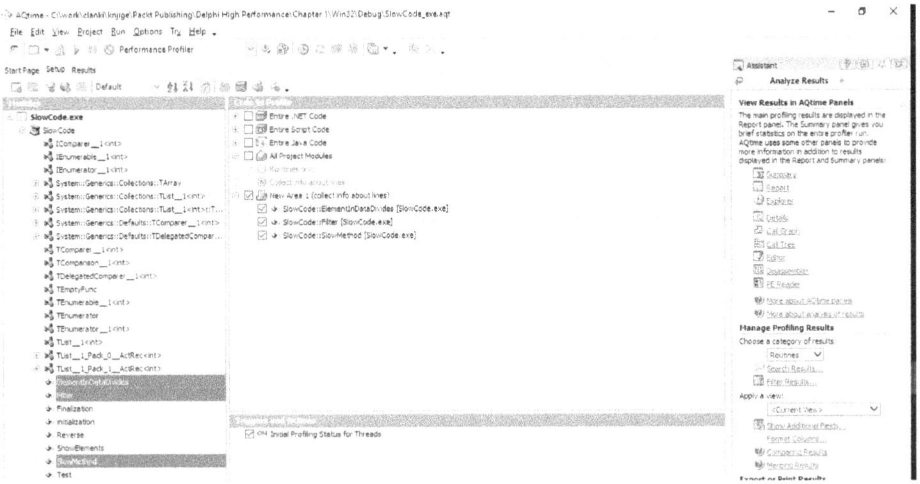

Figure 2.12 – Creating a new profiling area

Then start the program from AQTime—or select the **AQTime | Run with profiling** menu from Delphi, do the necessary steps you want to profile, and exit. AQTime will show the profiling results. Similar to all other profilers, it will show a grid with measured methods, the net time spent in each, time with children, and a **hit count** indicator—an indicator showing how many times the method executed.

More interesting info is hiding in the lower panel. There is a very detailed *call graph*, which displays a call tree for the selected method, and a very useful **Editor** panel, which shows the source together with the *hit count* information for each line:

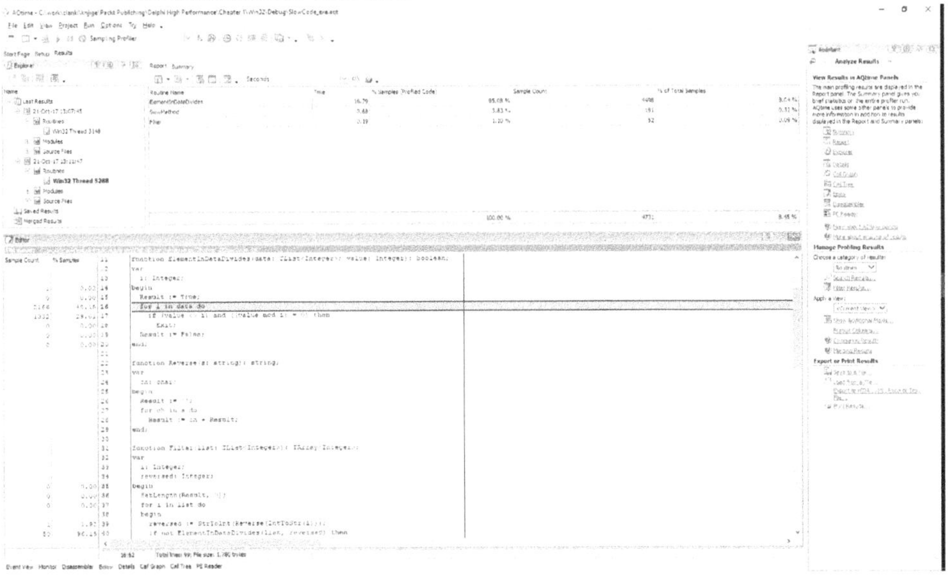

Figure 2.13 – The Editor view showing the hit count for instrumented methods

AQTime is a great tool, provided that you stay away from the very limited Standard edition and go directly for Professional.

The Nexus Quality Suite

NQS is a successor to the long-defunct TurboPower's SleuthQA, published by NexusQA Pty Ltd. It supports 32- and 64-bit applications written in Delphi from 5 to 11.3 Alexandria. You can find it at `www.nexusdb.com`.

The trial version has fewer limitations than AQTime's. Some functions are disabled, and some are limited in the quantity of collected data. Still, the program is not so limited that you wouldn't be able to test it out.

NQS integrates into Delphi's **Tools** menu and extends it with all the profilers it brings to Delphi. Of the most interest to us are **Method Timer**, an instrumenting profiler working at a method level, and **Line Timer**, an instrumenting profiler working at a line level. There is also **Block Timer**, an instrumenting profiler working on a block level (a `for` loop, for example), which was not working correctly at the time of writing this book and so I wasn't able to test it. That's really bad luck, as there are no other profilers for Delphi that work on a block level, and it would be really interesting to compare it with more standard approaches.

A few other tools are also interesting from the profiling viewpoint. **Coverage Analyst** will help you analyze code coverage, which is an important part of unit testing. After all, you definitely want to know whether your unit tests test all of the methods in a unit or not.

Also interesting is **CodeWatch**, which hunts for bugs in code by looking for memory and resource leaks.

All these profilers, with the exception of CodeWatch, are available in 32-bit and 64-bit versions, although the 64-bit operation is not as stable as in the 32-bit counterparts. I was only able to use Line Timer in 32-bit mode, for example, while Method Timer worked flawlessly in 32-bit and 64-bit modes.

Both Method Timer and Line Timer require no special preparation. You just have to have debug information turned on in the linker options.

When you start Method Timer, a profiler window opens. Click on the **Routines** button to select methods to profile. To change the profiling status of a method, double-click its name or right-click and select **Profile Status | Enable Profile Status For Selected**.

When you are done, press *F9* to start the program, go through the steps that you want to profile, and exit the program.

The program will then display basic timing information, including net time per method and *gross time* (what other programs call "time with children"). If you click on a method, the lower two panes will display information about the methods that called the current method and the methods that were called from the current method.

If you double-click on a method, another window will appear showing the source code for the selected method but without any information about the profiling results:

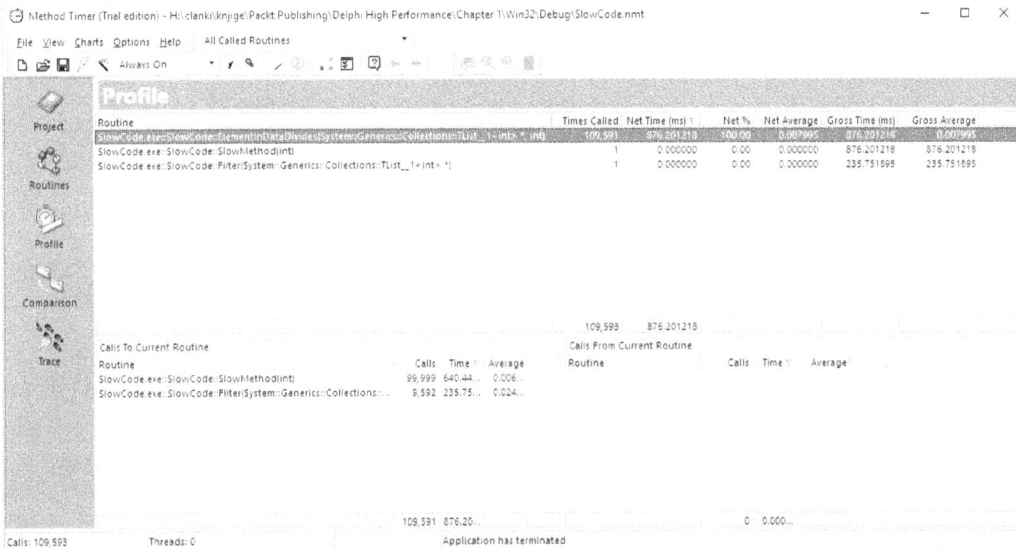

Figure 2.14 – The Method Timer results window

Line Timer has a similar user interface. First, you select the methods to be profiled in the **Routines** view, then you run the program, and at the end, examine the results in the **Line Times** window.

This profiler has a display that is a bit different from other profilers that support line-level profiling. It is not grouped by methods but by line numbers. This gives us an immediate overview of the most critical part of the code but is hard to integrate into a bigger picture.

As in Method Timer, a double-click on a line in the results grid opens up an editor window that displays the source code, together with the time spent in each profiled line and the number of times this line was executed:

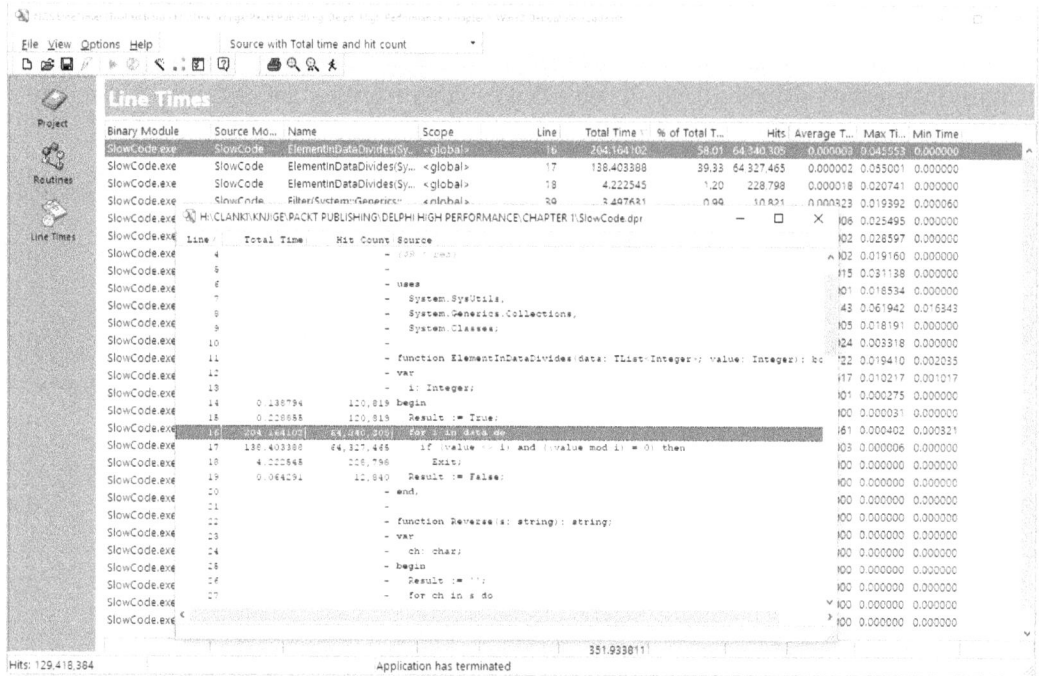

Figure 2.15 – Line Timer with built-in code display

NQS is a nice set of tools, and we can only hope that its stability improves with future releases.

Summary

In this chapter, we looked into the topic of profiling. We used a manual approach and specialized tools—profilers—to find the slowest part of a simple program. With this, we have confirmed the results of our analysis of the SlowCode program from the previous chapter.

In the next chapter, I'll briefly return to the topic of selecting the correct algorithm for the job. With a few examples, I'll show you how an algorithm can make or break a program's performance.

3
Fixing the Algorithm

In the previous chapter, we explored the concept of *performance* and looked at different scenarios where we would like to make a program faster. The previous chapter was largely theoretical, but now is the time to look at it in a more practical way.

There are two main approaches to speeding up a program, as follows:

- Replace the algorithm with a better one
- Fine-tune the code so that it runs faster

I spent lots of time in the previous chapter discussing *time complexity* simply to make it clear that a difference between two algorithms can result in an impressive speed-up. It can be much more than a simple constant factor (such as a 10-times speed-up). If we go from an algorithm with bad time complexity (say, $O(n^2)$) to an algorithm with better behavior ($O(n \log n)$, for example), then the difference in speed becomes more and more noticeable when we increase the size of the data.

Saying all that, it should not be surprising that I prefer the first approach (fixing the algorithm) to the second one (fine-tuning the code). To continue this point, this chapter will deal with several practical examples of speeding up the code by changing the algorithm.

Firstly, we will look at user interfaces. As we can't speed up the **Visual Component Library** (VCL) or Windows by fine-tuning the code (simply because we cannot modify that code), any speed-up in user interface responsiveness can only be a result of a better algorithm.

After that, I'll introduce the concept of *caching*. This approach can help you when everything else fails. Maybe it is not possible to change an algorithm, but introducing a cache of frequently calculated values can still improve the code speed drastically.

As implementing a fast cache is quite a tricky business, I'll also present a generic cache class that you can use freely in your own code.

Before concluding the chapter, we'll look at Mr. Smith's `SlowCode` example and try to speed it up by implementing a better algorithm.

We will cover the following topics in this chapter:

- Writing responsive user interfaces
- Updating VCL and **FireMonkey** (**FMX**) controls without waiting
- Speeding up functions by introducing caching
- Implementing a reusable generic caching class
- Analyzing and improving an unknown algorithm

Technical requirements

All code in this chapter was written with Delphi 11.3 Alexandria. I was not using the latest additions to the language so that most of the code could still be executed on Delphi XE and newer. You can find all the examples on GitHub at `https://github.com/PacktPublishing/Delphi-High-Performance---Second-Edition/tree/main/ch3`.

Writing responsive user interfaces

A user's first contact with any program is always the user interface. A good user interface can make or break a program. Leaving the user interface design aside (as I am not qualified to speak about that), I will focus on just one fact. Users hate user interfaces that are not *responsive*.

In other words, every good user interface must react quickly to a user's input, be that a keyboard, mouse, touchpad, or anything else.

What are the tasks that can make a user interface unresponsive? Basically, they all fall into one of two categories, as follows:

- A program is running a slow piece of code. While it is running, the user interface is not responding.
- Updating the user interface itself takes a long time.

The problems from the first category fall into two subsets—functions that have nonblocking (asynchronous) alternatives and functions that don't.

Sometimes we can replace the slow function with another one that runs *asynchronously*. For example, instead of using a standard function for reading from a file, we can use the Windows asynchronous file I/O API. File reading will then occur in parallel with our program and we will be notified of successful or unsuccessful reads via some other mechanism.

Often, however, this is not possible (there is no alternative for the slow function), or the asynchronous version is too complicated to implement. In that case, we can execute the slow code in a thread. This will be covered in *Chapters 7 to 10*.

An excellent candidate for the multithreaded approach is the file reading example. True—Windows offers an asynchronous API for that, but it is quite complicated and tricky to implement correctly. Pushing a read operation into a thread is simpler *and* cross-platform compatible if that is important for your application.

The second category (updating the user interface takes a long time) is typically solved with a simple solution: *Do less updating!* Delphi/Windows controls are typically very fast, and they only fail when we want them to do too many operations in a short time.

We will now focus on that category. I'll show a few examples of overabundant updates and give solutions for faster code.

Updating a progress bar

Almost any change to VCL controls can cause one or more Windows messages to be sent to the operating system. That takes time, especially as the program waits for the operating system to process the message and return an answer. This can happen even if nothing has changed in the user interface.

The next demonstration will show how an abundance of messages can slow down the execution speed. I must admit that the code in the `ProgressBar` demo is a bit contrived. Still, I assure you that I have seen similar code running in production.

This demo simulates reading a large file block by block. (The code doesn't really open and read the file; it just runs a loop that would do the reading.) For each block, a progress bar is updated.

For speed comparison, this progress bar update is done in two ways. In the first, slow approach, the `Max` property of the progress bar is set to the size of the file we are reading. After each block, the progress bar's `Position` property is set to the number of bytes read so far:

```
function TfrmProgressBar.Test0To2G: Integer;
var
  total: Integer;
  block: Integer;
  sw: TStopwatch;
begin
  sw := TStopwatch.StartNew;
  ProgressBar1.Max := CFileSize;
  ProgressBar1.Position := 0;

  total := 0;
  while total < CFileSize do begin
    block := CFileSize - total;
    if block > 1024 then
      block := 1024;
    // reading 'block' bytes
```

```
    Inc(total, block);

    ProgressBar1.Position := total;
    ProgressBar1.Update;
  end;
  Result := sw.ElapsedMilliseconds;
end;
```

This code runs slowly for two reasons. Firstly, setting the Position property sends a PBM_SETPOS message to Windows, and that is a relatively slow operation when compared with non-graphical program code. Secondly, when we call Update, the UpdateWindow Windows API function gets called. This function repaints the progress bar even if its position doesn't change, and that takes even more time. As all this is called 1,953,125 times, it adds up to a considerable overhead.

The second, faster approach sets Max to 100. After each block, the progress is calculated as a *percentage* with currPct := Round(total / CFileSize * 100).

The Position property is updated only if this percentage differs from the current progress bar's position.

As reading the Position property also sends one message to the system, the current position is stored in a local variable, lastPct, and a new value is compared to it:

```
function TfrmProgressBar.Test0To100: Integer;
var
  total: Integer;
  block: Integer;
  sw: TStopwatch;
  lastPct: Integer;
  currPct: Integer;
begin
  sw := TStopwatch.StartNew;
  ProgressBar1.Max := 100;
  ProgressBar1.Position := 0;
  lastPct := 0;

  total := 0;
  while total < CFileSize do begin
    block := CFileSize - total;
    if block > 1024 then
      block := 1024;
    // reading 'block' bytes
    Inc(total, block);

    currPct := Round(total / CFileSize * 100);
    if currPct > lastPct then
    begin
```

```
      lastPct := currPct;
      ProgressBar1.Position := currPct;
      ProgressBar1.Update;
    end;
  end;
  Result := sw.ElapsedMilliseconds;
end;
```

The file size is set to 2,000,000,000 bytes or 1.86 GB. That is a lot, but completely reasonable for a video file or database storage. The block is set to 1024 to amplify the problem. With a more reasonable size of 65,536 bytes, the difference is less noticeable.

As you can see, the second example contains more code and is a bit more complex. The result of those few additional lines is, however, more than impressive.

When you start the test program and click on a button, it will first run the slow code followed by the fast code. You will see a noticeable difference in display speed. This is also confirmed by the measurements:

Figure 3.1 – Progress bar demonstration program

If you ran the demo code, you would have probably noticed that the message showing the timing results is displayed *before* the progress bar finishes its crawl toward the full line. This is caused by a feature introduced in Windows Vista.

In Windows XP and before, updates to the progress bar were immediate. If you set `Max` to 100, `Position` to 0, and then a bit later updated `Position` to 50, the progress bar display would jump immediately to the middle. Since Vista, every change to the progress bar is animated. Changing `Position` from 0 to 50 results in an animation of the progress bar going from the left side to the middle of the control.

This makes for a nicer display but sometimes—as in our example—causes weird program behavior. As far as I know, there is only one way to work around the problem.

When I said that *every* change results in an animation, I lied a bit. In reality, the animation is only triggered if you *increase* the position. If you *decrease* it, the change is displayed immediately.

We can therefore use the following trick. Instead of setting the `Position` property to the desired value—ProgressBar1.Position := currPct—we do that in two steps. Firstly, we set it a bit too high (this causes the animation to start) and then we position it to the correct value (this causes an immediate update):

```
ProgressBar1.Position := currPct+1;
ProgressBar1.Position := currPct;
```

This leaves us with the problem of forcing the progress bar to display a full line when all processing is done. The simplest way I could find is to decrease the `Max` property so that it is lower than the current `Position` value. That also causes an immediate update to the progress bar:

```
ProgressBar1.Position := ProgressBar1.Max;
ProgressBar1.Max := ProgressBar1.Max - 1;
```

This technique of minimizing the number of changes also applies to multithreaded programs, especially when they want to update the user interface. I will return to this in *Chapter 9*, *Exploring Parallel Practices*.

Bulk updates

Another aspect of the same problem of overabundance of messages occurs when you want to add or modify multiple lines in `TListBox` or a `TMemo`. The demonstration program, `BeginUpdate`, demonstrates the problem and shows possible solutions.

Let's say we have a listbox and we want to populate it with lots of lines. The demo program displays 10,000 lines, which is enough to show the problem.

A naive program would solve the problem in two lines, like so:

```
for i := 1 to CNumLines do
  ListBox1.Items.Add('Line ' + IntToStr(i));
```

This, of course, works. It is also unexpectedly slow. 10,000 lines are really not much for modern computers, so we would expect this code to execute very quickly. In reality, it takes 3.3 seconds on my test machine!

We could do the same with a TMemo. The result, however, is a lot worse:

```
for i := 1 to CNumLines do
  Memo1.Lines.Add('Line ' + IntToStr(i));
```

With the memo, we can see lines appear one by one. The total execution time on my computer is a whopping 18 seconds!

This would make both the listbox and memo unusable for displaying a large amount of information. Luckily, this is something that the original VCL designers anticipated, so they provided a solution.

If you look into the code, you'll see that both `TListBox.Items` and `TMemo.Lines` are of the same type, `TStrings`. This class is used as a base class for `TStringList` and also for all graphical controls that display multiple lines of text.

The `TStrings` class also provides a solution. It implements the `BeginUpdate` and `EndUpdate` methods, which turn the visual updating of a control on and off. While we are in *update* mode (after `BeginUpdate` was called), the visual control is not updated. Only when we call `EndUpdate` will Windows redraw the control to the new state.

`BeginUpdate` and `EndUpdate` calls can be nested. Control will only be updated when every `BeginUpdate` method is paired with an `EndUpdate` method:

```
ListBox1.Items.Add('1'); // immediate redraw
ListBox1.BeginUpdate;     // disables updating visual
                          // control
ListBox1.Items.Add('2'); // display is not updated
ListBox1.BeginUpdate;     // does nothing, we are already in
                          //*update* mode
ListBox1.Items.Add('3;); // display is not updated
ListBox1.EndUpdate;       // does nothing, BeginUpdate was
                          // called twice
ListBox1.Items.Add('4'); // display is not updated
ListBox1.EndUpdate;       // exits update mode, changes to
                          // ListBox1 are
                          // displayed on the screen
```

Adding `BeginUpdate`/`EndUpdate` to existing code is very simple. We just have to wrap them around existing operations:

```
ListBox1.Items.BeginUpdate;
for i := 1 to CNumLines do
  ListBox1.Items.Add('Line ' + IntToStr(i));
ListBox1.Items.EndUpdate;

Memo1.Lines.BeginUpdate;
for i := 1 to CNumLines do
  Memo1.Lines.Add('Line ' + IntToStr(i));
Memo1.Lines.EndUpdate;
```

If you click on the second button in the demo program, you'll see that the program reacts much faster. Execution times on my computer were 285 ms for TListBox and 1,189 ms for TMemo. This second number seems suspicious. Why does TMemo need 1.2 seconds to add 10,000 lines if changes are not painted on the screen?

To find an answer to that we have to dig into the VCL code, in the TStrings.Add method:

```
function TStrings.Add(const S: string): Integer;
begin
  Result := GetCount;
  Insert(Result, S);
end;
```

Firstly, this method calls GetCount so that it can return the proper index of appended elements. Concrete implementation in TMemoStrings.GetCount sends two Windows messages even when the control is in the *updating* mode:

```
Result := SendMessage(Memo.Handle, EM_GETLINECOUNT, 0, 0);
if SendMessage(Memo.Handle, EM_LINELENGTH, SendMessage(Memo.Handle,
  EM_LINEINDEX, Result - 1, 0), 0) = 0 then Dec(Result);
```

After that, TMemoStrings.Insert sends three messages to update the current selection:

```
if Index >= 0 then
begin
  SelStart := SendMessage(Memo.Handle, EM_LINEINDEX,
    Index, 0);
  // some code skipped ... it is not executed in our case
  SendMessage(Memo.Handle, EM_SETSEL, SelStart, SelStart);
  SendTextMessage(Memo.Handle, EM_REPLACESEL, 0, Line); end;
```

All that causes five Windows messages to be sent for each appended line, and that slows the program down. Can we do this better? Sure!

To speed up TMemo, you have to collect all updates in some secondary storage—for example, in TStringList. At the end, just assign the new memo state to its Text property, and it will be updated in one massive operation.

The third button in the demo program does just that:

```
sl := TStringList.Create;
for i := 1 to CNumLines do
  sl.Add('Line ' + IntToStr(i));
Memo1.Text := sl.Text;
FreeAndNil(sl);
```

This change brings execution speed closer to the listbox. My computer needed only 1,178 ms to display 10,000 lines in a memo with this code.

An interesting comparison can be made by executing the same code in the `FireMonkey` framework. Graphical controls in `FireMonkey` are not based directly on Windows controls, so the effects of `BeginUpdate/EndUpdate` may be different.

The `BeginUpdateFMX` program in the code archive does just that. I will not go through the whole process again but just present the measurements. All times are in ms:

Framework	Update method	TListBox	TMemo
VCL	Direct update	3,272	17,694
	BeginUpdate	285	1,178
	Text	N/A	56
FireMonkey	Direct update	18,333	61
	BeginUpdate	71	28
	Text	N/A	27

Table 3.1 – Comparison between various TListBox and TMemo update methods

We can see that `BeginUpdate/EndUpdate` are also useful in the `FireMonkey` framework, especially with a `TListBox`.

> **Tip**
> If a class implements `BeginUpdate/EndUpdate`, use it when doing bulk updates. This will speed up the program.

In the next section, we'll see how we can replace a `TListBox` with an even faster solution if we are programming with the VCL framework.

Virtual display

Let's take a look at our problem again. We are adding 10,000 lines to a listbox. Simple job for a programmer, but what about our users? From their viewpoint, we are filling a scrolling control with 10,000 lines of data. 10,000! Will anyone scroll through them all and read them all? I don't think so!

Maybe we have a legitimate reason to display 10,000 lines. Still, we can be pretty sure that no one will ever look at all of them. If so, why should we even generate all that data? Wouldn't it be better if we only generated the lines that are actually visible on the screen? Yes, it would, and that is exactly the idea behind *virtual display controls*.

Virtual TListBox

We can easily convert standard VCL TListBox into a virtual control. We have to do the following:

- Set its Style property to lbVirtual
- Write an OnData event handler
- Optionally write OnDataFind and OnDataObject event handlers

Adding 10,000 lines to a listbox is then just a matter of setting its Count property to that value:

```
procedure TfrmVirtualListbox.Button1Click(Sender: TObject);
var
  stopwatch: TStopwatch;
begin
  stopwatch := TStopwatch.Create;
  ListBox1.Count := CNumLines;
  stopwatch.Stop;
  StatusBar1.SimpleText := Format('ListBox: %d ms',
    [stopwatch.ElapsedMilliseconds]);
end;
```

This is extremely fast. My test computer needs just a few ms to do it.

To actually display the data, VCL will call the OnData event handler for each line visible on the screen (and only for those lines). The input into this handler is a line number (starting with 0). Our input is the line itself, just as in the following code:

```
procedure TfrmVirtualListbox.ListBox1Data(
  Control: TWinControl; Index: Integer; var Data: string);
begin
  Data := 'Line ' + IntToStr(Index + 1);
end;
```

In this event handler, we can generate the data (as in this example), or we can use the Index parameter as a lookup key into some other structure.

If we want to find a specific string inside a virtual listbox by using the Items.IndexOf function, we have to write an OnDataFind event handler. This handler can again look for the data in some external structure, or it can—as in the next example—parse the line and determine its position from its contents:

```
function TfrmVirtualListbox.ListBox1DataFind(
  Control: TWinControl; FindString: string): Integer;
begin
  if Copy(FindString, 1, Length('Line ')) <> 'Line ' then
    Exit(-1);
```

```
    Delete(FindString, 1, Length('Line '));
    Result := StrToIntDef(FindString, 0) - 1;
  end;
```

The `OnDataObject` handler works the same as the `OnDataFind` handler. It is used to access objects associated with listbox items (the `Items.Objects[]` property).

Here's a recap:

- Each use of `Items[]` is translated into an `OnData` event handler
- Each use of `Items.IndexOf` is translated into an `OnDataFind` event handler
- Each use of `Items.Objects[]` is translated into an `OnDataObject` event handler

A virtual listbox is very practical but quite limited. If you need more functionality, you have to look around and find a better component. We'll look into one of them in the next section.

Virtual TreeView

A long time ago, Mike Lischke wrote a great Delphi component—Virtual TreeView. He stopped supporting it a long time ago, but the component found a new sponsor and a place on GitHub at `https://github.com/JAM-Software/Virtual-TreeView`.

Virtual TreeView supports the VCL framework in Delphi XE3 and later. There are also versions of code that support older versions of Delphi, but you'll have to look around for them. To use it, you have to download the source and recompile two included packages. That will add three new components to Delphi. This example will use the most useful of them, `TVirtualStringTree`.

> **Tip**
> Alternatively, you can install Virtual TreeView with Delphi's GetIt Package Manager if you are using Delphi XE8 or newer.

The `VirtualTree` demo compares `TListBox` with `TVirtualStringTree`, where the latter is used in different ways. Although `TVirtualStringTree` is very flexible and designed to display tree structures, we can also use it as a very fast listbox, as I will do in this example. To use it as a listbox, you should *remove* the `toShowRoot` and `toShowTreeLines` options from the component's `TreeOptions.PaintOptions` property.

> **Tip**
> Before you start using Virtual TreeView, go through the included demos. They show all the power of this component.

This demo compares two different modes of operation. One is adding lots of lines in one operation. We already know that we should use `BeginUpdate`/`EndUpdate` in this case.

The other mode is adding just one line to the list. As this is hard to measure precisely, the operation is repeated 100 times.

Virtual TreeView is different from other components included with Delphi. It operates on a *view/model* principle. The component itself just displays the data (presents a *view* of the data) but doesn't store it internally. Data itself is stored in a storage that we have to maintain (a *model*). Virtual TreeView only keeps a short reference that helps us access the data.

The first thing that we must do before we can use the component is to decide how large this *reference to data* is and set the `NodeDataSize` property accordingly.

Typically, we'll use an integer index (4 bytes), or an object or interface (*4 bytes on Win32* and *8 bytes on Win64*). We can also store larger quantities of data in this area, but that kind of defeats the view/model separation principle. In this example, I'll use a simple `integer` property so that `NodeDataSize` is set to 4.

The simplest way to use `TVirtualStringTree` is to use the `AddChild` method to add a new *node* (display line) and pass *user data* (a reference to a model) as a parameter:

```
VirtualStringTree1.BeginUpdate;
for i := 1 to 10000 do begin
  idx := FModel1.Add('Line ' + IntToStr(i));
  VirtualStringTree1.AddChild(nil, pointer(idx));
end;
VirtualStringTree1.EndUpdate;
```

The code uses global `FModel1: TStringList` for data storage. It firstly adds the data to the `FModel1.Add` model and sets the index of this data (`idx`) as user data for the newly created node (`AddChild`).

The first parameter to `AddChild` is a reference to the node's parent. As we are not displaying a tree structure but a list, we simply set it to `nil` (meaning that there is no parent). The second parameter represents user data. `AddChild` supports only `pointer` user data, so we have to cast our parameter accordingly.

We also have to take care of retrieving data from the model so that it can be displayed on the screen. For that, we have to create the `OnGetText` event handler. This event is called once for each column of each *visible* line. It is not called for lines that are not visible on the screen.

The code must first call `Node.GetData` to get the *user data* associated with the node. To be more precise, `GetData` returns a *pointer* to user data. As we know that our user data is just an integer, we can cast this pointer to `PInteger` to access the value (index into the `FModel1` string list):

```
procedure TfrmVTV.VirtualStringTree1GetText(
  Sender: TBaseVirtualTree; Node: PVirtualNode;
```

```
    Column: TColumnIndex; TextType: TVSTTextType;
    var CellText: string);
  begin
    CellText := FModel1[PInteger(Node.GetData)^];
  end;
```

If you run the demonstration program and click the **Add 10,000 lines** button, you'll see that the listbox needs more time for the operation (68 ms, in my case) than Virtual TreeView (16 ms). Quite nice!

Clicking the second button, **Add 1 line 100 times**, shows something completely different. In this case, the listbox is a lot faster (17 ms) than the Virtual TreeView (184 ms).

I must admit that I didn't expect that, so I did some digging around. As it turns out, TVirtualStringTree *sorts* its data on each call to AddChild (unless we called BeginUpdate before that!) This may be useful if you are using a multi-column view where data can be sorted on the selected column, but in our case, it only destroys the performance.

The fix for that is very simple. Just remove the toAutoSort option from the component's TreeOptions.AutoOptions property.

I cleared this flag in the second TVirtualStringTree instance, and the result is obvious. Adding 100 lines now takes 17 ms instead of 184 ms and is on par with the listbox.

The third TVirtualStringTree instance pushes the whole virtual aspect to the max. Instead of initializing nodes in code (by calling AddChild), we just tell the component how many nodes it must display by setting the RootNodeCount property, and the component will call the OnInitNode event handler for each node that is currently *visible on the screen*. In this mode, not only the painting but the initialization is executed on demand!

```
for i := 1 to 10000 do
  FModel3.Add('Line ' + IntToStr(i));
VirtualStringTree3.RootNodeCount :=
  VirtualStringTree3.RootNodeCount + 10000;
```

This approach is only feasible if we can somehow determine the part of the model that is associated with the given node in the OnInitNode handler. The only information we have at that moment is the node's index in the list (the Node.Index property). The first node gets index 0, the second gets index 1, and so on. Luckily, that is exactly the same as the index into TStringList, so we can just use the SetData method to set the user data:

```
procedure TfrmVTV.VirtualStringTree3InitNode(
  Sender: TBaseVirtualTree; ParentNode, Node: PVirtualNode;
  var InitialStates: TVirtualNodeInitStates);
begin
  Node.SetData(pointer(Node.Index));
end;
```

> **Note**
>
> In addition to `SetData` and `GetData`, there are also generic versions of those functions, `GetData<T>` and `SetData<T>`, which are of tremendous use if we want to use user data as storage for an object or an interface. The code implements all necessary reference counting so that interfaces are correctly managed.

This on-demand initialization approach speeds the program up even more. The addition of 10,000 lines now takes only 3 ms. The speed when adding lines one by one is not affected, though:

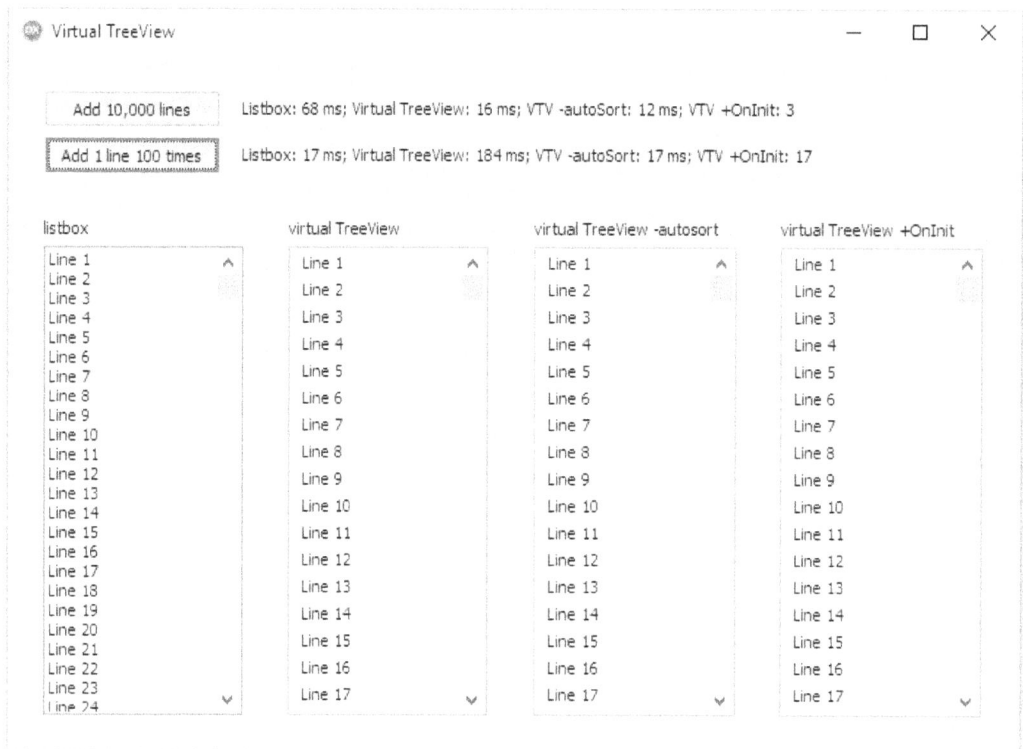

Figure 3.2 – VirtualTree demo showing different approaches to managing the TVirtualStringTree component

This concludes my foray into user interface land. The second part of this chapter will deal with an aspect of algorithmic improvement that we usually ignore—caching. And at the very end, I will return to Mr. Smith's code from *Chapter 1, About Performance*, and make it run much faster.

Caching

Our good friend Mr. Smith has improved his programming skills considerably. Currently, he is learning about recursive functions, and he programmed his first recursive piece of code. He wrote a simple seven-liner that calculates the nth element in the *Fibonacci* sequence:

```
function TfrmFibonacci.FibonacciRecursive(element: int64):
  int64;
begin
  if element < 3 then
    Result := 1
  else
    Result := FibonacciRecursive(element - 1) +
              FibonacciRecursive(element - 2);
end;
```

I will not argue with him—if you look up the definition of a Fibonacci sequence it really looks like it could be perfectly solved with a recursive function.

A sequence of Fibonacci numbers, *F*, is defined with two simple rules, as follows:

- The first two numbers in the sequence are both 1 ($F_1 = 1$, $F_2 = 1$)
- Every other number in the sequence is the sum of the preceding two ($F_n = F_{n-1} + F_{n-2}$)

You will also find a different definition of the Fibonacci sequence in the literature, starting with values 0 and 1, but it only differs from our definition in the initial zero. Our definition will produce the sequence 1, 1, 2, 3, 5, 8 ... while the second definition will produce 0, 1, 1, 2, 3, 5, 8.

As I said, a naive approach to writing a Fibonacci function is to write a recursive function, as Mr. Smith did. This way, however, leads to an exceedingly slow program.

Try it yourself. A program, *Fibonacci*, from the code archive for this book implements Mr. Smith's functions in the `FibonacciRecursive` method. Enter a number up to 3 0 in the **Element number** edit field, click **Recursive**, and you'll get your answer in a few ms. Increase this value to 4 0, and you'll have to wait about a second!

The numbers show that if you increase the element number just by one, the calculation time goes up by about 50%. That accumulates quickly. As we can see here, my computer needs more than 13 seconds to calculate element number 46, and 96 seconds—a minute and a half—for element number 50. What is going on?

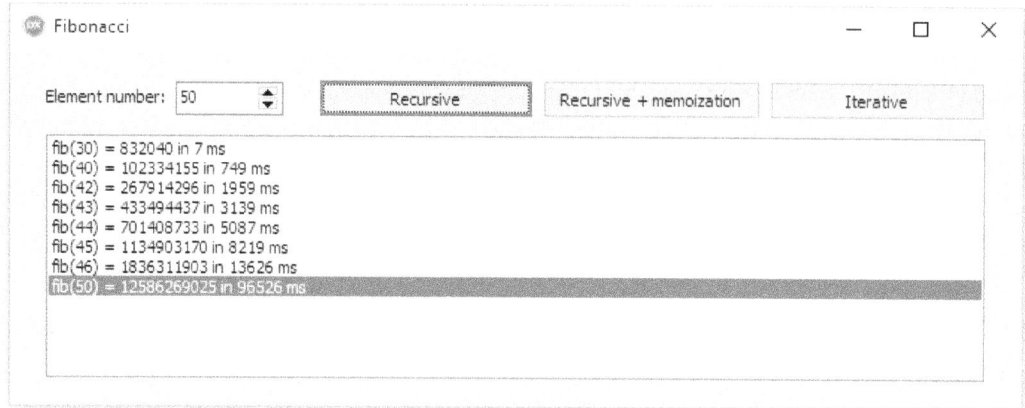

Figure 3.3 – Calculating Fibonacci numbers the recursive way

The problem with the naive implementation is that it has $O(2^n)$ complexity. When you calculate a value of the function, the first call to `FibonacciRecursive` causes two new calls. These cause two more calls each ($2*2 = 4$), and these four cause two more calls each ($2*4 = 8$), and so on. We have, therefore, 2^0 (= 1) calls initially, $2^1 = 2$ on the first level of recursion, $2^2 = 4$ on the second level, $2^3 = 8$ on the third, and that goes on and on until we need to calculate elements 1 or 2.

A careful examination of these recursive calls shows us something interesting. Let's say we want to calculate the 10th element in the sequence, F_{10}. That causes recursive calls to calculate F_9 and F_8. F_9 causes recursive calls for F_8 and F_7, while F_8 causes recursive calls for F_7 and F_6. In the next level of recursion, we have to calculate F_7 and F_6, F_6 and F_5, F_6 and F_5 again, and F_5 and F_4. We see that we are calculating the same values over and over again, and that definitely doesn't help the execution speed.

A solution to this type of problem is *caching* (sometimes also called *memoization*). Simply put, whenever we calculate a result corresponding to some inputs, we put it into storage—a *cache*. When we have to calculate a value corresponding to the *same* inputs, we just pull it from the cache.

This is a very powerful technique that has its own collection of problems. We have to decide the size of the cache. Maybe we could cache every already calculated value, but they will often be too abundant to store. We also have to create and destroy this cache somewhere.

Memoization

When dealing with the Fibonacci sequence, the size of the cache is not really a problem. If we want the nth Fibonacci number, we will only need to store values of elements from 1 to n. We can therefore represent the cache with a simple array:

```
var
  FFibonacciTable: TArray<int64>;
```

```
function TfrmFibonacci.FibonacciMemoized(element: int64):
  int64;
var
  i: Integer;
begin
  SetLength(FFibonacciTable, element+1);
  for i := Low(FFibonacciTable) to High(FFibonacciTable) do
    FFibonacciTable[i] := -1;
  Result := FibonacciRecursiveMemoized(element);
end;
```

The main function, `FibonacciMemoized`, first creates a `FFibonacciTable` cache. As dynamic arrays start counting elements from 0, the code allocates one element more than needed, simply to be able to address this array with the `[element]` index instead of `[element-1]`. One little `int64` type more for better code clarity; not a bad compromise!

Secondly, the code initializes all cache elements to -1. As the elements of the Fibonacci sequence are always positive numbers, we can use -1 to represent an uninitialized cache slot.

At the end, the code calls the memoized version of the recursive function.

As dynamic arrays are managed by the compiler, we don't have to destroy the cache after the calculation.

The `FibonacciRecursiveMemoized` recursive method is very similar to Mr. Smith's code, except that it adds cache management. At the beginning, it will check the cache to see if the value for the current element was already calculated. If it was, the result is simply taken from the cache. Otherwise, the value is calculated recursively and the result is added to the cache:

```
function TfrmFibonacci.FibonacciRecursiveMemoized(
  element: int64): int64;
begin
  if FFibonacciTable[element] >= 0 then
    Result := FFibonacciTable[element]
  else
  begin
    if element < 3 then
      Result := 1
    else
      Result := FibonacciRecursiveMemoized(element - 1) +
                FibonacciRecursiveMemoized(element - 2);
    FFibonacciTable[element] := Result;
  end;
end;
```

In the *Fibonacci* program, you can use the **Recursive + memoization** button to test this version. You'll see that it will need very little time to calculate Fibonacci sequence elements up to number 92. On my computer, it always needs less than 1 ms.

Why 92, actually? Well, because the 93rd Fibonacci number exceeds the highest `int64` value and the result turns negative. Indeed, F_{92} is larger than 2^{63}, which is a really big number in itself!

So, is this memoized solution a good way of calculating Fibonacci numbers? No, it is not! A much better way is just to start with the first two values (1 and 1), then add them to get F_3 (1 + 1 = 2), add F_2 and F_3 to get F_4 (1 + 2 = 3), and so on. This *iterative* solution is much faster than the cached recursive solution, although our timing doesn't show that. The precision is simply not good enough, and both versions will show a running time of 0 ms:

```
function TfrmFibonacci.FibonacciIterative(element: int64):
  int64;
var
  a,b: int64;
begin
  a := 1;
  b := 0;
  repeat
    if element = 1 then
      Exit(a);
    b := b + a;
    if element = 2 then
      Exit(b);
    a := a + b;
    Dec(element, 2);
  until false;
end;
```

A static cache is usually all that you need to speed up the program. Sometimes, however, you will need to write a more complicated caching solution. In the next section, we'll look at a more flexible approach.

Dynamic cache

A static cache that can only grow has limited use. We could, for example, use it to store **runtime type information** (RTTI) properties of types, as access to the RTTI is relatively slow. In most cases, however, a limited-size cache that stores only N most recently used values is more appropriate.

Writing such a cache is quite a tricky operation. First and foremost, we need quick access to a value associated with a specific input (let's call that input a *key*). In *Chapter 1, About Performance*, we found out that the best way to do that is a hash table (or, in Delphi terms, `TDictionary`), which has $O(1)$ lookup time.

> **Note**
>
> As the old joke goes, there are only two hard problems in computer science—naming things, cache invalidation, and off-by-one errors.

So, that's solved—our cache will store data in a dictionary. But we also have to remove the values from the dictionary (as this cache has a limited size), and there lies the big problem!

When a cache is full (when it reaches some pre-set size) we would like to remove the *oldest* element from the cache. We could, for example, replace the value part of the dictionary with a pair (date/time, value) where *date/time* would contain the last modification time of this key, but finding the oldest key in the cache then becomes an $O(n)$ operation (in the worst case, we have to scan all (date/time, value) pairs), which is something we would like to avoid.

> **Keep in mind**
>
> Removing the *oldest* elements from the cache is an implementation detail, not a requirement. We could also remove the element that was not *accessed* for the longest time or that was least recently *updated*.

Another alternative would be a sorted list of all keys. When we need to update a key, we can find it with $O(\log n)$ steps and insert it at the beginning in $O(1)$ steps (but this is an expensive step as it needs to move other elements around in memory). To get the oldest element, we can just check the last element in the list (very fast $O(1)$). However, we can do even better than that.

Let's look again at the requirements. We need a data structure that satisfies the following criteria:

- Data is kept in *modification* order. In other words, when you insert an item into the structure, that item becomes the first element in the structure.

- When you update an item, it moves to the front (it becomes the first element).

Whenever you need a data structure with these conditions, the answer is always a *doubly linked list*. This is a list of items where each item contains some value, plus two pointers pointing to the previous and next elements. The first element doesn't have a predecessor and the last element doesn't have a successor. These two links point to a special value (for example, `nil`):

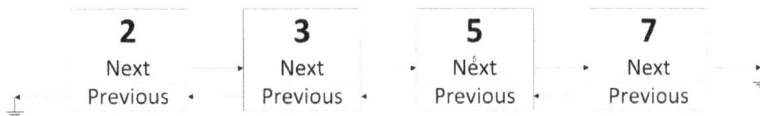

Figure 3.4 – A doubly linked list

Removing an element from such a list is very fast. You just have to change a few pointers. The same goes for insertion into such a list.

> **Tip**
>
> Keep in mind that *pointer* in this context doesn't necessarily mean Delphi `pointer`. One can also build a list in an array, and a *pointer* would then simply be an integer index into this array.

I have implemented a very fast cache, `TDHPCache<K,V>`, based on a hash table and doubly linked list. You can find it in the `DHPCache` unit, together with a simple test app, `CacheDemo`.

`TDHPCache` is a generic cache class with two type parameters—the K key and the V value. Its public interface implements only a few functions:

```
TDHPCache<K,V> = class
public
  constructor Create(ANumElements: Integer;
    AOwnsValues: boolean = false);
  destructor Destroy; override;
  function TryGetValue(const key: K; var value: V):
    boolean;
  procedure Update(const key: K; const value: V);
end;
```

The `Create` constructor creates a cache with the specified maximum size. The cache can optionally own values that you insert into the cache, which allows you to cache objects.

The `TryGetValue` function tries to retrieve a value associated with the specified key. Success or failure is indicated by the Boolean result.

The `Update` procedure stores the value in the cache and associates it with the specified key. It also makes sure that the key is now first in the list of the most recently used keys.

Internally, all keys and values are stored in a doubly linked list. As we know the maximum size of this list (`ANumElements` parameter passed to the constructor), we can create it as a simple array of elements of (internal) type `TListElement`:

```
strict private type
  TListElement = record
    Next : Integer;
    Prev : Integer;
    Key : K;
    Value: V;
  end;
var
  FCache : TDictionary<K,Integer>;
  FKeys : TArray<TListElement>;
  FFreeList : Integer;
  FHead : Integer;
  FTail : Integer;
```

This list is not using Delphi pointers to link elements. Instead, each `Next` and `Prev` field simply contains an index into the `FKeys` array or -1 if there is no next/previous item. (-1 in this case corresponds to a `nil` pointer.)

Two lists are actually stored in this array. One element is always in exactly one of those two lists, so parallel storage is not a problem.

The first list contains all unused items. The `FFreeList` field points to the first element in this list. When a `TDHPCache` class is created, the `BuildLinkedList` method (not shown here) adds all elements of `FKeys` to this list.

The second list contains all used (cached) items. The `FHead` field points to the first (most recently modified) element in this list, and the `FTail` field points to the last (oldest) element in this list. In the beginning, this list is empty, and both `FHead` and `FTail` contain -1. The cache also uses an `FCache` dictionary, which maps keys into array indexes.

To retrieve a value, `TryGetValue` calls `FCache.TryGetValue` to get the array index associated with the key. If that function returns `True`, the associated value is read from the `FKeys` array. Both operations execute in $O(1)$ time, which makes the whole function $O(1)$:

```
function TDHPCache<K, V>.TryGetValue(const key: K; var value: V):
boolean;
var
  element: Integer;
begin
  Result := FCache.TryGetValue(key, element);
  if Result then
    value := FKeys[element].Value;
end;
```

Updating a value is a bit more complicated. The function first tries to find the key in the cache. If it is found, the value in the `FKeys` array is updated. Both operations are $O(1)$.

The code looks a bit more complicated because it must handle the destruction of the old value when the cache owns its values:

```
procedure TDHPCache<K, V>.UpdateElement(element: Integer;
  const key: K; const value: V);
var
  oldValue: V;
begin
  if not FOwnsValues then
    FKeys[element].Value := value
  else
  begin
    oldValue := FKeys[element].Value;
    if PObject(@value)^ <> PObject(@oldValue)^ then
```

```
    begin
      FKeys[element].Value := value;
      PObject(@oldValue)^.DisposeOf;
    end;
  end;
  MoveToFront(element);
end;

procedure TDHPCache<K, V>.Update(const key: K; const value: V);
var
  element: Integer;
begin
  if FCache.TryGetValue(key, element) then
    UpdateElement(element, key, value)
  else
    AddElement(key, value);
end;
```

AddElement gets executed when a new (key, value) pair is added to the cache. First, it checks whether the list is full. This can be done simply by checking the FFreeList pointer—if it points to an array element, the list is not yet full. Then, the code either removes the oldest element from the list (discussed next) or allocates a new element from the free list. The latter is done by moving a few pointers around in *O(1)* time.

Next, a new element is inserted at the beginning of the list. Again, just a few pointers are moved and the code runs in *O(1)*.

Finally, key and value are updated and (key, index) mapping is inserted into the dictionary, which is again an *O(1)* operation:

```
procedure TDHPCache<K, V>.AddElement(const key: K; const value: V);
var
  element: integer;
begin
  if IsFull then
    element := RemoveOldest
  else
    element := GetFree;
  InsertInFront(element);
  FKeys[element].Key := key;
  FKeys[element].Value := value;
  FCache.Add(key, element);
end;
```

The only unsolved part is the `RemoveOldest` function. It will first remove the last element from the list (`Unlink(FTail)`), which is a simple *O(1)* operation. Then, it will remove the (key, index) mapping from the cache (*O(1)*) and destroy the old value if the cache owns its values:

```
function TDHPCache<K, V>.RemoveOldest: Integer;
var
  element: Integer;
begin
  if FTail < 0 then
    raise Exception.Create('TDHPCache<K, V>.RemoveOldest:' +
      'List is empty!');
  Result := FTail;
  Unlink(FTail);
  FCache.Remove(FKeys[Result].Key);
  if FOwnsValues then
    PObject(@FKeys[Result].Value)^.DisposeOf;
end;
```

As you can see, we have created a data structure that can insert, update, delete (when it is full), and retrieve elements all in *O(1)*. In *Chapter 1*, *About Performance*, however, I stated that there is always a trade-off and that not all operations can be fast in one data structure. What is going on here?

The answer is that we are gaining speed by duplicating the data, namely the keys. Each key is stored twice—once in the doubly linked list and once in the dictionary. If you remove one copy of this key, some operations will slow down.

Speeding up SlowCode

For the last practical example, we can try speeding up Mr. Smith's `SlowCode` from *Chapter 1*, *About Performance*. Here, we immediately ran into a problem. To fix or change an algorithm, we must understand what the code does. This happens a lot in practice, especially when you inherit some code. Reading and understanding code that you didn't write is an important skill.

Let's try to understand the first part of `SlowCode`. The `for` loop in `SlowMethod` starts counting with 2. Then, it calls `ElementInDataDivides`, which does nothing as the `data` list is empty. Next, `SlowMethod` adds 2 to the list.

Next, i takes the value of 3. `ElementInDataDivides` checks if 3 is divisible by 2. It is not, `SlowMethod` adds 3 to the list.

In the next step, *i = 4*, it is divisible by 2, and 4 is *not* added to the list. 5 is then added to the list (it is not divisible by 2 or 3), 6 is not (divisible by 2), 7 is added (not divisible by 2, 3, or 5), 8 is not (divisible by 2), 9 is not (divisible by 3), and so on:

```
function ElementInDataDivides(data: TList<Integer>;
  value: Integer): boolean;
var
  i: Integer;
begin
  Result := True;
  for i in data do
    if (value <> i) and ((value mod i) = 0) then
      Exit;
  Result := False;
end;

function SlowMethod(highBound: Integer): TArray<Integer>;
var
  i: Integer;
  temp: TList<Integer>;
begin
  temp := TList<Integer>.Create;
  try
    for i := 2 to highBound do
      if not ElementInDataDivides(temp, i) then
        temp.Add(i);
    Result := Filter(temp);
  finally
    FreeAndNil(temp);
  end;
end;
```

We now understand what the first part (before calling the `Filter` method) does. It calculates prime numbers with a very simple test. It tries to divide each candidate by every already generated prime. If the candidate number is not divisible by any known prime number, it is also a prime number and we can add it to the list.

Can we improve this algorithm? With a bit of mathematical knowledge, we can do exactly that! As it turns out (the proof is presented next, and you can safely skip it), we don't have to test divisibility with every generated prime number. If we are testing number *value*, we only have to test divisibility with a prime number smaller or equal to *Sqrt(value)* (square root of *value*).

Proof by contradiction

My assumption is this: If we have a number *n* that is not divisible by any number smaller than or equal to *Sqrt(n)*, then it is also not divisible by any number larger than *Sqrt(n)*. The simplest way to prove that is by establishing a contradiction.

Let's say that we can divide *n* by *p*. We can therefore write $n = p * k$. As *n* is not divisible by numbers smaller than or equal to *Sqrt(n)*, both *p* and *k* must be strictly larger than *Sqrt(n)*. If that is true, $p * k$ must be strictly larger than $Sqrt(n) * Sqrt(n)$, or *n*.

On one side, we have $p * k = n$, and on the other, $p * k > n$. These cannot both be true at the same time, so we have run into a contradiction. Therefore, the initial assumption is correct.

We can therefore rewrite `ElementInDataDivides` so that it will stop the loop when the first number larger than `Sqrt(value)` is encountered. You can find the following code in the `SlowCode_v2` program:

```
function ElementInDataDivides(data: TList<Integer>;
  value: Integer): boolean;
var
  i: Integer;
  highBound: integer;
begin
  Result := True;
  highBound := Trunc(Sqrt(value));
  for i in data do begin
    if i > highBound then
      break;
    if (value <> i) and ((value mod i) = 0) then
      Exit;
  end;
  Result := False;
end;
```

If we compare the original version with time complexity $O(n^2)$ with this improved version with time complexity $O(n * sqrt\ n)$, we can see a definite improvement. Still, we can do even better!

Try to search *prime number generation* on the internet, and you'll quickly find a very simple algorithm that was invented way before computers. It is called the *sieve of Eratosthenes* and was invented by a Greek scholar who lived in the third century BC.

I will not go into details here. For the theory, you can check *Wikipedia*, and for a practical example, see the `SlowCode_Sieve` program in the code archive. Let me just say that the sieve of Eratosthenes has a time complexity of $O(n\ log\ log\ n)$, which is very close to $O(n)$. It should therefore be quite a bit faster than our *Sqrt(n)* improvement.

The following table compares `SlowCode`, `SlowCode_v2`, and `SlowCode_Sieve` for different inputs. The highest input was not tested with the original program as it would run for more than an hour. All times are in ms:

Highest number tested	SlowCode	SlowCode_v2	SlowCode_Sieve
10,000	19	2	0
100,000	861	29	7
1,000,000	54,527	352	77
10,000,000	Not tested	5,769	1,072

Table 3.1 – Execution times in milliseconds for three versions of the SlowCode

As we have demonstrated, you can sometimes speed up a program by a considerable factor, but to do that, you have to understand the code so that you can find a better solution.

Summary

In this chapter, I touched on two very different topics. First, I looked at responsive user interfaces and how to make them faster, and then I continued with the concept of *caching*. That topic was introduced with a simple example, followed by a fully featured cache implementation that you can use in your programs.

At the end, I returned to the example `SlowCode` program from *Chapter 1, About Performance*. First, I used some detective work to find out what it really does, which allowed me to improve it a lot. The code now processes a million numbers in mere ms and not *minutes* as the original program did.

In the next chapter, I'll continue pointing out that the best way to improve program speed is to use a better algorithm. Most programs need some way to effectively organize internal data, which makes this part a great candidate for optimization, and since we would rather use great existing code than reinvent a mediocre version, we'll look into Spring for Delphi support for **collections**.

4

Don't Reinvent, Reuse

After dedicating one chapter to a discussion on how to fix an algorithm, it is now time to … do it again! Truth be told, we merely scraped the surface in the previous chapter. You could write an entire book about improving algorithms but I don't have that much space dedicated to the topic, so you will have to be satisfied with one (more) chapter.

In the previous chapter, we looked at some examples of improving speed by doing less work. This time, we'll take a different approach and run better, more optimized code. We will, however, not optimize our code – as that is a job for the next chapter – but take the smarter way and use a well-written external library. It will not only be better than our code but will also perform better than Delphi's **Run-Time Library (RTL)**.

There is, of course, no silver bullet and we cannot expect that one external library will solve all our problems and help us write programs twice as fast. I can, however, ensure you that it will improve your code in a way that matters a lot for all non-trivial applications – managing data collections.

If you are like me, your code is full of `TObjectList` and `TStringList`, maybe with a `TDictionary` thrown in, and has a `TQueue` here and there if you feel really brave. In this chapter, I'll show how you can replace them with better alternatives and I'll also show how to use collection types that have no counterpart in the Delphi RTL. You will learn the following:

- How to work with Spring's powerful `IEnumerable` support
- Working with all kinds of lists
- Leveraging the power of stacks, queues, and deques
- How to use trees in your code
- All about sets, multisets, dictionaries, and multimaps

Technical requirements

All code in this chapter was written with Delphi 11.3 Alexandria. I did not use the latest additions to the language so that most of the code could still be executed on Delphi XE or above. You can find all the examples on Github: `https://github.com/PacktPublishing/Delphi-High-Performance---Second-Edition/tree/main/ch4`.

Spring for Delphi

Spring for Delphi (also known as **Spring4D**) is a collection of modules written for Delphi. It was originally written by Baoguan Zuo and is now maintained by Stefan Glienke. It contains multiple parts, which are mostly independent but share some common code (and design philosophy). Inside, you'll find the following:

- A collection of small helpers (smart pointers, nullable objects, multi-dispatch events, and so on)
- A set of interface-based collections, including a powerful way to access them
- A Dependency Injection container
- An interception and mocking library to help with unit testing
- A reflection library that extends Delphi's RTTI
- An object-relational mapping layer
- An encryption library
- And more…

All this is released under Apache License, Version 2.0 which in short means that you can do almost anything with the software, as long as you include the required notices. There are no problems using it in commercial programs.

Before we start, we have to add Spring (as I intend to call it for brevity) to Delphi, and that is mostly a manual process.

To get the latest version, open `https://bitbucket.org/sglienke/spring4d` in any browser and click on **Downloads**. This will bring you to the **Download** page, where you should click **Download repository** to get the latest stable version. Then, unzip the downloaded file in an appropriate place. Alternatively, click the **Clone** button at the top of the page and clone the library using Git.

After that, you can either add Spring folders to Delphi's library path, to the project's search path, or use the included `Build.exe` to precompile Spring sources to `.dcu` files and add them to the library path. The library's author recommends always using precompiled `.dcu` files, as that will result in your program being compiled more quickly.

This chapter focuses on Spring version 2, which was just released during the writing of this book. You may already be familiar with Spring version 1, where some details were implemented differently, and therefore you may experience some issues during the upgrade. This list will help you resolve the issues:

- Collections in Spring v1 were partially dependent on the `System.Generics.Collections` unit. In version 2, all functionality from that unit is reimplemented in `Spring.Collections` and you should be able to remove Delphi's unit from the code. If you have to use both from the same source, be aware that Spring v2 *redefines* the `TPair` type, which could lead to weird conflicts.

- Spring v2 also redefines Delphi's `TDictionaryOwnership` type (in the `Spring` unit).

- Spring v2 *filter* functions now accept `const` parameters, so you may have to update filter function definitions.

- The bidirectional dictionary, `IBidiDictionary`, has a simplified (and quite different) interface. For details, see the *Sets and dictionaries* section later in this chapter.

In this chapter, we will focus on the most important part of the Spring library – at least in my opinion – the rich and powerful *collections* support. All units related to the collections module can be found in the `Source\Base\Collections` subfolder.

Enumerations, collections, and lists

I have used the term *collection* quite a few times already but I (quite intentionally) "forgot" to define this quite a loose term. Let's fix this now.

We will define a **collection** as a bunch of data items of the same type combined with some kind of an API that allows us to access (read, write, and delete) those data items in an organized manner. Collections can be further classified by the API they provide (we'll return to that in a moment). Delphi's `TList`, for example, is a collection (more specifically, a *list*) of pointers and `TList<T>` is a collection of items of type `T`.

Spring collections are implemented using interfaces (an API from the previous paragraph is Delphi's interface). The core of Spring collections is built on generic types and interfaces, which are carefully crafted so that they don't result in slow compilation times and EXE bloat when you are using generic collections of many different data types. They are also at least as fast as the Delphi implementation, while some operations can be many times faster.

Some interfaces (`IEnumerable` and `ICollection`) are also available as non-generic versions, which store TValue items. They are based on generic implementations and were designed for data serialization where you don't know the type of stored data in advance.

The supported APIs are organized into multiple interfaces, which all derive from IEnumerable (TValue versions) or IEnumerable<T> (generic versions). A full tree of generic collection interfaces is reproduced here:

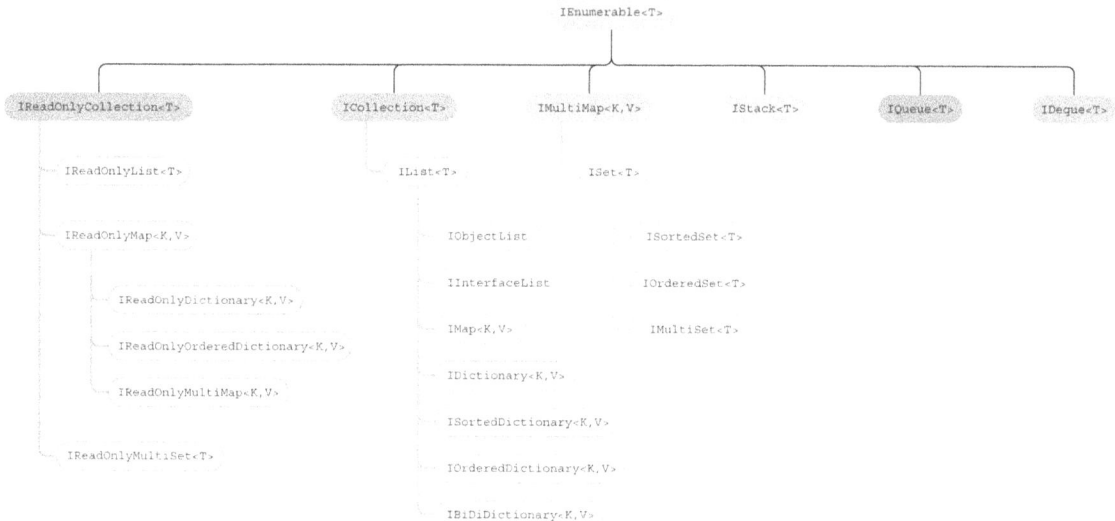

Figure 4.1 – Inheritance diagram of Spring4D collections interfaces

For example, this tree tells us that we can treat every ISortedDictionary<K,V> also as an IMap<K,V>, an ICollection<T> (where T is actually TPair<K,V>), and – as holds for all collections – as an IEnumerable<T>. This allows us to treat a collection at the level of abstraction that is required for the current use case. For example, if we need a function that walks through all the elements in a set and processes the elements in some manner, we don't have to work with the ISet<T> interface. As we will only walk through the collection, enumerating all elements, we can write this code so that it accepts an IEnumerable<T> parameter. This will also allow us to replace the set with any other collection and the code will still compile.

An interesting branch of this tree is *read-only collections*. They implement a simplified interface that allows only read-only access, thus protecting the data from unwanted changes. This also allows for an optimized implementation that doesn't have to take care of possible modifications. Spring collections often return read-only collections as a function result.

We will now take a brief look at some interesting interfaces from this hierarchy, starting with the most important of them all, IEnumerable<T>.

IEnumerable<T>

At the most basic level, the IEnumerable<T> interface implements support for read-only *sequences* of data. As part of that, it also implements support for Delphi's for .. in loop statement. This allows any Spring collection to participate in this extremely useful construct.

As an example, the following code from the IEnumerable program lists numbers from 1 to 5. (All the examples in this section come from the same program.) The code first creates an IEnumerable<integer> by calling TEnumerable.Range. This function (which we'll cover later in this section) returns an enumerable sequence of numbers. A for .. in construct then *enumerates* the collection and displays all elements:

```
procedure TfrmEnumerable.btnForInClick(Sender: TObject);
var
  enum: IEnumerable<integer>;
  i: integer;
begin
  enum := TEnumerable.Range(1, 5);
  for i in enum do
    ListBox1.Items.Add(IntToStr(i));
end;
```

The interface by itself offers no guarantee about the order of the elements. The order is enforced by the specific implementation of a collection. If a collection is *sorted* or the order is enforced in some other way (for example, in a *queue*), then the enumeration will return elements in the sort order, as we could expect. Otherwise, the order is not defined.

> **Tip**
> While the order of unsorted collections in Spring v1 was really undefined – adding an element to a *set*, for example, could reorder the elements – Spring v2 keeps unsorted collections in *insertion order*. You should, however, not depend on that. As the **dependency inversion** principle says, you should depend on abstraction and not on concrete implementation.

As IEnumerable<T> is a very rich interface, I'll present its functionality in smaller segments, starting with functions for querying and filtering data.

Querying sequences

The for .. in support is actually just a small part of the IEnumerable<T> interface. Its full power lies in its extensive support for *querying* a collection.

If you look up this interface in the Spring.Collections unit, you'll see that it implements a large number of functions returning itself (that is, an IEnumerable<T> interface). These *query* functions allow you to filter the original collection into a new collection containing only selected entries. While we will look at some of them in detail, here is the full list for reference:

```
function Concat(const second: IEnumerable<T>):
  IEnumerable<T>;
function Memoize: IEnumerable<T>;
function Ordered: IEnumerable<T>; overload;
function Ordered(const comparer: IComparer<T>):
  IEnumerable<T>; overload;
function Ordered(const comparer: TComparison<T>):
  IEnumerable<T>; overload;
function Reversed: IEnumerable<T>;
function Shuffled: IEnumerable<T>;
function Skip(count: Integer): IEnumerable<T>;
function SkipLast(count: Integer): IEnumerable<T>;
function SkipWhile(const predicate: Predicate<T>):
  IEnumerable<T>; overload;
function SkipWhile(
  const predicate: Func<T, Integer, Boolean>):
  IEnumerable<T>; overload;
function Take(count: Integer): IEnumerable<T>;
function TakeLast(count: Integer): IEnumerable<T>;
function TakeWhile(const predicate: Predicate<T>):
  IEnumerable<T>; overload;
function TakeWhile(
  const predicate: Func<T, Integer, Boolean>):
  IEnumerable<T>; overload;
function Where(const predicate: Predicate<T>):
  IEnumerable<T>; overload;
function Where(const predicate: Func<T, Integer, Boolean>):
  IEnumerable<T>; overload;
```

As you can see in this list, some functions accept `Func<T>` and `Predicate<T>` parameters. These types are functionally equivalent to Delphi's `TFunc<T>` and `TPredicate<T>` except that they accept `const` parameters. This allows for faster execution of the code when `T` represents a string, an array, a record, or an interface. The following code fragment shows Delphi and Spring definitions of those types:

```
// System.SysUtils
  TFunc<T,TResult> = reference to function (Arg1: T): TResult;
  TPredicate<T> = reference to function (Arg1: T): Boolean;
// Spring
  Func<T, TResult> = reference to function(const arg: T): TResult;
  Predicate<T> = reference to function(const arg: T): Boolean;
```

Let us now look at some examples of querying an IEnumerable<T> sequence. The code under the **Where** button shows two ways of using the Where query. The following code creates a sequence containing integer values from 1 to 20 and then outputs all values that are divisible by 3 and all values that are divisible by 5:

```
function TfrmEnumerable.IsDivisibleBy3(const i: integer): boolean;
begin
  Result := (i mod 3) = 0;
end;

procedure TfrmEnumerable.btnIEnumerable1Click(Sender: TObject);
var
  enum: IEnumerable<integer>;
begin
  enum := TEnumerable.Range(1, 20);
  ListBox1.Items.Add(
    'Where/3: '+ Join(',', enum.Where(IsDivisibleBy3)));
  ListBox1.Items.Add('Where/5: ' + Join(',',
    enum.Where(function (const i: integer): boolean
               begin Result := (i mod 5) = 0 end)));
end;
```

The filter function is used in two different ways. The first example uses the IsDivisibleBy3 method, which is automatically converted into an anonymous function (when passed to the Where query) by the compiler. The second example uses an anonymous function defined at the place of the call.

The RTL Join function takes a sequence and converts it into a string. Let's ignore it for now, as it uses concepts that will be covered later in the chapter in the *TEnumerable* section. We'll revisit its implementation at that point.

As query functions return IEnumerable<T> interfaces, we can *chain* them by calling another query function on the result. This allows code that can express its intention in very little space, without any loops. The following example, attached to the **Chaining** button, goes through the following steps:

1. It generates numbers from 1 to 10.

2. Next, it skips the first three elements (resulting in numbers from 4 to 10).

3. The code then skips the last three elements (resulting in numbers from 4 to 7).

4. It then reverses the order (resulting in numbers from 7 to 4).

5. At the end of the code, it outputs the resulting collection.

As you can see in the following example, all that happens in few lines of code and without any loops:

```
procedure TfrmEnumerable.btnIEnumerable2Click(Sender: TObject);
var
  enum: IEnumerable<integer>;
begin
  enum := TEnumerable.Range(1, 10);
  ListBox1.Items.Add('Skip: ' + Join(',',
    enum.Skip(3).SkipLast(3).Reversed));
end;
```

The main power of queries comes from the fact that they are only *executed when needed*. This can greatly speed up the program while also allowing us to prepare filters in advance and execute them later. The following code fragment (attached to the **Deferred execution** button) demonstrates that.

The code first creates a sequence of numbers from 1 to 200. Then, it prepares a `filtered` variable, which queries this sequence with a `Where` filter (keeping only odd numbers) and a chained `Take` filter (returning only the first three numbers). The code then runs over `filter` with the `for ..  in` statement and displays all values:

```
procedure TfrmEnumerable.btnIEnumerable4Click(Sender: TObject);
var
  enum: IEnumerable<integer>;
  filtered: IEnumerable<integer>;
  i: integer;
begin
  enum := TEnumerable.Range(1, 200);
  filtered := enum.Where(
              function (const value: integer): boolean
              begin
                Result := Odd(value);
                ListBox1.Items.Add('? ' + IntToStr(value));
              end)
              .Take(3);
  ListBox1.Items.Add('Start');
  for i in filtered do
    ListBox1.Items.Add('=> ' + IntToStr(i));
end;
```

A naïve filter implementation would filter all 200 numbers immediately when `Where` is called and then take first three elements from the resulting list of 100 numbers. All that would happen before the `for ..  in` statement was even executed. The Spring implementation, however, is much smarter.

As we can see in the following figure, filter construction does not execute any filtering code. The first log entry is Start, just before for .. in begins its work. Only then is the Where filter executed (lines starting with ?) and it is only executed five times. As soon as the Take filter has its three elements, enumeration stops and the procedure exits:

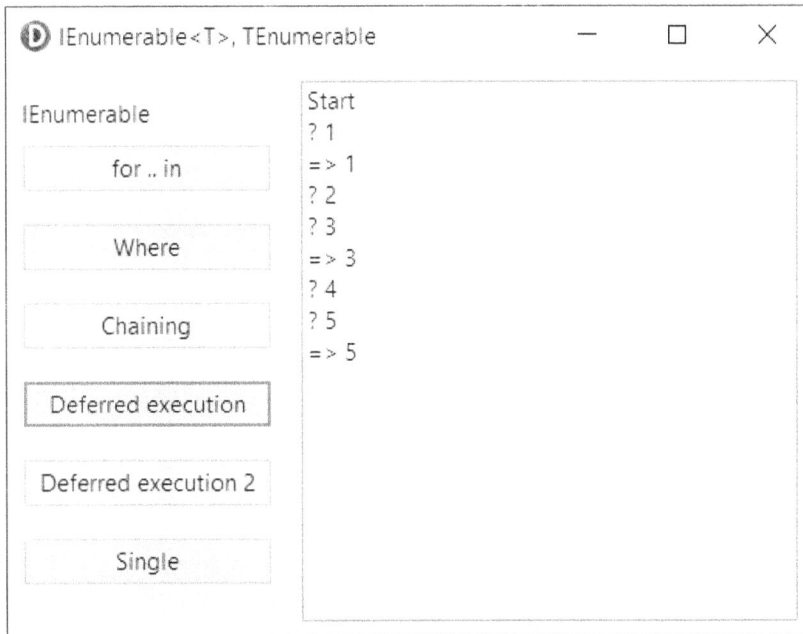

Figure 4.2 – A demonstration of a deferred query execution

Similar care for performance can be found throughout the Spring code and is the main reason why I'm recommending using it over home-grown custom code. To give another example, the TEnumerable.Range function, which I'm using in the examples, doesn't actually create a collection of numbers (200 in the last example). It creates a TRangeIterator class (defined in the Spring. Collections.Extensions unit) which implements for .. in support and calculates return values when requested.

An interesting after-effect of deferred execution is that you can modify the sequence after the filter was created but before it is executed and the filter will be applied to the new collection content. To demonstrate this, I had to cheat a little (as IEnumerable<T> collections cannot be modified) and skip ahead. The next demo uses a *list* (IList<T>). You should, however, have no problems understanding the code, as IList<T> implements the same interface as Delphi's TList<T>.

The code (attached to the **Deferred execution 2** button) creates a list and a filter, similarly to the previous example. The filter simply returns all values that are divisible by 3. The code then adds three elements to the list, displays the content of the filter, adds another three elements, and displays the content of the filter again:

```
procedure TfrmEnumerable.btnEnumerable5Click(Sender: TObject);
var
  list: IList<integer>;
  filter: IEnumerable<integer>;
begin
  list := TCollections.CreateList<integer>;
  filter := list.Where(IsDivisibleBy3);
  list.AddRange([1, 2, 3]);
  ListBox1.Items.Add(Join(',', filter));
  list.AddRange([4, 5, 6]);
  ListBox1.Items.Add(Join(',', filter));
end;
```

This short fragment results in two lines being output:

```
3
3,6
```

With this example, I will stop talking about query functions and briefly cover other parts of the IEnumerable<T> interface. You can look at the documentation for other query functions in the code (the Spring.Collections unit).

Accessing single elements

The next set of IEnumerable<T> functions returns one element that satisfies some special conditions. We can use them to get the first or last element (First or Last), find the smallest or largest value in the sequence (Min or Max), run an aggregation function over the sequence (Aggregate or Sum), or verify that the sequence has exactly one element (Single). The following list shows all functions from this category:

```
function Aggregate(const func: Func<T, T, T>): T;
function ElementAt(index: Integer): T;
function ElementAtOrDefault(index: Integer): T; overload;
function ElementAtOrDefault(index: Integer;
  const defaultValue: T): T; overload;
function First: T; overload;
function First(const predicate: Predicate<T>): T; overload;
function FirstOrDefault: T; overload;
function FirstOrDefault(const defaultValue: T): T; overload;
function FirstOrDefault(const predicate: Predicate<T>): T; overload;
function FirstOrDefault(const predicate: Predicate<T>;
```

```
    const defaultValue: T): T; overload;
function Last: T; overload;
function Last(const predicate: Predicate<T>): T; overload;
function LastOrDefault: T; overload;
function LastOrDefault(const defaultValue: T): T; overload;
function LastOrDefault(const predicate: Predicate<T>): T; overload;
function LastOrDefault(const predicate: Predicate<T>;
    const defaultValue: T); overload;
function Max: T; overload;
function Max(const comparer: IComparer<T>): T; overload;
function Max(const comparer: TComparison<T>): T; overload;
function Min: T; overload;
function Min(const comparer: IComparer<T>): T; overload;
function Min(const comparer: TComparison<T>): T; overload;
function Single: T; overload;
function Single(const predicate: Predicate<T>): T; overload;
function SingleOrDefault: T; overload;
function SingleOrDefault(const defaultValue: T): T; overload;
function SingleOrDefault(const predicate: Predicate<T>): T; overload;
function SingleOrDefault(const predicate: Predicate<T>;
    const defaultValue: T): T; overload;
function Sum: T; overload;
function TryGetFirst(var value: T): Boolean; overload;
function TryGetFirst(var value: T;
    const predicate: Predicate<T>): Boolean; overload;
function TryGetLast(var value: T): Boolean; overload;
function TryGetLast(var value: T;
    const predicate: Predicate<T>): Boolean; overload;
function TryGetSingle(var value: T): Boolean; overload;
function TryGetSingle(var value: T;
    const predicate: Predicate<T>): Boolean; overload;
```

You can see that most functions in this list appear in multiple alternative forms. Some accept custom functions to compare values (IComparer<T> or TComparison<T>), while others can optionally run a filter first (Predicate<T>). Some functions have ...OrDefault variations, which return default values if a collection is empty (or if the filtering predicate returns an empty collection). If the default value is not provided, Spring uses a commonly accepted default value for the type (0, ' ', nil, and so on). Some of the functions also exist in the TryGet... form, which doesn't raise an exception but returns False if the value cannot be returned.

The example program shows how the Single function can be used. This interesting function converts a collection into its only element. It raises an exception if the collection is empty or if it contains more than one element. (The SingleOrDefault variant raises an exception only if a collection has more than one element and returns the default value if a collection is empty.) The code, which is attached to the **Single** button, creates a sequence of numbers from 1 to 20, removes all elements that are not divisible by 3 and by 5, and displays the only remaining number (15) in a message box:

```
function TfrmEnumerable.DivisibleBy(const num: integer):
  Predicate<integer>;
begin
  Result :=
    function (const i: integer): boolean
    begin
      Result := (i mod num) = 0;
    end;
end;

procedure TfrmEnumerable.btnIEnumerable3Click(
  Sender: TObject);
begin
  ShowMessage(
    TEnumerable.Range(1, 20)
      .Where(DivisibleBy(3))
      .Where(DivisibleBy(5))
      .Single.ToString);
end;
```

> **Tip**
>
> This example also demonstrates another way to create filter functions. Previously, we passed a method as a filter (`IsDivisibleBy3`) or we constructed an anonymous function on the fly. This example uses a *function factory* and *variable-capturing* to create filter functions on demand.

Tests

The last big group of `IEnumerable<T>` functions contains *tests*. All the functions in this group return a `Boolean` answer. Here, you can find functions that test the size of the collection (`AtLeast`, `AtMost`, `Between`, `Exactly`, or `IsEmpty`), test whether elements satisfy a condition (`All` or `Any`), look for an element (`Contains`), and compare this collection with another (`EqualsTo`):

```
function All(const predicate: Predicate<T>): Boolean;
function Any: Boolean; overload;
function Any(const predicate: Predicate<T>): Boolean; overload;
function AtLeast(count: Integer): Boolean;
function AtMost(count: Integer): Boolean;
function Between(min, max: Integer): Boolean;
function Contains(const value: T): Boolean; overload;
function Contains(const value: T;
```

```
  const comparer: IEqualityComparer<T>): Boolean; overload;
function Contains(const value: T;
  const comparer: TEqualityComparison<T>): Boolean; overload;
function EqualsTo(const values: array of T): Boolean; overload;
function EqualsTo(const values: IEnumerable<T>): Boolean; overload;
function EqualsTo(const values: IEnumerable<T>;
    const comparer: IEqualityComparer<T>): Boolean; overload;
function Exactly(count: Integer): Boolean;
property IsEmpty: Boolean read GetIsEmpty;
```

As with the previously described functions, this set also implements various overloads accepting comparers (`IEqualityComparer<T>` or `TEqualityComparer<T>`) and filters (`Predicate<T>`).

> **Tip**
>
> The speed of the `Contains` function depends on the implementation. For example, in an unsorted list, `Contains` will be $O(n)$ while in a sorted list, it will be $O(log\ n)$.

Other functions

The last set of functions contains various helper functions:

```
function ToArray: TArray<T>;
procedure ForEach(const action: Action<T>);
property Count: Integer read GetCount;
function ElementAt(index: Integer): T;
function ElementAtOrDefault(index: Integer): T; overload;
function ElementAtOrDefault(index: Integer;
  const defaultValue: T): T; overload;
function TryGetElementAt(var value: T; index: Integer): Boolean;
```

`ToArray` converts a collection into a dynamic array. It is useful when interfacing with RTL functions.

`ForEach` executes an *action* (an anonymous procedure) for each element in the collection. The `Action<T>` type is similar to Delphi's `TProc<T>` type except that parameters are passed as `const` for performance reasons. The following code fragment compares the two implementations:

```
// System.SysUtils
TProc<T> = reference to procedure (Arg1: T);
// Spring
Action<T> = reference to procedure(const arg: T);
```

`Count` returns the number of elements in the collection.

`ElementAt` returns an element at a specific index. While it is implemented for every collection, its execution time differs according to the specific implementation. The default implementation, used in most collections, iterates over all elements in the collection looking for the specific one, and has a complexity of *O(n)*. Only specific implementations (hash-based sets and dictionaries, lists) will execute this function in a constant time, *O(1)*. `TryGetElementAt` and `ElementAtOrDefault` variations can be useful if you're not sure whether the element is actually present in the collection.

This concludes our tour of the most important interface in the Spring collection hierarchy. In the next pages, we'll take a quicker look at other interfaces.

ICollection<T>

An `ICollection<T>` interface represents a "bag" of items. They can be put into the bag and removed, but there's no implied order. As with `IEnumerable<T>`, the order depends on the specific implementation. A slightly shortened definition of the interface is displayed here:

```
ICollection<T> = interface(IEnumerable<T>)
  function Add(const item: T): Boolean;
  procedure AddRange(const values: array of T); overload;
  procedure AddRange(const values: IEnumerable<T>); overload;
  function Extract(const item: T): T;
  procedure Clear;
  function CopyTo(var values: TArray<T>; index: Integer): Integer;
  function MoveTo(const collection: ICollection<T>):
    Integer; overload;
  function MoveTo(const collection: ICollection<T>;
    const predicate: Predicate<T>): Integer; overload;
  function Remove(const item: T): Boolean;
  function RemoveAll(const predicate: Predicate<T>): Integer;
  function RemoveRange(const values: array of T): Integer; overload;
  function RemoveRange(const values: IEnumerable<T>):
   Integer; overload;
  function ExtractAll(const predicate: Predicate<T>): TArray<T>;
  procedure ExtractRange(const values: array of T); overload;
  procedure ExtractRange(const values: IEnumerable<T>); overload;
  property OnChanged: ICollectionChangedEvent<T> read GetOnChanged;
end;
```

This interface contains methods for adding elements (`Add` or `AddRange`), removing elements (`Extract, Remove, RemoveAll, RemoveRange, ExtractAll`, or `ExtractRange`), clearing the collection (`Clear`), and moving/copying data to another collection or an array (`CopyTo` or `MoveTo`).

The `OnChange` event handler will notify you of changes in the collection (the addition or removal of elements).

The code archive for this book contains the `ICollection` program, which demonstrates some operations on a collection.

IList<T>

An `IList<T>` interface represents an *ordered* collection. Each item is stored at a specific location and can be referenced by that location. Items can be added or inserted at specific locations. A list as a whole can be sorted (which may, of course, change the association between locations and items).

The `IList<T>` interface is very similar to the public part of Delphi's `TList<T>` class. This allows you to replace `TList<T>` with `IList<T>` without much work. In most cases, you will just change the list creation and destruction.

The `IList` program shows how `TList<T>` and `IList<T>` are similar but not completely the same. The differences between the two code fragments are marked in **bold**:

```
procedure TfrmIList.btnListClick(Sender: TObject);
var
  i: integer;
  list: TList<integer>;
  loc: integer;
begin
  list := TList<integer>.Create;
  for i := 1 to 5 do list.Add(i);
  list.AddRange(list);
  list.Insert(5, 6);
  ListBox1.Items.Add('TList: ' + Join(' ', list.ToArray));
  list.Sort;
  ListBox1.Items.Add('Sorted: ' + Join(' ', list.ToArray));
  ListBox1.Items.Add('Pos(5): ' +
    IntToStr(list.IndexOf(5)));
  if list.BinarySearch(5, loc) then
    ListBox1.Items.Add('Search(5): ' + IntToStr(loc));
  list.Free;
end;

procedure TfrmIList.btnIListClick(Sender: TObject);
var
  i: integer;
  list: IList<integer>;
  loc: integer;
begin
  list := TCollections.CreateList<integer>;
  for i := 1 to 5 do list.Add(i);
```

```
list.AddRange(list);
list.Insert(5, 6);
ListBox1.Items.Add('IList: ' + Join(' ', list.ToArray));
list.Sort;
ListBox1.Items.Add('Sorted: ' + Join(' ', list.ToArray));
ListBox1.Items.Add('Pos(5): ' +
   IntToStr(list.IndexOf(5)));
ListBox1.Items.Add('LastPos(5): ' +
   IntToStr(list.LastIndexOf(5)));
if TArray.BinarySearch<integer>(list.ToArray, 5, loc) then
   ListBox1.Items.Add('Search(5): ' + IntToStr(loc));
list.Reverse;
ListBox1.Items.Add('Reversed: ' + Join(' ', list.ToArray));
end;
```

To construct a Delphi list, you would call the `TList<integer>.Create` *constructor* while a Spring list is constructed by calling a `TCollections.CreateList<integer>` *factory* (more on that in a moment). The code then works the same up to the line that outputs the `Pos(5)` string. The Spring code then demonstrates the use of the `LastIndexOf` function, which has no equivalent on the Delphi side, and finds the last occurrence of a value.

The Delphi code then calls the `BinarySearch` function to look for a value (we know that the list is sorted because we sorted it by calling the `Sort` method a few lines before that). Spring has no `BinarySearch` function – if you want a fast search, you would start with a *sorted list* – but you can work around this by calling the `ToArray` function and passing data to Delphi's `TArray.BinarySearch`. This is slower than the `TList` version because `ToArray` creates a copy of all the data.

At the end, we have to destroy the `TList<integer>` object explicitly by calling `Free` (or `FreeAndNil`), while we can leave `IList<integer>` untouched, and it will be automatically destroyed when the method exits.

All Spring collections should be created by calling an appropriate function of the `TCollections` class, which is defined in the `Spring.Collections` unit. The following list shows different factories that we can use to create a list (I have removed the `overload` and `static` directives for clarity):

```
class function CreateList<T>: IList<T>;
class function CreateObjectList<T: class>: IList<T>;
class function CreateInterfaceList<T: IInterface>:IList<T>;
class function CreateStringList: IList<string>;
class function CreateSortedList<T>: IList<T>;
class function CreateSortedObjectList<T>: IList<T>;
class function CreateSortedInterfaceList<T: IInterface>: IList<T>;
```

You can create a "normal" list, a list of objects (which can optionally own the objects, just like Delphi's `TObjectList<T>`), a list of interfaces, and a list of strings (equivalent to `TStringList`). All those lists except the last can also be created in the Sorted version, which enforces the sort order during insertion and speeds up searches.

> **Tip**
>
> It is worth mentioning that the sorted list doesn't implement all the functions of the IList<T> interface, as they would break the sort order of the elements. The Insert method, for example, raises an exception instead of corrupting the data.

The TCollections class also implements many overloads for factory functions, which I have removed from the previous example for brevity. You can look up the full list in the code. For example, here are all the versions of the CreateList factory (again, I have removed the overload and static directives; I have also removed functions that are deprecated):

```
class function CreateList<T>: IList<T>;
class function CreateList<T>(const comparer: IComparer<T>): IList<T>;
class function CreateList<T>(
  const comparer: TComparison<T>): IList<T>;
class function CreateList<T>(const values: array of T): IList<T>;
class function CreateList<T>(const values: IEnumerable<T>): IList<T>;
```

These overloaded versions allow you to set a custom comparison function (IComparer<T>) and optionally pass in initial values as an array of values or as an IEnumerable<T> sequence.

To help you select an appropriate tool for solving a problem, I'll present a short overview of the most important time complexities for each Spring collection type. Those are as follows:

- **Access** describes the behavior when directly accessing elements in the collection by using an integer index (element := list[index])

- **Search** tells us about the speed when trying to find an element in a collection by its value (index := list.IndexOf(element))

- **Insert** complexity tells us how costly it is to add elements to the collection (list.Add(element))

- **Delete** describes how the collection behaves when we remove elements (list.Remove(element))

- Most of the time, I will only list *average* behavior. In cases where it's important information, I'll list the *best-case-scenario* or *worst-case-scenario* behavior too.

- Sometimes, an operation results in different behavior depending on specific conditions. If that is the case, I'll list the different possibilities.

> **Algorithm complexity**
>
> All lists, sorted and unsorted, are implemented as arrays and have the following complexities:
>
> - **Access**: *O(1)*
> - **Search**: *O(n)* for an unsorted list, *O(log n)* for a sorted list
> - **Insert**: *O(n)* as the average case for an unsorted list, *O(1)* best-case scenario for an unsorted list, *O(log n)* for a sorted list
> - **Delete**: *O(n)*

Other interfaces

Other interfaces will be covered in detail later in this chapter, so here's just a quick overview.

`IDictionary<TKey,TValue>` stores key-value pairs of any type. The key can be used to look up its value. It is equivalent to Delphi's `TDictionary<TKey,TValue>`. The key and the value can be of any type. An `IMultiMap<TKey, TValue>` maps one key to multiple values.

`ISet<T>` is similar to a Delphi `set` type. It stores items in no particular order – an item is either stored in the set or not. Each value can only be stored once. `IMultiSet<T>` is similar but can store multiple copies of the same value. This type of collection is sometimes also called a *bag*.

`IStack<T>` is a **Last In, First Out** (**LIFO**) data structure (a *stack*) with standard `Push` and `Pop` operations. `IQueue<T>` is a **First In First Out** (**FIFO**) data structure (a *queue*) with standard `Enqueue` and `Dequeue` operations. There's also an `IDeque<T>` collection, which implements a *double-ended queue* (also called a *deque*). It is similar to `IQueue<T>` but allows additions and removals at both ends of the structure.

TEnumerable

Before we can start exploring other data structures, I have to mention the `TEnumerable` class from the `Spring.Collections` unit.

The `TEnumerable` class implements some useful operations that are not part of the `IEnumerable<T>` interface (and repeats some operations that are included in `IEnumerable<T>`). Here's a slimmed-down version, including some interesting tidbits (for a full list of supported functions, see the code):

```
TEnumerable = class
  class function Chunk<T>(const source: IEnumerable<T>;
    size: Integer): IEnumerable<TArray<T>>;
  class function Distinct<T>(const source: IEnumerable<T>):
    IEnumerable<T>;
  class function DistinctBy<T, TKey>(
    const source: IEnumerable<T>;
```

```
      const keySelector: Func<T, TKey>): IEnumerable<T>;
  class function Empty<T>: IReadOnlyList<T>; static
  class function From<T>(const values: array of T): IReadOnlyList<T>;
  class function Intersect<T>(const first,
    second: IEnumerable<T>): IEnumerable<T>;
  class function Range(start, count: Integer): IReadOnlyList<Integer>;
  class function Repeated<T>(const element: T;
    count: Integer): IEnumerable<T>;
  class function Select<T, TResult>(
    const source: IEnumerable<T>;
    const selector: Func<T, TResult>):IEnumerable<TResult>;
  class function Union<T>(const first,
    second: IEnumerable<T>): IEnumerable<T>;
```

I have already used the `Range` function in multiple examples. There's also `Repeated<T>`, which repeats a single element, and `Empty<T>`, which returns an empty read-only list (and is a practical example of a **Null object** design pattern). A few of the other functions are demonstrated in the `TEnumerable` program.

The **Range, From** button shows how `TEnumerable` can be used to pass data from Spring's `IEnumerable<T>` to Delphi's `TList<T>` and back:

```
procedure TfrmTEnumerable.btnTEnumerable1Click(
  Sender: TObject);
var
  list: TList<integer>;
begin
  list := TList<integer>.Create(
             TEnumerable.Range(11, 9).ToArray);
  ListBox1.Items.Add('Range(11,9): ' +
    Join(' ', TEnumerable.From<integer>(list)));
  list.Free;
end;
```

> **Keep in mind**
>
> The `TEnumerable.Range` function parameters represent a starting value and a number of values, not a starting value and an end value, as in the `for` statement!

To initialize a `TList<integer>` object, the code creates `IEnumerable<integer>` by calling `TEnumerable.Range(11, 9)` and converts it into `TArray<integer>` by calling the `ToArray` function of the `IEnumerable<T>` interface. This dynamic array is then used to initialize `TList`. As interfaces are dynamic and reference-counted, we don't have to take any care to properly destroy the collection and free the data.

In the next line, the code wraps Delphi's TList<T> into an interface implementing IEnumerable<T> by calling the TEnumerable.From<integer> function. This allows us to pass a Delphi object to a function that is expecting a Spring interface. The code then uses the Join function to convert the resulting collection into a string object.

I have used the Join function multiple times in the code in this chapter without explaining how it works. Now is the perfect time to rectify that.

Join accepts a delimiter string and an IEnumerable<integer>. It then uses TEnumerable.Select to convert the collection of integers into a collection of strings. Select accepts an IEnumerable sequence of one type and a conversion function and returns an IEnumerable sequence of another type. The resulting IEnumerable<string> is then converted into an array and passed to Delphi's Join helper function:

```
function TfrmTEnumerable.Join(const delim: string;
  const enum: IEnumerable<integer>): string;
begin
  Result := string.Join(delim,
              TEnumerable.Select<integer,string>(
                enum, IntToString).ToArray);
end;

function TfrmTEnumerable.IntToString(const val: integer): string;
begin
  Result := IntToStr(val);
end;
```

The conversion function is defined as Func<T, TResult>, which expects a const parameter. That's why I had to write a proxy function, IntToString, which accepts a const value and passes it to the RTL IntToStr function (which accepts a non-const value).

Although I have barely scratched the surface, I have to wrap up this introductory part. In the rest of the chapter, we'll examine other collection types implemented in Spring.

Stacks and queues

Spring collections implement three types of collections that are foundational to computer science – a stack (**LIFO**), a queue (**FIFO**), and a double-ended queue (deque). Although both queues (and sometimes the stack) are frequently implemented with a *linked list*, Spring collections implement them with dynamic arrays. This approach is much faster as it doesn't require frequent allocations of small memory blocks.

> **Tip**
>
> Spring collections include a linked list implementation, ILinkedList<T>, where each
> node is represented by a TLinkedListNode<T> object, but it is better to ignore it. The
> implementation in Spring is perfectly fine but this approach – allocating many small blocks of
> memory – doesn't work well on modern CPU architectures. Allocated memory is usually not
> sequential, which leads to poor cache performance when walking over such structures. Use
> other collections if at all possible.

IStack<T>

A stack is a collection with the following properties:

- Elements are organized in a sequence that cannot be reordered

- The only two operations are *Push* (add an element to the end of the sequence) and *Pop* (remove
  an element from the same end)

A stack is usually represented as a vertically oriented stack of boxes. We can only add a box at the
top (*Push*) or remove it from the top (*Pop*). Spring's implementation is richer than that, as it also
implements the IEnumerable<T> interface. We can therefore walk over the stack with for ..
in without disturbing the content, which is not possible with the traditional implementation (but
can be very practical, especially for debugging).

Spring's implementation is defined by the IStack<T> interface, which is shown here in a slightly
shortened form:

```
IStack<T> = interface(IEnumerable<T>)
  procedure Clear;
  function Push(const item: T): Boolean;
  function Pop: T;
  function Extract: T;
  function Peek: T;
  function PeekOrDefault: T;
  function TryPop(var item: T): Boolean;
  function TryExtract(var item: T): Boolean;
  function TryPeek(var item: T): Boolean;
  procedure TrimExcess;
  property Capacity: Integer
    read GetCapacity write SetCapacity;
  property OnChanged: ICollectionChangedEvent<T>
    read GetOnChanged;
end;
```

The interface is very simple. We have one method that pushes an item onto the stack (Push) and two methods to remove the item from the top of the stack (Pop and Extract). The difference between the latter two only takes effect when the stack contains objects. If that is the case, Pop always returns nil (which is kind of non-intuitive), while Extract returns the item. We can also use the Peek function to "look at" the top element in the stack without removing it.

All the methods for removing items from the stack raise an exception if the stack is empty. If that is a possibility and you don't want to check the number of elements in the stack (Count or IsEmpty) before popping an item, you can use the "safe" versions–PeekOrDefault, TryPop, TryExtract, and TryPeek.

Spring implements two stack variations – normal and *bounded*. The only difference is that the latter is limited in size. On a normal stack, the Push function will always succeed. On a bounded stack, it will work as expected until the stack is at the limit, after which Push will silently throw the data away and return False.

The following code fragment shows the two most basic factories for stack creation (you can look up other factories in the code):

```
class function CreateStack<T>: IStack<T>;
class function CreateBoundedStack<T>(size: Integer): IStack<T>;
```

The StackQueue program in the code archives includes a simple demonstration of a stack of strings. When you click on the **Stack** button, the code shown here creates a stack, pushes characters from '1' to '7', shows the content of the stack as seen through the IEnumerable<string> interface, and then removes (and logs) all the strings from the stack:

```
procedure TfrmStackQueueMain.btnStackClick( Sender: TObject);
var
  ch: char;
  stack: IStack<string>;
  s: string;
begin
  stack := TCollections.CreateStack<string>;
  for ch := '1' to '7' do
    stack.Push(ch);
  ListBox1.Items.Add('Stack: ' + ''.Join(' ', stack.ToArray));

  s := 'Stack remove: ';
  while not stack.IsEmpty do
    s := s + stack.Pop + ' ';
  ListBox1.Items.Add(s);
end;
```

As we can see in the following figure, this code emits two identical sequences. This shows that the "native" order in a stack (the order in which the `for .. in` enumerator operates) is the same as the order in which the `Pop` operation works:

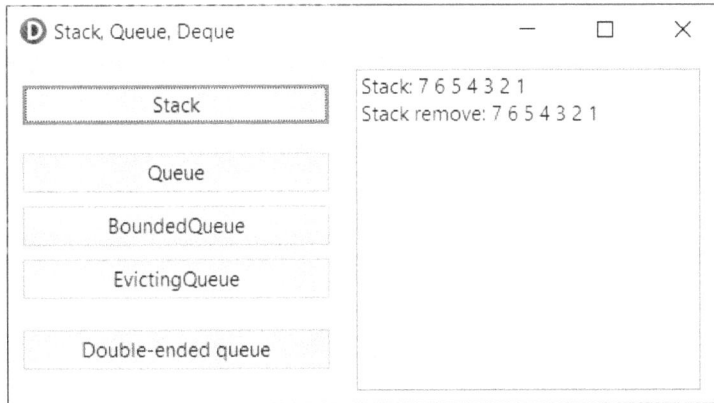

Figure 4.3 – Demonstration of basic operations on a stack

Accessing the stack via the `IEnumerable<T>` interface will therefore return items in the order they will be removed, not in the order they were added.

If you need to access the items in the order they were added, you can simply use the `Reversed` method from the `IEnumerable<T>` interface, as follows:

```
for item in stack.Reversed do
  Process(item);
```

IQueue<T>

A queue is a collection with the following properties:

- Elements are organized in a sequence that cannot be reordered
- The only two operations are *enqueue* (add an element on one end) and *dequeue* (remove it from the other end)

The end of the sequence at which the elements are added is usually called the *tail* (or *rear* or *back*) of the queue. The other end (from which the elements are removed) is usually called the *head* (or *front*) of the queue. A queue works just the same as any other queue you will encounter in real life. You join a queue at the *back*, then wait for some (typically much too long) time, and when you reach the *front* of the queue, you can finally be processed and exit the queue.t

As with the stack, the Spring queue also implements the `IEnumerable<T>` interface, which allows us to walk over the elements without removing them. The queue interface is called `IQueue<T>` and is shown here in a slightly shortened form:

```
IQueue<T> = interface(IEnumerable<T>)
  procedure Clear;
  function Enqueue(const item: T): Boolean;
  function Dequeue: T;
  function Extract: T;
  function Peek: T;
  function PeekOrDefault: T;
  function TryDequeue(var item: T): Boolean;
  function TryExtract(var item: T): Boolean;
  function TryPeek(var item: T): Boolean;
  procedure TrimExcess;
  property Capacity: Integer
    read GetCapacity write SetCapacity;
  property OnChanged: ICollectionChangedEvent<T>
    read GetOnChanged;
end;
```

This interface is quite similar to `IStack<T>`. We add data to the queue by calling the `Enqueue` function and remove it with `Dequeue` or `Extract`. The latter behave differently if a queue owns objects, where `Dequeue` returns `nil` and `Extract` returns the actual value. We can also use the `Peek` function to inspect the value at the head of the queue without removing it.

If there's a possibility that the queue is empty and you don't want to raise an exception (or check the queue length beforehand), you can call one of the "safe" versions–`PeekOrDefault`, `TryDequeue`, `TryExtract`, and `TryPeek`.

Spring implements three queue variations – normal, *bounded*, and *evicting*. The factory methods reflect that, as can be seen in the following fragment (without showing all possible overloaded versions):

```
class function CreateQueue<T>: IQueue<T>;
class function CreateBoundedQueue<T>(size: Integer): IQueue<T>;
class function CreateEvictingQueue<T>(size: Integer): IQueue<T>;
```

A normal queue is unlimited in size, while both bounded and evicting queues have a maximum size. When this size is reached and the code calls the Enqueue function, a bounded queue will throw the enqueued value away while the evicting queue will throw away the value at the head. This behavior is demonstrated in the `StackQueue` program.

The **Queue** button shows basic operations on a normal queue. The code creates a queue, enqueues elements from `'1'` to `'7'`, shows the content of the queue by using the `IEnumerable<string>` interface, and removes (and logs) all the elements from the queue:

```
procedure TfrmStackQueueMain.btnQueueClick(Sender: TObject);
var
  ch: char;
  queue: IQueue<string>;
  s: string;
begin
  queue := TCollections.CreateQueue<string>;
  for ch := '1' to '7' do
    queue.Enqueue(ch);
  ListBox1.Items.Add('Queue: ' + ''.Join(' ', queue.ToArray));

  s := 'Queue remove: ';
  while not queue.IsEmpty do
    s := s + queue.Dequeue + ' ';
  ListBox1.Items.Add(s);
end;
```

As you can see in the following figure, the enumeration returns elements in the same order that they will be dequeued (similar to the behavior of `IStack<T>`). This picture also shows the behavior of bounded and evicted queues, which we'll cover next:

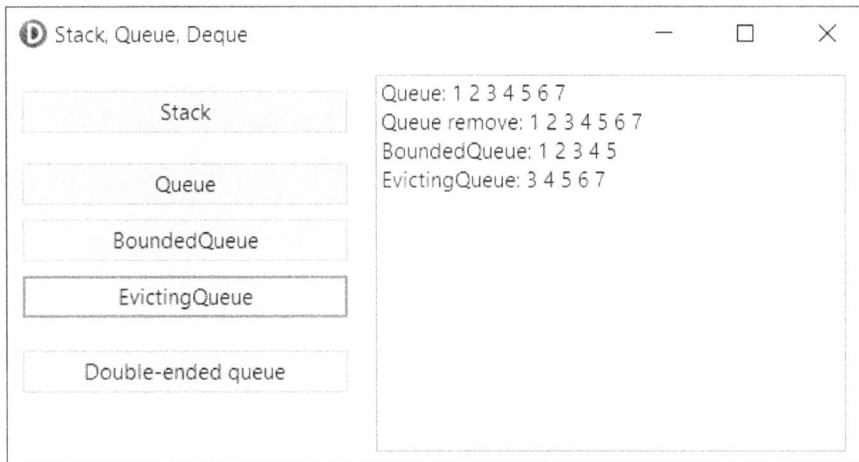

Figure 4.4 – Demonstration of all three queue variations

The bounded queue behavior is tested by clicking on the **BoundedQueue** button. This executes the code shown here, which creates a bounded queue with a maximum size of 5, enqueues strings from '1' to '7', and displays the content of the queue:

```
procedure TfrmStackQueueMain.btnBoundedQueueClick(Sender: TObject);
var
  ch: char;
  queue: IQueue<string>;
begin
  queue := TCollections.CreateBoundedQueue<string>(5);
  for ch := '1' to '7' do
    queue.Enqueue(ch);
  ListBox1.Items.Add('BoundedQueue: ' +
    ''.Join(' ', queue.ToArray));
end;
```

As we saw in the preceding figure, the queue reaches its maximum size when element '5' is added, after which elements '6' and '7' are thrown away.

The evicting queue behavior is demonstrated with the following code, attached to the **EvictingQueue** button. The code creates an evicting queue with a maximum size of 5, enqueues strings from '1' to '7', and displays the content of the queue:

```
procedure TfrmStackQueueMain.btnEvictingQueueClick(Sender: TObject);
var
  ch: char;
  queue: IQueue<string>;
begin
  queue := TCollections.CreateEvictingQueue<string>(5);
  for ch := '1' to '7' do
    queue.Enqueue(ch);
  ListBox1.Items.Add('EvictingQueue: ' +
    ''.Join(' ', queue.ToArray));
end;
```

The output was shown in *Figure 4.3*. When element '6' is added, the implementation throws away (*evicts*) the element at the head of the queue ('1'). Similarly, '2' is evicted when '7' is added.

IDeque<T>

The last collection in this section, the **double-ended queue** (**deque** for short), is similar to a queue except that we can add and remove elements from both ends. In Spring, it is defined by the IDeque<T> interface, which is – in a slightly shortened form – shown here:

```
IDeque<T> = interface(IEnumerable<T>)
  procedure Clear;
  function AddFirst(const item: T): Boolean;
  function AddLast(const item: T): Boolean;
  function RemoveFirst: T;
  function RemoveLast: T;
  function ExtractFirst: T;
  function ExtractLast: T;
  function TryRemoveFirst(var item: T): Boolean;
  function TryRemoveLast(var item: T): Boolean;
  function TryExtractFirst(var item: T): Boolean;
  function TryExtractLast(var item: T): Boolean;
  procedure TrimExcess;
  property Capacity: Integer
    read GetCapacity write SetCapacity;
  property OnChanged: ICollectionChangedEvent<T>
    read GetOnChanged;
end;
```

This interface should feel quite familiar by now. Elements are added to the deque by calling AddFirst or AddLast. They are removed by calling RemoveFirst, RemoveLast, ExtractFirst, or ExtractLast. As with the previous two collections, RemoveXXX returns nil and ExtractXXX returns an actual object if a deque owns objects. There is also a set of TryXXX functions, which don't raise exceptions.

To look ("peek") at the first or last element in the deque, you can use functions from the IEnumerable<T> interface – First, Last, FirstOrDefault, TryGetFirst, and so on.

Spring implements three variations of a deque – normal, *bounded*, and *evicting*. This is reflected in the factory functions, which are shown here in shortened form (without overloads):

```
class function CreateDeque<T>: IDeque<T>;
class function CreateBoundedDeque<T>(size: Integer):
  IDeque<T>;
class function CreateEvictingDeque<T>(size: Integer):
  IDeque<T>;
```

A normal deque is unlimited in size, while both bounded and evicting deques have a maximum size. When this size is reached and the code tries to add another element, a bounded deque will throw the new element away, while the evicting deque will remove the element at the other end (AddFirst will implicitly use RemoveLast and AddLast will implicitly execute RemoveFirst).

The deque operation is demonstrated by the code associated with the **Double-ended queue** button. The program creates a deque and adds the strings from '1' to '7' to it. Strings that represent odd values are added to the head (AddFirst) while the remaining strings are added to the end (AddLast). The code then displays all the elements in the deque by enumerating it with the IEnumerable<string> functionality. At the end, the code removes elements alternately from the head (RemoveFirst) and the tail (RemoveLast) and logs the removed values:

```
procedure TfrmStackQueueMain.btnDequeClick(Sender: TObject);
var
  i: integer;
  llDeque: IDeque<string>;
  s: string;
begin
  llDeque := TCollections.CreateDeque<string>;
  for i := 1 to 7 do
    if Odd(i) then
      llDeque.AddFirst(IntToStr(i))
    else
      llDeque.AddLast(IntToStr(i));
  ListBox1.Items.Add('Deque: ' + ''.Join(' ', llDeque.ToArray));

  s := 'Deque remove: ';
  for i := 1 to 7 do
    if Odd(i) then
      s := s + llDeque.RemoveFirst + ' '
    else
      s := s + llDeque.RemoveLast + ' ';
  ListBox1.Items.Add(s);
end;
```

In the following figure, we can see that the enumerator works from head to tail (the last value, '7', was inserted at the head and is displayed first), just as with the normal queue:

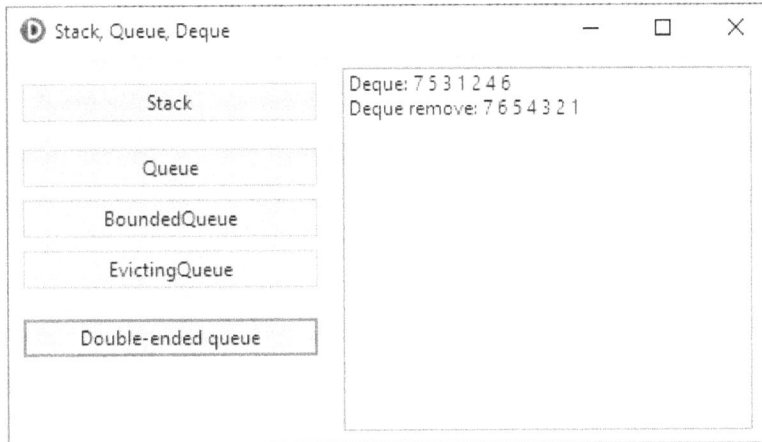

Figure 4.5 – Demonstrations of basic double-ended queue operations

This is not surprising as both `IQueue<T>` and `IDeque<T>` derive most of the functionality from the `TCircularArrayBuffer<T>` class (defined in the `Spring.Collections.Base` unit), which implements an array-based circular buffer with insertion and removal operations operating on both ends.

Algorithm complexity

Stacks and queues are implemented with arrays. Insertions can execute in linear time if there is room in an array (the best case) or can cause a potentially major memory reallocation and movement of existing items to a new array (the worst case). As Spring takes great care to manage arrays as well as possible, this worst case will only rarely occur.

- **Access**: $O(n)$
- **Search**: $O(n)$
- **Insert**: $O(1)$ average, $O(n)$ worst-case
- **Delete**: $O(1)$

Trees

All three structures we looked at in the previous section have something in common – they can be represented by a list. We can start at one side (the head of the queue, the top of the stack), follow a structure in which each item is connected to the next item only, and arrive at the end (the tail of the queue, the bottom of the stack). This simplicity makes them easy to implement but also causes them to not be very performant. Adding and removing elements at the end of the list is fast, while access and search operations are quite slow.

The structure we'll talk about now is quite different. It can be simple or it can be complex, but most importantly, it can be balanced – not too simple and not too complex, not extremely fast but also never very slow. As you may already have guessed from the section heading, this structure is called a **tree**.

A tree can be very easily defined by using recursion. A tree is either one of the following:

- An empty tree

- A *node* pointing to zero or more trees

Each non-empty tree has a starting node (a *root* node), which is connected to zero or more *child* nodes (inner nodes, inodes, or branch nodes), which are connected to zero or more nodes, and so on. Some nodes are not connected to any child nodes and we call them *leaves* (outer nodes or terminal nodes). Each child (except the root) has exactly one parent. (This rule prevents the tree from establishing cycles and turning into a **graph**.)

An important property of a node is its *height*. It is defined as the longest path from a node to any of the leaves that can be accessed from the node (without traveling up to its parent). The height of the tree is defined as the height of the root node. Leaves have a height of 0 by definition.

A height of a node can also be defined recursively. It is either one of the following:

- Zero, if the node has no children

- The maximum of its children's heights, plus one

In the following figure, you can find a small tree with a root, two inner nodes, and four leaves. The root node is marked with a thicker border while the leaves are bordered with a dashed line. A number in each node represents its height:

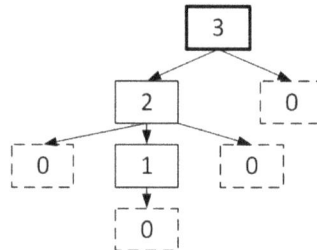

Figure 4.6 – A small tree with a height of 3

Another useful property of a node is its *level*. This is defined as a number of connections between the node and the root of the tree. In the previous figure, the root has level 0, its two children have level 1, the three children of node **2** have level 2, and the last node has level 3.

In computer science, we frequently deal with *binary* trees as they are the simplest to work with. A binary tree is just a simple normal tree with one constraint – each node can have only up to two children. We also say that in a binary tree, each node has a *left* and *right* subtree (any of which may be empty). The tree in the previous figure is not a binary tree as the inner node with height 2 has three children.

While being frequently used, binary trees are not the be-all and end-all. Other frequently encountered trees (which I will not cover in this chapter) are 2-3 trees, B-trees, and B+ trees.

As can be seen in *Figure 4.5*, trees can grow unevenly. In the previous example, the root node has two children – one leaf and another one with descendants of heights 2 and 4. In the worst case, a tree becomes a collection of nodes where each has only one child. In other words, a tree becomes a list. And we already know that lists have fast insertions and removals (at the end) but don't behave that well when inserting (in the middle) or searching. To get the best performance, we want the tree to be as even as possible. In other words, we want it to be *balanced*.

> **Definition**
> A balanced binary tree is a tree in which the left and right subtrees differ in height by no more than 1.

We can easily extend this definition to a non-binary tree. For each node, we calculate the height of all its children, and if the minimum and maximum height differ by no more than 1, the tree is balanced.

A useful tree is not just nodes and connections. In practical applications, each node will also hold a *value* that we are interested in. Actually, that value is the important part of the node; connections between parents and children are just the plumbing that allows us to quickly find and manage those values.

In the context of implementing optimal data structures, we also want our trees to implement some order. We say that a binary tree is *ordered* (this definition gets slightly complicated if a tree is not binary) when the value in each node in the left subtree is smaller than the value of the parent node and the value in each node in the right subtree is larger than the value of the parent node. In other words, smaller values are on the left and larger values are on the right, no matter at which node we are looking.

An ordered binary tree is also called a **binary search tree** (**BST**). Such a tree is great because it is simple to implement and all operations on it are of order *O(log n)*. (I will offer no proof of that. If you are interested, look it up.) BSTs also have one big problem. It is hard to keep them balanced.

The problem is demonstrated in the following picture. The tree on the left is ordered and balanced – node **3** has a height of 1 while node **6** has a height of 0 and the difference is 1, which is OK. When we add value 1 to the tree, a new node is created under node 2 (on the left, because 1 is smaller than 2). The height of the node **3** is now 2 and the tree is not balanced anymore. To fix this, we execute a sequence of *rotations* (the exact process is not important here), which makes node **3** a root of a new balanced tree:

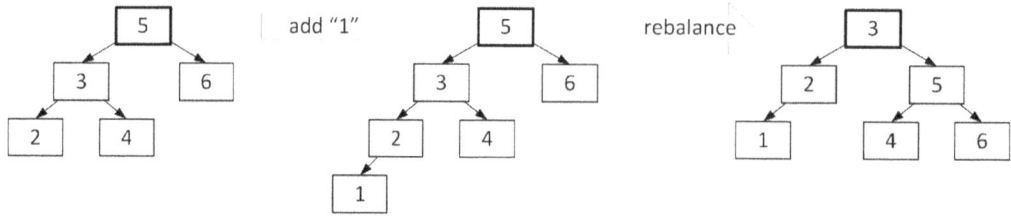

Figure 4.7 – Rebalancing a BST

The problem here is that keeping trees fully balanced is an expensive process. As you can see in the figure, two of the nodes (**3** and **5**) had their children changed in the process. In a larger tree, this number would typically be even higher.

A better idea is to keep the tree "somewhat balanced." We don't care whether it is fully balanced or not, but we also don't want it to morph into a list. There are multiple solutions to this problem (such as AVL trees), but I will ignore them and cover only the option implemented in the Spring collections – a red-black tree.

Red-black trees

A *red-black tree* is a BST with some additional information and constraints. In addition to holding a value, each node has an additional property called a *color*. A red-black tree is then defined by the following properties:

- Each tree node has a color (red or black)

- A red node can only have black children

- Every path from a node to any of its descendant leaves goes through the same number of black nodes

Such a tree is *self-balanced*. If we ensure that the rules are satisfied after every insertion or deletion, the tree will be balanced enough so the searching in the tree is an $O(log\ n)$ operation. It is also guaranteed that the balancing executes in $O(1)$ time on average. All in all, red-black trees are pretty awesome. It is no surprise that they became Mr. Smith's favorite data structure!

Red-black trees are rarely perfectly balanced, as that is too expensive to achieve. Because of that, we can have different trees that all represent the same data and are all valid red-black trees, as we can see in the following figure:

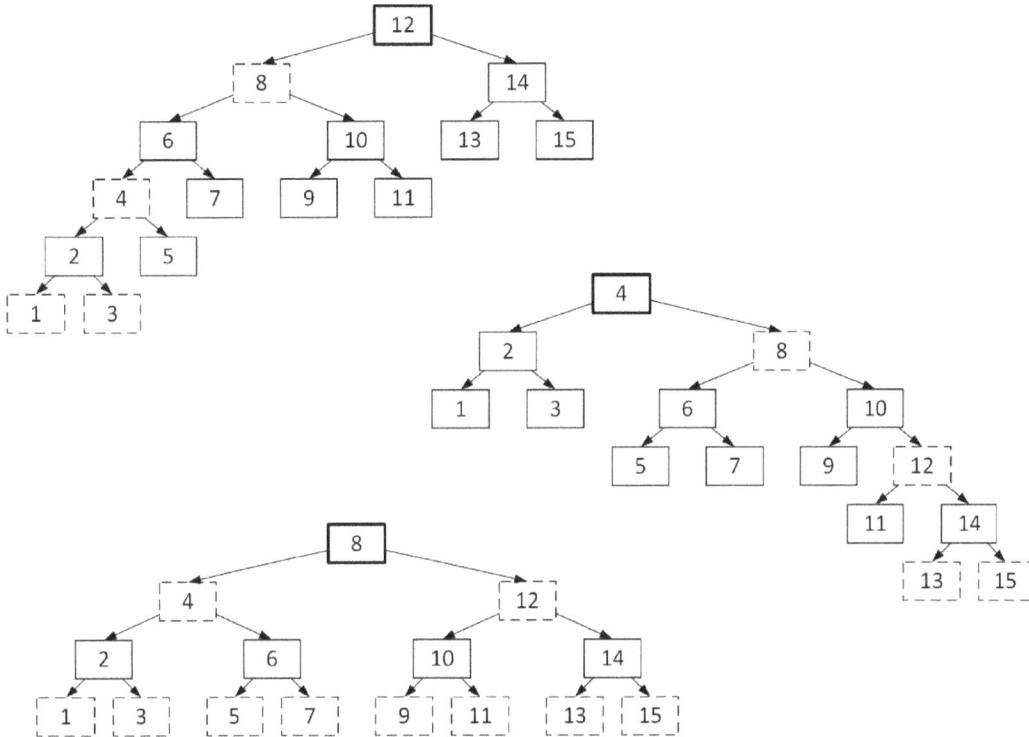

Figure 4.8 – Three ways of storing values from 1 to 15 in a red-black tree, where a solid border represents black, a dashed border represents red, and root nodes have thicker borders

The first tree was created by inserting numbers in descending order (from 15 to 1), the second by inserting numbers in ascending order (from 1 to 15), and the last one by inserting them in a specially tailored order. Later in the chapter, I will show you the code that generated these three trees.

From these examples, we can easily grasp how red-black trees achieve fast searching and not-too-expensive rebalancing. If we look only at the black nodes, they represent fully balanced (but not binary) trees (this comes from the third property). To save time when manipulating the tree, we can add red nodes, which don't affect the balancing criteria (the third property) and can therefore be added relatively quickly. The second property (no two red nodes in a row), on the other hand, prevents the tree from becoming too unbalanced (full of red nodes).

As the color has only two possible values (*red* and *black*), we can encode it with only one bit. This bit usually doesn't cause any growth of the structure size as we can hide it inside some other data. For example, Spring hides this information inside connections between nodes (inside pointers).

> **Space-saving tip**
>
> If we keep nodes 4-byte-aligned (we will talk about that in the next chapter), a pointer to each node contains a value of 0 in its bottom two bits. They can be reused to hold some data (for example, the *color*) if we don't mind doing some very fast operations before using the pointer (clearing those bits).

In `Spring.Collections.Trees`, the following two simple functions extract color and the actual pointer from the (pointer/color) combination:

```
const
  ColorMask = IntPtr(1);
  PointerMask = not ColorMask;

function TRedBlackTree.TNode.GetColor: TNodeColor;
begin
  Result := TNodeColor(IntPtr(fParent) and ColorMask);
end;

function TRedBlackTree.TNode.GetParent: PNode;
begin
  Result := Pointer(IntPtr(fParent) and PointerMask);
end;
```

Spring's implementation of a red-black tree is designed more to be a basis for other collection types than for general use. It is nonetheless a fully working tree, which we can use directly in code.

A red-black tree is defined by the two interfaces – `IRedBlackTree<T>` and `IRedBlackTree<TKey, TValue>`. The former defines a tree that stores values of type T while the latter stores (TKey, TValue) pairs but only uses the TKey part for comparison (for ordering elements). These two interfaces are derived from `IBinaryTree<T>` and `IBinaryTree<TKey, TVal>`, respectively.

Both implementations are functionally the same, so I'll focus on the simpler `IBinaryTree<T>` and `IRedBlackTree<T>` interfaces, which are shown here:

```
IBinaryTree = interface
  property Count: Integer read GetCount;
  property Height: Integer read GetHeight;
  property Root: PBinaryTreeNode read GetRoot;
end;
IBinaryTree<T> = interface(IBinaryTree)
  function Add(const key: T): Boolean;
  function Delete(const key: T): Boolean;
  function Exists(const key: T): Boolean;
  function Find(const key: T; var value: T): Boolean;
```

```
  procedure Clear;
  function GetEnumerator: IEnumerator<T>;
  function ToArray: TArray<T>;
  property Root: TNodes<T>.PRedBlackTreeNode read GetRoot;
end;

IRedBlackTree<T> = interface(IBinaryTree<T>)
  function FindNode(const value: T): TNodes<T>.PRedBlackTreeNode;
  procedure DeleteNode(node: TNodes<T>.PRedBlackTreeNode);
end;
```

As we can see, a red-black tree is a structure with a simple interface. We can add nodes (Add), remove them (Delete and DeleteNode), and look for them (Exists, Find, and FindNode). We can also extract all nodes as an array (ToArray) or examine them via for .. in enumeration or by manually walking around the tree starting at Root. While adding, searching for, and removing data is simple, walking around the tree is not. We will dig deeper into it in the following examples.

Traversing trees

In general, we use two different ways of traversing a tree – depth-first and breadth-first. With depth-first traversal, we visit all children of a node before visiting the next node at the same level. We can do this in three different ways – by visiting the current node before any of its children (*pre-order*), between visiting its children (*in-order*), or after visiting its children (*post-order*).

With breadth-first traversal, we visit the nodes "by level." We start at the root (the single node at level 0), then visit all children of the root (all nodes at level 1), then all grandchildren of the root (nodes at level 2), and so on.

All this is simpler to explain with an example, so imagine we want to traverse the tree in the following figure:

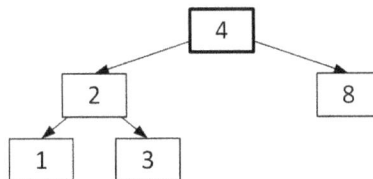

Figure 4.9 – A small tree with three levels

- A depth-first pre-order traversal would return 4, 2, 1, 3, 8.

- A depth-first in-order traversal would return 1, 2, 3, 4, 8. This is the traversal used to return elements in the sorted order.

- A depth-first post-order traversal would return 1, 3, 2, 8, 4.

- A breadth-first traversal would return 4, 2, 8, 1, 3.

The following code from the `Trees` program demonstrates this in practice. Activated with the **RB tree depth-first** button, the code first builds a tree by inserting numbers from 15 to 1 (resulting in the first tree from the first figure in this section) and then traverses it in four different ways – by using the `for .. in` statement, in in-order (`for node in tree.Root.InOrder do`), in pre-order (`for node in tree.Root.PreOrder`), and in post-order (`for node in tree.Root.PostOrder`):

```
procedure TfrmTrees.btnRBTree1Click(Sender: TObject);
var
  tree: IRedBlackTree<integer>;
  i: integer;
  node: TNodes<integer>.PRedBlackTreeNode;
  s: string;
begin
  tree := TRedBlackTree<integer>.Create;
  for i := 15 downto 1 do
    tree.Add(i);
  s := '';
  for i in tree do s := s + IntToStr(i) + ' ';
  ListBox1.Items.Add('Enumerator: ' + s);
  s := '';
  for node in tree.Root.InOrder do
    s := s + IntToStr(node.Key) + ' ';
  ListBox1.Items.Add('InOrder: ' + s);
  s := '';
  for node in tree.Root.PreOrder do
    s := s + IntToStr(node.Key) + ' ';
  ListBox1.Items.Add('PreOrder: ' + s);
  s := '';
  for node in tree.Root.PostOrder do
    s := s + IntToStr(node.Key) + ' ';
  ListBox1.Items.Add('PostOrder: ' + s);
end;
```

Traversing the tree with the `for .. in` statement is trivial, as it directly returns the node values. Traversing in other manners is more complicated, as the enumerator returns a pointer to the internal node representation (`TNodes<T>.PRedBlackTreeNode`) from which we have to extract the data (the `Key` field).

Executing this code results in the following output:

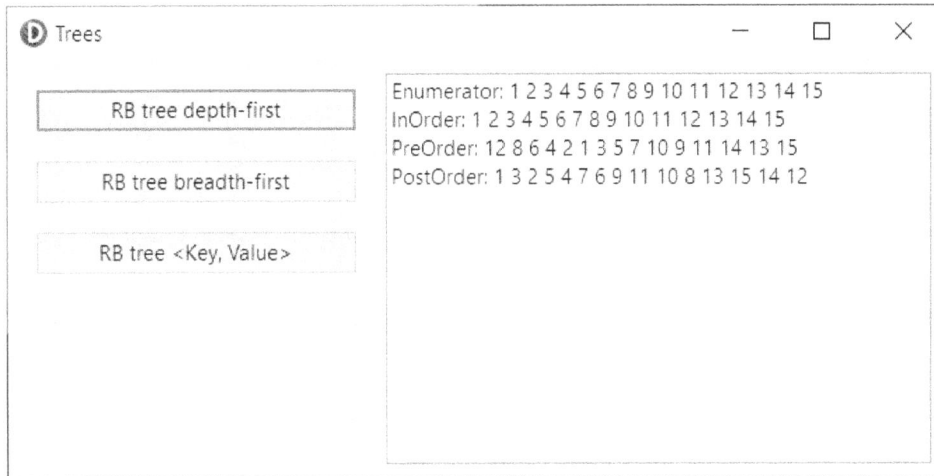

Figure 4.10 – The result of traversing a tree in depth-first order

We can see that the `for  ..  in` enumeration returns the values in sorted in-order, which was somehow expected.

The breadth-first traversal is not implemented in Spring. We can do this type of traversal with a slightly convoluted piece of code, demonstrated in the following code fragment from the `Trees` program. The code creates a tree as in the previous example and manages two queues, one containing the values from the current level that we have yet to traverse (`q`), and another containing the values from the next level to traverse (`q2`). It also logs the value and color (`r` and `b`) of each node plus an "end of data" marker (`x`) after each leaf:

```
procedure TfrmTrees.btnRBTree2Click(Sender: TObject);
var
  q, q2: IQueue<TNodes<integer>.PRedBlackTreeNode>;
  node: TNodes<integer>.PRedBlackTreeNode;
  t: IRedBlackTree<integer>;
  i: integer;
  s: string;
begin
  t := TRedBlackTree<integer>.Create;
//  for i := 1 to 15 do
  for i := 15 downto 1 do
//  for i in GenerateOptimalInsertionOrder(1, 15) do     t.Add(i);

  q := TCollections.CreateQueue<
        TNodes<integer>.PRedBlackTreeNode>;
```

```
  q.Enqueue(TNodes<integer>.PRedBlackTreeNode(t.Root));
  while not q.IsEmpty do begin
    s := '';
    q2 := TCollections.CreateQueue<
            TNodes<integer>.PRedBlackTreeNode>;
    while not q.IsEmpty do begin
      node := q.Dequeue;
      if not assigned(node) then
        s := s + 'x '
      else begin
        s := s + IntToStr(node.Key) +
                IfThen(node.Color = Black, 'b ', 'r ');
        q2.Enqueue(node.Left);
        q2.Enqueue(node.Right);
      end;
    end;
    ListBox1.Items.Add(s);
    q := q2;
  end;
end;
```

The code extracts information from the node data of type TNodes<T>.PRedBlackTreeNode. It gives us access to the node's value (Key), color (Color), and children (Left and Right). The only tricky part is the initial setup of the q queue. As the Root property returns a non-generic pointer, we have to cast it to TNodes<integer>.PRedBlackTreeNode before it is inserted into the queue.

> **Note**
>
> Although nodes are represented as pointers, they are not allocated individually. Spring stores node data in a dynamically allocated "array of arrays of node data" structure fStorage: TArray<TArray<TRedBlackTreeNode<T>>> and then subdivides this space into individual nodes.

This method was used to generate all three red-black tree examples shown earlier in this section. If you comment out the for i := 1 to 15 do loop at the top of the method and uncomment one of the other for loops, you will get the other two examples. The GenerateOptimalInsertionOrder method (which is part of the example code) returns elements in an "optimal" insertion order – that is, one that results in a perfectly balanced tree. Further analysis of this method is left as a practical example for the reader.

Let's conclude this section with a simple example of the IRedBlackTree<TKey, TValue>. The following code from the Tree program, attached to the **RB tree <Key, Value>** button, creates a tree with an integer key and a string value. Then it inserts four (key-value) pairs and logs the tree content:

```
procedure TfrmTrees.btnRBTree3Click(Sender: TObject);
var
  tree: IRedBlackTree<integer, string>;
  node: TPair<integer, string>;
begin
  tree := TRedBlackTree<integer, string>.Create;
  tree.Add(3, 'three');
  tree.Add(1, 'one');
  tree.Add(2, 'two');
  tree.Add(1, 'One');
  for node in tree do
    ListBox1.Items.Add(IntToStr(node.Key) + ': ' + node.Value);
end;
```

Even though the code inserts four different (key-value) pairs, there are only three different keys (**1**, **2**, and **3**), which results in a tree with three nodes, as shown in the following figure:

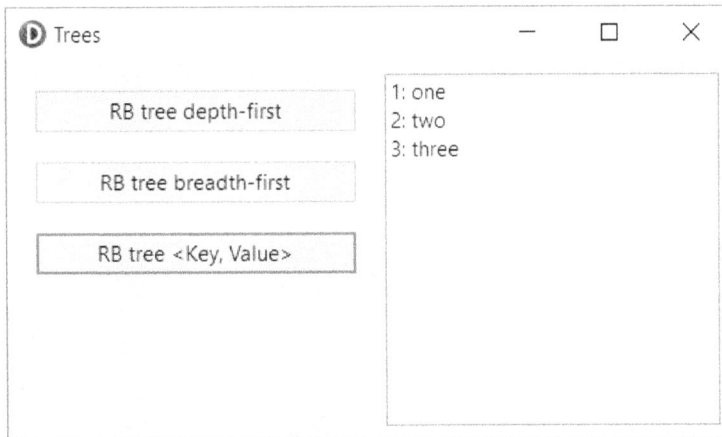

Figure 4.11 – Inserting data into a (key, value) tree

As we can see from the output, the last Add call is simply ignored. The new value, 'One', does not overwrite the original value, 'one'. The Add function returns True if the value was added and False if it was ignored.

> **Algorithm complexity**
>
> Rebalancing a red-black tree has `O(1)` complexity on average. All other operations execute in logarithmic time:
>
> * **Access**: *O(log n)*
> * **Search**: *O(log n)*
> * **Insert**: *O(log n)*
> * **Delete**: *O(log n)*

Sets and dictionaries

Our journey through Spring collections is almost at its end. Before we discover wonderful sets and dictionaries, I have to – albeit briefly – mention *hash tables*. They will appear extensively in this section, and although I don't have space to talk about them, I can at least describe them in general terms.

Hash tables

The simplest way to introduce hash tables is to start with arrays. An **array** is a linear structure containing multiple elements of the same type, where we can use an offset from the start (an *index*) to access one element of the array. We can, for example, write a simple code fragment to store some strings in an array:

```
var
  names: array [1..3] of string;
names[1] := 'Spring';
names[2] := 'February';
names[3] := 'Wednesday';
```

A **hash table** is a generalization of this concept. Instead of using integer values, we can use data of any type for the index (in hash tables, we call this index a *key*). A hash table where the key is a string and the value is an integer could, for example, be used as shown here (warning – this is a made-up example and would not compile):

```
var
  order: THashTable<string,integer>;
order['Spring'] := 1;
order['February'] := 2;
order['Wednesday'] := 3;
```

Hash tables do their magic through a process called *hashing*. They use a special function (*hash function*) to convert a key into a number (*hash*). Then, they apply some additional magic to convert this hash to an index into some internal array where values are stored. Additional complications are caused by *collisions* (where two keys are converted into the same hash) and by the fact that we also want to remove values, not just add them.

All this results in a pretty complicated mess with many competing implementations, but luckily you can ignore all that. Instead of using hash tables directly, you would just use one of Spring collections and everything else would be taken care of automatically.

One question springs up – why use hash tables at all? Well, they have some pretty awesome characteristics. We can add, remove, and look up elements in constant time *O(1)*! Admittedly, reading a value from a hash table is much slower than reading it from an array but the speed of access remains the same, however big the hash table gets.

The next question is – why do we need other types of storage if hash tables are so great? The most significant reason is that they cannot implement a sort order. If we need the data to be sorted, we have to use something else – either a sorted array or a tree. Another problem with most implementations is that enumerating hash tables is *unstable*. If we add an element to a hash table, we usually have no idea how it will be inserted into its internal array. Because of that, we don't know in what order `for .. in` will return elements from the table.

Actually, scratch that last problem. A Spring hash table (implemented in `Spring.HashTable`) is written so that the elements are always added one after another. This keeps enumeration predictable (elements will always be returned in the *insertion order*) and, even better, allows us to access the elements directly via an integer index. This makes it possible even to access elements (by index) as an `O(1)` operation.

> **Note**
>
> Accessing elements is an *O(1)* operation only if you don't remove elements from the hash table. Removed elements leave "holes" in the table, and those holes must be filled before accessing an element by index. This changes the complexity of the operation to *O(n)*.

Let that be enough of an introduction. If you want to know more about hash tables, you'll have to find information online – there are plenty of examples and explanations. Here is a link at which you can learn more: `https://www.hackerearth.com/practice/data-structures/hash-tables/basics-of-hash-tables/tutorial/`. I, however, will continue with more practical topics.

ISet<T>

It feels right to start this part of the journey with the simplest data structure. A Spring *set* represents an extension to Delphi's `set` type. It is similar to a mathematical concept of a *set*. We can put data in a set, or remove it from a set, with the limitation that a set will only keep one copy of any given value. For example, the following code fragment will result in a set of two values – 1 and 2:

```
set.Add(1); // set contains 1
set.Add(2); // set contains 1, 2
set.Add(3); // set contains 1, 2, 3
```

```
set.Add(3); // set contains 1, 2, 3
set.Remove(3); // set contains 1, 2
```

A set is represented by the `ISet<T>` interface, which is simply an `ICollection<T>` (that one already handles addition and removal of elements) with added functions to calculate operations on sets (intersection, union, difference, equality checks, and so on), as shown in the next code fragment:

```
ISet<T> = interface(ICollection<T>)
  procedure ExceptWith(const other: IEnumerable<T>);
  procedure IntersectWith(const other: IEnumerable<T>);
  procedure UnionWith(const other: IEnumerable<T>);
  function IsSubsetOf(const other: IEnumerable<T>): Boolean;
  function IsSupersetOf(const other: IEnumerable<T>): Boolean;
  function SetEquals(const other: IEnumerable<T>): Boolean;
  function Overlaps(const other: IEnumerable<T>): Boolean;
  procedure TrimExcess;
  property Capacity: Integer read GetCapacity write SetCapacity;
end;
```

Spring implements two variations of a set – a normal (unsorted) set and a sorted set. The following code fragment shows the two basic `TCollections` factories, without the additional overloads:

```
class function CreateSet<T>: IOrderedSet<T>;
class function CreateSortedSet<T>: ISet<T>;
```

Unlike the lists that also have unsorted and sorted versions, which are both basically the same, Spring implements unsorted and sorted sets in two completely different ways. An unsorted set is implemented with a hash table while the sorted set is implemented with a red-black tree. This is also the reason why the `CreateSet<T>` factory returns `IOrderedSet<T>` and not `ISet<T>`. This interface allows us to access set elements via integer indices, as shown here:

```
IOrderedSet<T> = interface(ISet<T>)
  function IndexOf(const item: T): Integer;
  property Items[index: Integer]: T read GetItemByIndex;
end;
```

The `SetDictionaryMultiMap` program implements a short demonstration of Spring sets. We will look at it later after we learn about multisets.

Algorithm complexity

As the unsorted set is implemented with a hash table, operations on it are faster than operations on a sorted set, which is implemented with a tree.

- **Access**: *O(1)* best case scenario for an unsorted set, *O(n)* worst case, not available in a sorted set

- **Search**: *O(1)* for an unsorted set, *O(log n)* for a sorted set

- **Insert**: *O(1)* for an unsorted set, *O(log n)* for a sorted set

- **Delete**: *O(1)* for an unsorted set, *O(log n)* for a sorted set

---

IMultiSet<T>

A *multiset* (in computer science usually called a *bag*) is a variation of a set that can store the same value multiple times. It is defined by the IMultiSet<T> interface, which is (in a slightly shortened form) shown here:

```
TMultiSetEntry<T> = packed record
  Item: T;
  Count: Integer;
end;

IMultiSet<T> = interface(ICollection<T>)
  function Add(const item: T; count: Integer): Integer;
  function Remove(const item: T; count: Integer): Integer;
  function SetEquals(const other: IEnumerable<T>): Boolean;
  function OrderedByCount: IReadOnlyMultiSet<T>;
  property Entries: IReadOnlyCollection<TMultiSetEntry<T>>
    read GetEntries;
  property Items: IReadOnlyCollection<T> read GetItems;
  property ItemCount[const item: T]: Integer
    read GetItemCount write SetItemCount; default;
end;
```

Unlike a normal set, a multiset doesn't implement standard operations on sets (unions, intersections, and so on). It can, however, return different *views* on the data. The Entries property returns a read-only collection of (value, count) pairs, the Items property returns a collection containing only values, and the ItemCount property returns or sets a number of occurrences of an element. We can even access elements ordered by the number of occurrences (OrderedByCount).

> **Tip**
>
> If you iterate over IMultiSet<T> with a for .. in construct, it will return each element as many times as it is contained in the collection. This is demonstrated later in this section. The ToArray function will also create an array containing multiple copies of the same element. The Items collection, however, contains each value only once (it is, in fact, a true *set*).

In addition to an unsorted version, Spring also implements a sorted multiset. The following code fragment shows the basic version of the corresponding factories, without other overloads:

```
class function CreateMultiSet<T>: IMultiSet<T>;
class function CreateSortedMultiSet<T>: IMultiSet<T>;
```

Although both factories return the same interface (there is no *ordered* access to a multiset), the implementation of the two differs, the same as in ISet<T>. An unsorted multiset is implemented with a hash table while the sorted version is implemented with a red-black tree.

The SetDictionaryMultiMap program demonstrates the basic multiset interface. A click on the **Set** button executes the following code, which tests all variations of sets and multisets. The code creates a collection, adds numbers from 4 to 1 in descending order, adds 1 again, and lists the content of the collection:

```
procedure TfrmSetMultiMap.TestSet(const name: string;
  const aset: ICollection<integer>);
var
  i: integer;
begin
  for i := 4 downto 1 do
    aset.Add(i);
  aset.Add(1);
  ListBox1.Items.Add(name + ': ' + Join(' ', aset));
end;

procedure TfrmSetMultiMap.btnSetClick(Sender: TObject);
begin
  TestSet('Set',
    TCollections.CreateSet<integer>);
  TestSet('SortedSet',
    TCollections.CreateSortedSet<integer>);
  TestSet('MultiSet',
    TCollections.CreateMultiSet<integer>);
  TestSet('SortedMultiSet',
    TCollections.CreateSortedMultiSet<integer>);
end;
```

In the next figure, we can see that both "normal" versions output elements in the insertion order while both "sorted" versions output elements in the sorted order. Additionally, both sets contain only one occurrence of value 1 while both multisets contain two occurrences of value 1:

Figure 4.12 – Comparing sorted and unsorted sets and multisets

---

Algorithm complexity

The unsorted multiset is implemented with a hash table and has better time complexity than the sorted multiset, which is implemented with a tree.

- **Access**: $O(1)$ unsorted multiset best-case, $O(n)$ worst-case, not available in sorted multiset
- **Search**: $O(1)$ unsorted multiset, $O(\log n)$ sorted multiset
- **Insert**: $O(1)$ unsorted multiset, $O(\log n)$ sorted multiset
- **Delete**: $O(1)$ unsorted multiset, $O(\log n)$ sorted multiset

---

IDictionary<TKey, TValue>

A dictionary is a natural extension of a set. While a set contains elements (*keys*) and nothing else, a dictionary stores keys associated with values. As with the set, a dictionary can store a specific key only once.

You are probably already familiar with dictionaries as the standard Delphi implementation has been with us for quite a long time. It is no wonder that the Spring interface, IDictionary<TKey, TValue>, closely mirrors the Delphi version, as shown here in a slightly shortened form:

```
IDictionary<TKey, TValue> = interface(IMap<TKey, TValue>)
  procedure AddOrSetValue(const key: TKey; const value: TValue);
    deprecated 'Use dict[key] := value instead';
  function Extract(const key: TKey): TValue;
  function GetValueOrDefault(const key: TKey): TValue;
  function GetValueOrDefault(const key: TKey;
```

```
      const defaultValue: TValue): TValue;
    function TryExtract(const key: TKey; var value: TValue): Boolean;
    function TryGetValue(const key: TKey; var value: TValue): Boolean;
    function TryUpdateValue(const key: TKey;
      const newValue: TValue; var oldValue: TValue): Boolean;
    function AsReadOnly: IReadOnlyDictionary<TKey, TValue>;
    procedure TrimExcess;
    property Capacity: Integer
      read GetCapacity write SetCapacity;
    property Items[const key: TKey]: TValue
      read GetItem write SetItem; default;
  end;
```

The simplest way to access a dictionary is through the Items property. Writing to Items[key] sets the value, associated with the key, or overwrites the previously associated value. The same thing can be achieved by calling the AddOrSetValue function. A similar function, TryUpdateValue, updates the value and returns True only if the key already exists in the dictionary. If the key is not found, the dictionary is not changed and the function returns False.

Reading from Items[key] returns the value associated with the key or raises an exception if the key is not found in the dictionary. If you want to avoid exceptions, use the TryGetValue function.

Extract and TryExtract can be used to remove a (key, value) pair from the dictionary and return the value.

IDictionary<TKey, TValue> defines only part of the interface. Some interesting functions can also be found in the IMap<TKey, TValue> interface, which is shown next:

```
  IMap<TKey, TValue> = interface(ICollection<TPair<TKey, TValue>>)
    procedure Add(const key: TKey; const value: TValue);
    function TryAdd(const key: TKey; const value: TValue): Boolean;
    function Remove(const key: TKey): Boolean;
    function Remove(const key: TKey; const value: TValue): Boolean;
    function RemoveRange(const keys: array of TKey): Integer;
    function RemoveRange(const keys: IEnumerable<TKey>): Integer;
    function Extract(const key: TKey; const value: TValue):
      TPair<TKey, TValue>;
    function Contains(const key: TKey; const value: TValue):
      Boolean; overload;
    function ContainsKey(const key: TKey): Boolean;
    function ContainsValue(const value: TValue): Boolean;
    property Keys: IReadOnlyCollection<TKey> read GetKeys;
    property Values: IReadOnlyCollection<TValue>
      read GetValues;
    property OnKeyChanged: ICollectionChangedEvent<TKey>
```

```
    read GetOnKeyChanged;
  property OnValueChanged: ICollectionChangedEvent<TValue>
    read GetOnValueChanged;
end;
```

This interface contains (besides other things) functions to check for the presence of a key (ContainsKey) or a value (ContaintsValue) and gives us access to all Keys and Values. (Be careful with ContainsValue, as it executes in *O(n)*. If you need quick access to values, see IBiDiDirectionary<TKey, TValue> later in this chapter.) As with Delphi's TDictionary, we can iterate over all (key, value) pairs (for kv in dictionary), over keys (for key in dictionary.Keys), or over values (for value in dictionary.Values).

It is worth repeating that Spring redefines the TPair<TKey, TValue> type in the Spring.Collections unit. This may result in conflicts if you use System.Generics.Collections and Spring.Collections in the same unit, so beware! Most problems can be solved by explicitly prefixing the TPair definition in the code with the appropriate unit, for example, as in the following code fragment:

```
var
   kv1: System.Generics.Collections.TPair<integer, string>;
   kv2: Spring.Collections.TPair<string, integer>;
```

As with sets, a dictionary can be either unsorted or sorted. The basic factories for creating both kinds of a dictionary are shown here:

```
class function CreateDictionary<TKey, TValue>:
  IOrderedDictionary<TKey, TValue>;
class function CreateSortedDictionary<TKey, TValue>:
  IDictionary<TKey, TValue>;
```

It probably won't surprise you to learn that an unsorted dictionary is implemented with a hash table while the sorted dictionary is implemented with a red-black tree. This allows the unsorted dictionary to implement ordered access with the IOrderedDictionary<TKey, Value> interface, as shown next:

```
IOrderedDictionary<TKey, TValue> = interface(IDictionary<TKey,
TValue>)
   function IndexOf(const key: TKey): Integer;
   function AsReadOnly:
     IReadOnlyOrderedDictionary<TKey, TValue>;
   property Items[index: Integer]: TPair<TKey, TValue>
     read GetItemByIndex;
end;
```

The SetDictionaryMultiMap program demonstrates the difference between unsorted and sorted dictionaries with the code attached to the **Dictionary** button. It is very similar to the code demonstrating the *set* collections, so I won't show it in the book. I will rather move to an interesting dictionary derivation – a bidirectional dictionary.

Algorithm complexity

Algorithmic complexities for dictionaries are based on the underlying implementation –hash tables for unsorted dictionaries and red-black trees for sorted dictionaries.

- **Access**: $O(1)$ best-case scenario for an unsorted dictionary, $O(n)$ worst case, not available in sorted dictionary

- **Search**: key – $O(1)$ for an unsorted dictionary, $O(\log n)$ for a sorted dictionary; value: $O(n)$

- **Insert**: $O(1)$ for an unsorted dictionary, $O(\log n)$ for a sorted dictionary

- **Delete**: $O(1)$ for an unsorted dictionary, $O(\log n)$ for a sorted dictionary

---

IBiDiDictionary<TKey, TValue>

A bidirectional dictionary is a dictionary that works "both ways." We can use a key to look up a value, but also a value to look up a key. Both operations execute with the same algorithmic complexity.

A bidirectional dictionary is implemented as a pair of interconnected dictionaries. One maps keys to values and the other maps values to keys. Whatever modification we apply to one of them is automatically reflected in the other. The defining interface, shown next, is reduced to the minimum:

```
IBidiDictionary<TKey, TValue> = interface(IDictionary<TKey, TValue>)
  property Inverse: IBidiDictionary<TValue, TKey>
    read GetInverse;
end;
```

An `IBidiDictionary<TKey, TValue>` is an unsorted dictionary with an additional function, `Inverse`, which returns the associated "inverse" dictionary mapping `TValue` to `TKey`.

As an example, instead of using `bidi.ContainsValue(value)` (which is slow, $O(n)$), we would use `bidi.Inverse.ContainsKey(value)` (which is fast, $O(1)$).

There are no sorted bidirectional dictionaries, so there's only one class of factory functions to create such a dictionary (with a bunch of overloads, which are not shown here, as usual):

```
class function CreateBidiDictionary<TKey, TValue>:
  IBidiDictionary<TKey, TValue>;
```

The demonstration program, `SetDictionaryMultiMap`, contains a short demo for the bidirectional dictionary, which I will not reproduce in this book.

> **Algorithm complexity**
>
> The algorithmic complexity for bidirectional dictionaries is the same as for an unsorted hash table-based dictionary. (It is actually approximately twice slower than the dictionary, but we are ignoring constant factors when determining algorithmic complexity.)
>
> - **Access**: Not available
> - **Search**: *O(1)*
> - **Insert**: *O(1)*
> - **Delete**: *O(1)*

IMultiMap<TKey, TValue>

To conclude this chapter, we'll look into the most complicated Spring collection, a *multimap*. This is a more complicated version of a dictionary where each key can be associated with multiple values. This collection is defined by the `IMultiMap<TKey, TValue>` interface, which is shown here:

```
IMultiMap<TKey, TValue> = interface(IMap<TKey, TValue>)
  function Add(const key: TKey; const value: TValue): Boolean;
  procedure AddRange(const key: TKey; const values: array of TValue);
  procedure AddRange(const key: TKey;
    const values: IEnumerable<TValue>);
  function Extract(const key: TKey): ICollection<TValue>;
  function TryGetValues(const key: TKey;
    var values: IReadOnlyCollection<TValue>): Boolean;
  function AsReadOnly: IReadOnlyMultiMap<TKey, TValue>;
  property Items[const key: TKey]:
    IReadOnlyCollection<TValue> read GetItems; default;
end;
```

Some of the functionality is defined by the `IMultiMap<TKey, TValue>` interface, which was covered in this chapter's section on `IDictionary<TKey, TValue>`. Other methods are similar to the `IDictionary<TKey, TValue>` versions except that they support adding multiple values for one key in one call (`AddRange`) and can return all values associated with a key as a collection (`Items` and `TryGetValues`).

The complexity of this collection arises from the implementation details. Multimaps can be unsorted (based on a hash table) or sorted (based on a red-black tree). In addition, we have three options for the collection of values associated with a key. It can support multiple values, in which case, it will be implemented as a list, or it can support unique values and be unsorted (hash table) or sorted (red-black tree). This gives us six combinations, which are reflected in the TCollections factories. The six simplest multimap factories are listed in the next code fragment:

```
class function CreateMultiMap<TKey, TValue>(
  ownerships: TDictionaryOwnerships = []):
  IMultiMap<TKey, TValue>;
class function CreateHashMultiMap<TKey, TValue>(
  ownerships: TDictionaryOwnerships = []):
  IMultiMap<TKey, TValue>;
class function CreateTreeMultiMap<TKey, TValue>(
  ownerships: TDictionaryOwnerships = []):
  IMultiMap<TKey, TValue>;
class function CreateSortedMultiMap<TKey, TValue>(
  ownerships: TDictionaryOwnerships = []):
  IMultiMap<TKey, TValue>;
class function CreateSortedHashMultiMap<TKey, TValue>(
  ownerships: TDictionaryOwnerships = []):
  IMultiMap<TKey, TValue>;
class function CreateSortedTreeMultiMap<TKey, TValue>(
  ownerships: TDictionaryOwnerships = []):
  IMultiMap<TKey, TValue>
```

The SetDictionaryMultiMap program demonstrates the difference between all six variations in the code associated with the **MultiMap** button. The code runs the same procedure, TestMultiMap, for each multimap type. This code maps between numbers 3, 2, and 1 and characters c, b, and a, respectively. Each mapping is added twice. The code then iterates over the Keys collection, and for each key, through the associated values (via the default Items property). At the end, it logs the contents of the multimap:

```
procedure TfrmSetMultiMap.TestMultiMap(const name: string;
  const mmap: IMultiMap<integer, string>);
var
  i: integer;
  ch: char;
  sLog,s: string;
begin
  for i := 3 downto 1 do
    for ch := 'c' downto 'a' do begin
      mmap.Add(i, ch);
      mmap.Add(i, ch);
```

```
    end;

  sLog := '';
  for i in mmap.Keys do begin
    if sLog <> '' then
      sLog := sLog + ', ';
    sLog := sLog + IntToStr(i) + ':';
    for s in mmap[i] do
      sLog := sLog + s;
  end;
  ListBox1.Items.Add(name + ': ' + sLog);
end;

procedure TfrmSetMultiMap.btnTestMultimapClick(Sender: TObject);
begin
  TestMultiMap('MultiMap',
    TCollections.CreateMultiMap<integer,string>);
  TestMultiMap('HashMultiMap',
    TCollections.CreateHashMultiMap<integer,string>);
  TestMultiMap('TreeMultiMap',
    TCollections.CreateTreeMultiMap<integer,string>);
  TestMultiMap('SortedMultiMap',
    TCollections.CreateSortedMultiMap<integer,string>);
  TestMultiMap('SortedHashMultiMap',
    TCollections.CreateSortedHashMultiMap<integer,string>);
  TestMultiMap('SortedTreeMultiMap',
    TCollections.CreateSortedTreeMultiMap<integer,string>);
end;
```

The results of this code are shown in the following figure:

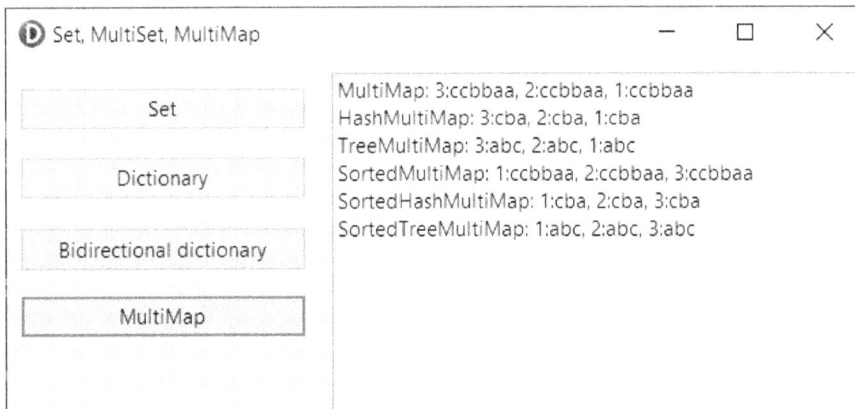

Figure 4.12 – Comparing six multimap variations

We can see that unsorted multimaps return keys in insertion order (3, 2, 1) while sorted multimaps return keys in sorted order (1, 2, 3). Additionally, "standard" versions return all values (ccbbaa) while "hash" versions return values in insertion order (cba), and "tree" versions return values in sorted order (abc).

Algorithm complexity

As there are so many different implementations, an algorithmic complexity table also becomes quite messy:

- **Access**: Not available

- **Search**: Key – $O(1)$ normal multimap, $O(\log n)$ sorted multimap; value – $O(n)$

- **Insert**: Key – $O(1)$ normal multimap, $O(\log n)$ sorted multimap; value – $O(1)$ list, hash, $O(\log n)$ tree

- **Delete**: Key – $O(1)$ normal multimap, $O(\log n)$ sorted multimap; value – $O(n)$ list, $O(1)$ hash, $O(\log n)$ tree

The complexities listed for inserting and deleting values do not include the cost of accessing the key. For example, in a `SortedHashMultiMap`, we can remove a value in $O(1)$, but as we have to access the key in $O(\log n)$ time first, the total complexity will be $O(\log n)$.

Summary

This chapter provided an overview of Spring4D's extensive support for data collection. I introduced the Spring4D library and discussed some potential incompatibilities with the previous version. Then, I spent some time looking at the `IEnumerable<T>` interface, which is common to all Spring collections and implements a rich set of functionalities. After that, I looked into Spring's implementation of the basic computer science collections – stacks, queues, and deques.

In the second half of the chapter, I introduced the concept of *trees*, which is largely unfamiliar to Delphi programmers, and then continued with a description of the Spring implementation of a red-black tree, which is used in many collection types, but also available for direct use in code.

I completed the chapter with a description of extremely useful sets, multisets, dictionaries, and multimaps, which all appear in multiple sub-implementations. This chapter compared their implementations, which will help the reader select the right tool for the job.

In the next chapter, I'll focus on optimizing the code at a smaller level. We will look at various Delphi language and RTL elements, and you'll learn which parts of the language and system libraries can behave slower than one would expect and what you can do to improve that.

5
Fine-Tuning the Code

I have stated repeatedly that the best way to improve code speed is to change the algorithm. Sometimes, however, this is just not possible because you already use the best possible algorithm. It may also be completely impractical, as changing the algorithm may require rewriting a large part of the program.

If this is the case and the code is still not running quickly enough, we have multiple available options. We can start *optimizing* the program, by which I mean that we can start changing small parts of the program and replacing them with faster code. This is the topic of this chapter.

Another option is to rewrite a part of the program in an *assembler*. Delphi offers quite decent support for this, and we'll look into that possibility near the end of this chapter. We can also find a *library* that solves our problem. It could be written in an assembler or created by a compiler that produces faster code than Delphi – usually, we don't care. We only have to know how to *link* it to our program, which will be the topic of *Chapter 11*, *Using External Libraries*.

The last option is to somehow split our problem into multiple parts and solve them in parallel. We'll dig into this approach from *Chapter 7* to *Chapter 10*.

But first, let's look at fine-tuning code, or how to make a program run faster with small changes. In this chapter, you'll discover the following:

- How Delphi compiler settings can affect the program speed
- Why it is good to use local variables to store intermediate results
- How to work with the CPU window
- What happens behind the scenes when you work with different data types
- How to speed up code by using RTL functions
- How to use an assembler to speed up code
- How we can make `SlowCode` run even faster

Technical requirements

All code in this chapter was written with Delphi 11.3 Alexandria. Most of the examples, however, can also be executed on Delphi XE and newer versions. You can find all the examples on GitHub: `https://github.com/PacktPublishing/Delphi-High-Performance---Second-Edition/tree/main/ch5`.

Delphi compiler settings

First things first – before you start meddling with the code, you should check the Delphi compiler settings for your project. In some situations, they can affect code speed quite a lot.

To check and possibly change compiler settings, open your project, and then select **Project | Options** from the menu or press *Ctrl + Shift + F11*. Relevant options can be found in **Building | Delphi Compiler | Compiling**, as shown in the following screenshot:

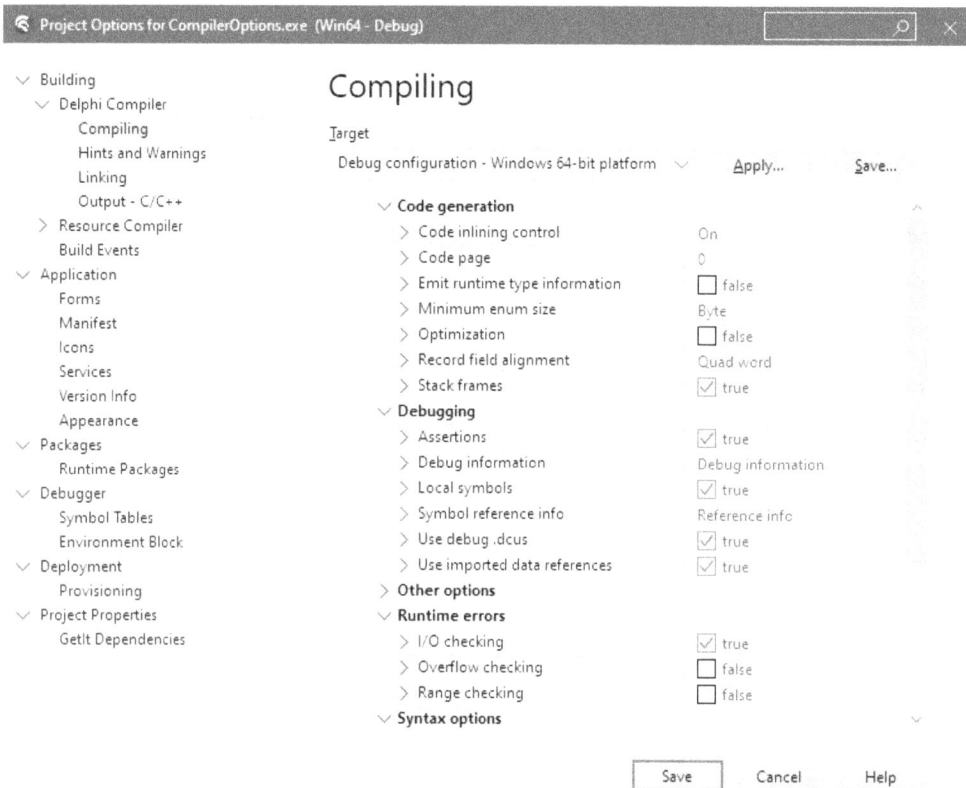

Figure 5.1 – Compiler settings that influence program speed

We will look at the following options:

- **Code inlining control**
- **Optimization**
- **Record field alignment**
- **Assertions**
- **Overflow checking**
- **Range checking**

All of these settings can be enabled/disabled for an entire project, and they can also be turned on/off in the code by using *compiler directives* (comments that start with {$).

Code inlining control

We will look into the concept of **code inlining** later in this chapter, in the *Optimizing method calls* section, so this section will serve just as an introduction.

By setting the **Code inlining control** value, you define the default behavior for the complete project. The possible values for that setting are **On**, **Off**, and **Auto**. This default value can be changed in the code by inserting {$INLINE ON}, {$INLINE OFF}, or {$INLINE AUTO} into the source.

The INLINE state can be set to a different value at a place where the inlined method is defined and a place where it is used (called). This creates six possible combinations.

When we look at the method definition, INLINE has the following meaning:

- INLINE ON: If the method is marked with the inline directive (more on that later), it will be marked as *inlinable*
- INLINE AUTO: This is the same as INLINE ON, except that any routine not marked with inline will still be marked *inlinable* if its code size is less than or equal to 32 bytes
- INLINE OFF: The routine will not be marked as *inlinable*, even if it is marked with inline

Depending on its place of use, INLINE has a different meaning:

- INLINE ON: If a called method is marked as *inlinable*, it will be expanded inline (at the place of call) if possible
- INLINE AUTO: This is the same as INLINE ON
- INLINE OFF: The routine will not be expanded inline, even if it is marked *inlinable*

Optimization

The **Optimization** option controls code optimization. The possible values are **true** and **false**. This setting can also be controlled with compiler directives {$OPTIMIZATION ON} and {$OPTIMIZATION OFF} or with the respective {$O+} and {$O-} equivalents.

When optimization is enabled, the compiler optimizes generated code for speed. Optimizations are guaranteed not to change the behavior of the program. The only reason you may want to turn optimization off is to improve debugging. When optimization is turned on, debugging – especially stepping through code and evaluating variables – may not work correctly in all cases.

> Tip
> My approach is to turn optimization off during development and on when building the release version.

The $O (or $OPTIMIZATION) directive can only turn optimization on or off for an entire method. You cannot turn optimization on or off for selected lines within a method.

To show the effect of different compiler options, I have written a CompilerOptions demo. When you click the **Optimization** button, the following code will be run twice – once with optimization turned on and once off.

The code initializes an array and then does some calculations on all of its elements. All of this is repeated 1,000 times because, otherwise, it is hard to measure differences. Modern CPUs are fast!

```
function NonOptimized: int64;
var
  arr1: array [1..50000] of Integer;
  i,j: Integer;
begin
  for j := 1 to 1000 do
  begin
    for i := Low(arr1) to High(arr1) do
      arr1[i] := i;
    for i := Low(arr1) to High(arr1)-1 do
      arr1[i] := arr1[i] + arr1[i+1];
  end;
end;
```

The results will be different, depending on the CPU speed and model, but on all computers, the optimized version should be significantly faster. On my test machine, the optimized version runs approximately six times faster than the non-optimized version! Here are the results:

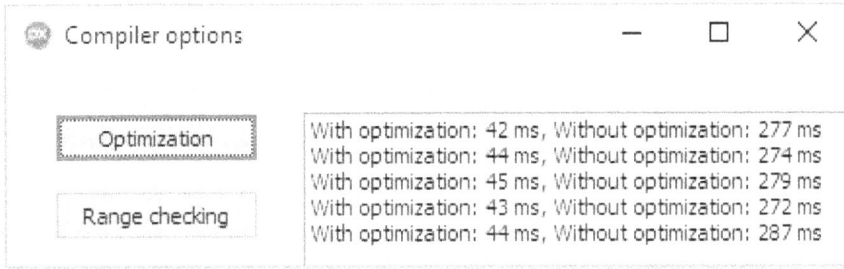

Figure 5.2 – The results of changing the Optimization setting

Although you can turn compilation options on and off with compiler directives, Delphi provides no direct support to switch back to the previous state. You can, however, check whether a compilation option is currently turned on or off by using the {$IFOPT} compiler directive.

Then, you can define or undefine some conditional compilation symbol that will tell you later what the original state of the compilation option was:

```
{$IFOPT O+}{$DEFINE OPTIMIZATION}{$ELSE}{$UNDEF OPTIMIZATION}{$ENDIF}
{$OPTIMIZATION ON}
```

When you want to restore the original state, use this conditional to turn the compilation option off if it was not enabled initially:

```
{$IFNDEF OPTIMIZATION}{$OPTIMIZATION OFF}{$ENDIF}
```

This technique is used throughout the CompilerOptions demo with different compilation options.

Record field alignment

The third compiler option I want to discuss regulates the alignment of fields in Delphi record and class types. It can be set to the following values – Off, Byte, Word, Double Word, and Quad Word. Settings are a bit misleading, as the first two values actually result in the same behavior.

You can use the {$ALIGN 1}, {$ALIGN 2}, {$ALIGN 4}, {$ALIGN 8}, and {$ALIGN 16} compiler directives to change record field alignment in code, or the equivalent short forms – {$A1}, {$A2}, {$A4}, {$A8}, and {$A16}, respectively. There are also two directives that exist only for backward compatibility. {$A+} means the same as {$A8} (which is also a default for new programs), and {$A-} is the same as {$A1}.

> **Note**
> You can set *double-quadword alignment* in the code ({$A16}) but not in the compiler options settings.

Field alignment controls exactly how fields in records and classes are laid out in memory.

Let's say that we have the following record, and let's say that the address of the first field in the record is simply 0:

```
type
  TRecord = record
    Field1: byte;
    Field2: int64;
    Field3: word;
    Field4: double;
  end;
```

With the {$A1} alignment, each field will simply follow the next one. In other words, Field2 will start at address 1, Field3 at 9, and Field4 at 11. As the size of double is 8 (as we'll see later in this chapter), the total size of the record is 19 bytes.

Tip

The Pascal language has a syntax that enforces this behavior without the use of compiler directives. You can declare a record as a packed record, and its fields will be packed together as with the {$A1} alignment, regardless of the current setting of this directive. This is very useful when you have to interface with libraries written in other languages.

With the {$A2} alignment, each field will start on a word boundary. In layman's terms, the address of the field (offset from the start of the record) will be divisible by 2. Field2 will start at address 2, Field3 at 10, and Field4 at 12. The total size of the record will be 20 bytes.

With the {$A4} alignment, each field will start on a double word boundary, so its address will be divisible by 4. (You can probably see where this is going.) Field2 will start at address 4, Field3 at 12, and Field4 at 16. The total size of the record will be 24 bytes.

Finally, with the {$A8} alignment, each field will start on a quad word boundary, so its address will be divisible by 8. Field2 will start at address 8, Field3 at 16, and Field4 at 24. The total size of the record will be 32 bytes.

Saying all this, I have to add that the $A directive doesn't function exactly as I described it. Delphi knows how simple data types should be aligned (for example, it knows that an integer should be aligned on a double-word boundary) and will not move them to higher alignment, even if it is explicitly specified by a directive. For example, the following record will use only 8 bytes, even though we explicitly stated that fields should be quad-word aligned:

```
{$A8}
TIntegerPair = record
  a: integer;
```

```
    b: integer;
  end;
```

If you need to exactly specify the size and alignment of all fields (for example, if you pass records to some API call or an external library), it is best to use the packed record directive and insert unused padding fields into the definition. The next example specifies a record containing two quad-word aligned integers:

```
TIntegerPair = packed record
  a: integer;
  filler: integer;
  b: integer;
end;
```

The following diagram shows how this record is laid out in memory with different record field alignment settings. Fields are renamed F1 to F4 for brevity. X marks unused memory:

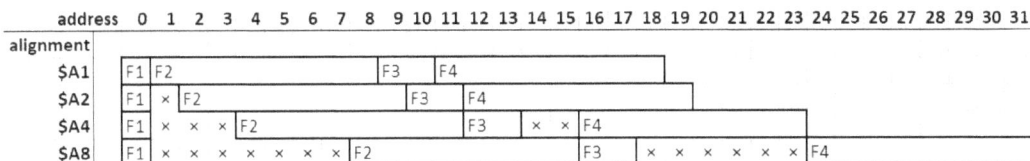

Figure 5.3 – Different ways of record fields packing

Why is all this useful? Why don't we always just pack fields together so that the total size of a record or class is as small as possible? Well, that is an excellent question!

As traditional wisdom says, CPUs work faster when the data is correctly aligned. Accessing 4-byte data (an `integer`, for example) is faster if its address is double-word aligned (i.e., is divisible by 4). Similarly, 2-byte data (`word`) should be word-aligned (i.e., the address is divisible by 2) and 8-byte data (`int64`) should be quad-word aligned (i.e., the address is divisible by 8). This will significantly improve performance in your program.

But will it really? Does this traditional wisdom make any sense in the modern world?

The `CompilerOptions` demo contains sets of measurements done on differently aligned records. It is triggered with the **Record field align** button.

Running the test shows something surprising – all four tests (for A1, A2, A4, and A8) run at almost the same speed. Actually, the code operating on the best-aligned record (A8) is the slowest! I must admit that I didn't expect this while preparing the test.

A little detective work has shown that sometime around 2010, Intel did a great job optimizing the data access on its CPUs. If you manage to find an older machine, it will show a big difference between unaligned and aligned data. However, all Intel CPUs produced after that time will run on unaligned and aligned data at the same speed. Working on unaligned (packed) data may actually be faster, as more data will fit into the processor cache.

What is the moral lesson of all this? **Guessing is nothing; hard numbers are everything!** Always measure. There is no guarantee that your changes will actually speed up the program.

Assertions

The **Assertions** compiler option enables or disables code generation for `Assert` statements. You can use the `{$ASSERTIONS ON}` and `{$ASSERTIONS OFF}` compiler directives to turn this option on or off in the code. The short forms of these directives are rather cryptic – `{$C+}` and `{$C-}`, respectively.

Delphi allows us to use runtime checks in the code in the form of `Assert` statements. `Assert` will check whether the first parameter evaluates to `false` and will raise an exception if it is `true`. The second (optional) parameter will be used for the exception message.

The following statement will raise an exception if the i variable is smaller than zero:

```
Assert(i >= 0, 'Expecting a positive value here');
```

If you turn code generation for assertions off, the compiler will just skip such statements, and they will not generate any code. You can, for example, use this to remove assertions from the release code. Even if any argument to `Assert` contains a function call, this function will not be called when assertions are turned off.

Overflow checking

The **Overflow checking** option regulates whether the compiler checks whether certain arithmetic operations (+, -, *, Abs, Sqr, Inc, Dec, Succ, and Pred) produce a number that is too large or too small to fit into the data type. You can use the `{$OVERFLOWCHECKS ON}` and `{$OVERFLOWCHECKS OFF}` compiler options (or the respective short forms, `{$Q+}` and `{$Q-}`) to turn this option on and off in a specific part of a program.

For example, the following program will silently increment `$FFFFFFFF` to 0 in the first `Inc` statement but will raise the `EIntOverflow` exception in the second `Inc` statement:

```
procedure TfrmCompilerOptions.btnOverflowErrorClick(Sender: TObject);
var
  i: cardinal;
begin
  {$Q-}
  i := $FFFFFFFF;
  // Without overflow checks, Inc works and i will be 0
```

```
  Inc(i);

  {$Q+}
  i := $FFFFFFFF;
  // With overflow checks, Inc will raise an exception
  Inc(i);
end;
```

Enabling overflow checking doesn't make a big change in the program speed in almost all Delphi programs, so you can freely use it whenever it is needed. The only exception is if you are doing a lot of numerical calculations, in which case you would want to profile the code running with and without the overflow checking to see whether turning it on makes a big difference for you.

I would suggest turning it on at least in debug configuration, as that will help to find errors in the code.

Range checking

The last compiler option I want to discuss, **Range checking**, tells the compiler whether to check indices to array elements and string characters. In other words, it will check whether an expression such as s[idx+1], where idx+1 represents a valid index.

You can turn range checking on and off with the {$RANGECHECKS ON} and {RANGECHECKS OFF} compiler directives (or {$R+} and {$R-}, respectively).

Let's take a look at an example. In the method shown here, the second for loop accesses the arr[101] element without any error, although the maximum array element is arr[100]. The third for loop, however, will raise an ERangeCheck exception when accessing arr[101]:

```
procedure TfrmCompilerOptions.btnRangeErrorClick(Sender: TObject);
var
  arr: array [1..100] of Integer;
  i: Integer;
begin
  for i := Low(arr) to High(arr) do
    arr[i] := i;

  {$R-}
  for i := Low(arr) to High(arr) do
    arr[i] := arr[i] + arr[i+1];

  {$R+}
  for i := Low(arr) to High(arr) do
    arr[i] := arr[i] + arr[i+1];
end;
```

This kind of checking is so important that I always leave it in my code, even in the release version. Accessing a nonexistent array or string element may not seem so dangerous, but what if you are writing into that element? If this is not caught, your code just overwrites some other data with nonsense values. This leads to extremely hard-to-find problems!

> **Warning**
> In older Delphi versions, range checking is turned off even in the debug build. I would strongly recommend turning it on!

What about the "cost" of this checking? As the `CompilerOptions` program shows, it can be significant. In this example, turning range checking on slows down code by 50%:

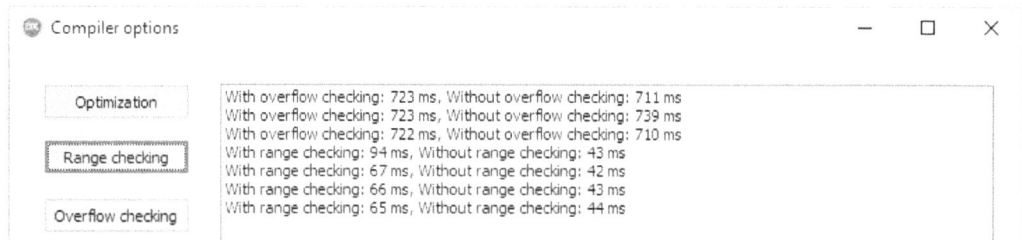

Figure 5.4 – Comparing the code running with or without overflow checking and range checking

In such cases, turning range checking off can speed up the program. I would still recommend that you do that just for critical parts of code, not the whole program. It is also recommended to do this only for the release version of the program.

This brings us to the end of this very long but necessary section, as understanding developer tools is always a good thing. Let us now (finally!) switch to something more interesting – real programming.

Extracting common expressions

This next tip will sound obvious, but it will nicely introduce us to the next topic. Plus, it is a real problem frequently found in production code.

The `ExtractCommonExpression` demo creates a list box with a mere 1,000 entries, all in the form of *author–title*. A click on the **Complicated expression** button runs a short piece of code that reverses the order of author and title in the list box so that it shows entries in the form of *title–author*:

```
procedure TfrmCommonExpression.Button1Click(
  Sender: TObject);
var
  i: Integer;
  sw: TStopwatch;
```

```
begin
  ListBox1.Items.BeginUpdate;
  try
    sw := TStopwatch.StartNew;
    for i := 0 to ListBox1.Count - 1 do
      ListBox1.Items[i] :=
        Copy(ListBox1.Items[i], Pos('-', ListBox1.Items[i])
             + 1, Length(ListBox1.Items[i]))
      + '-'
      + Copy(ListBox1.Items[i], 1,
             Pos('-', ListBox1.Items[i]) - 1);
    sw.Stop;
    Button1.Caption := IntToStr(sw.ElapsedMilliseconds);
  finally ListBox1.Items.EndUpdate; end;
end;
```

The code goes over the list and, for each entry, finds the '-' character, extracts the first and second parts of the entry, and puts them back together, reversed. It does that, however, in a terrible copy-and-paste way. The code refers to ListBox1.Items[i] five times while calculating the result. It also calls Pos('-', ListBox1.Items[i]) twice.

In a really good compiler, you could expect that both subexpressions mentioned in the previous paragraph would be calculated only once. Not with Delphi's compiler, though. It has some optimization built in, but Delphi's optimization is far from the level required for such tricks to work. That leaves the burden of optimization on us, the programmers.

The second button in this demo executes the code shown next. This implementation of the same algorithm is not only more readable but also accesses ListBox1.Items[i] only once. It also calculates the position of '-' inside the string only once:

```
procedure TfrmCommonExpression.Button2Click(
  Sender: TObject);
var
  i: Integer;
  s: string;
  p: Integer;
  sw: TStopwatch;
begin
  ListBox1.Items.BeginUpdate;
  try
    sw := TStopwatch.StartNew;
    for i := 0 to ListBox1.Count - 1 do begin
      s := ListBox1.Items[i];
      p := Pos('-', s);
```

```
        ListBox1.Items[i] := Copy(s, p + 1, Length(s)) + '-'
                             + Copy(s, 1, p - 1);
      end;
      sw.Stop;
      Button2.Caption := IntToStr(sw.ElapsedMilliseconds);
   finally ListBox1.Items.EndUpdate; end;
end;
```

Comparing both approaches shows a definite improvement in the second case. The first method uses around 40 ms and the second one around 30 ms, which is 25% faster. The code only times the inner `for` loop, not the update of the list box itself, which takes the same time in both cases.

I can recap all this with a simple statement – *calculate every subexpression only once.* This is good advice, but, still, don't exaggerate. Simple expressions, such as `i+1` or `2*i`, are so *cheap* (in computing time) that extracting them in a subexpression won't speed up the code.

The helpful CPU window

In situations similar to the previous example, it doesn't hurt if you can look at the generated assembler code to check what is going on behind the scenes. Luckily, the Delphi IDE provides a great way to do just that.

I will be the first to admit that examining assembler code is not for everyone. You can be a great Delphi programmer even though you have no idea how to read assembler instructions.

If you recognized yourself in the previous statement, don't worry. This section is included just for people who want to know everything. You can safely skip it knowing that you'll still be able to understand everything else in the book. However, if you're still interested, then, by all means, read on!

The Delphi IDE gives us a few different tools to view the low-level state of the code and computer. They are hidden away in the not-so-obvious **View | Debug Windows | CPU Windows** submenu. The most useful view is called **Entire CPU** and encompasses all other views in that submenu. It is displayed as an editor tab named **CPU**.

Be aware, though, that the CPU window is only useful while you are debugging a program. The best way to use it is to pause the program (either by inserting a breakpoint, clicking the *pause* icon, or selecting the **Run | Program Pause** menu) and then switch to the **CPU** view.

> Tip
>
> You can quickly access the **Entire CPU** view with the *Ctrl + Alt + C* key combination. If that doesn't work, check whether you have installed some add-on expert that overrides this key combination. For example, GExperts assigns this combination to the `Copy Raw Strings` command.

This view is composed of five parts. The **Disassembly** panel (at the top left) shows your code as assembler instructions intermixed with Pascal source lines for easier orientation. If you don't see Pascal lines, right-click into that view and verify whether the menu item, **Mixed source**, is checked.

To the right of that (in the top-right corner) are two small panels showing processor registers and flags. Below them (in the bottom-right corner) is the current thread's stack. Finally, a panel in the bottom-left corner shows the process's memory at some address:

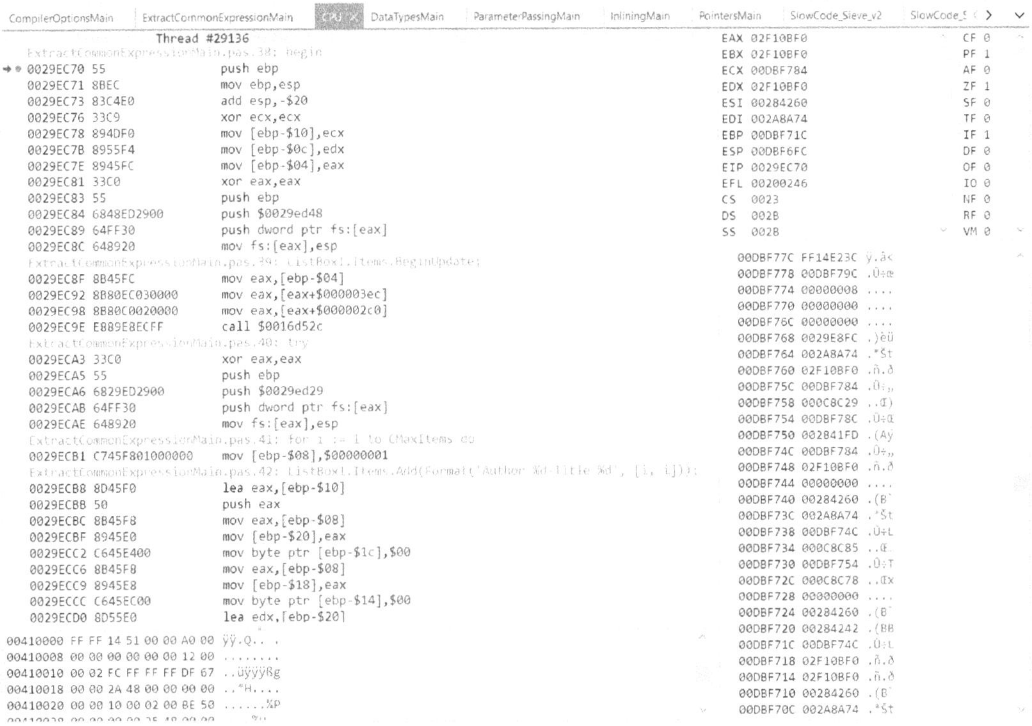

Figure 5.5 – The CPU window in Delphi 11.3 Alexandria

If you use the *F8* key to step to the next command while the **Entire CPU** view is active, Delphi will not step to the next Pascal instruction but to the next assembler instruction instead. If you want to go to the next Pascal instruction, use the *Shift + F7* combination or the **Run | Trace to Next Source Line** menu.

We can use the **CPU** view to verify that the first code example from the previous section really compiles to much worse code than the second example.

The following figure shows part of the code from the original (on the left) and the hand-optimized method (on the right). Green rectangles mark the code that accesses `ListBox1.Items[i]`. We can clearly see that the original code does that twice, once before calling the `Pos` function (`call Pos`) and once immediately after that, while the hand-optimized method executes the same sequence of instructions only once:

```
ExtractCommonExpressionMain.pas.55: ListBox1.Items[i] :=
→ 005DEE0F 8D45D0        lea eax,[ebp-$30]
  005DEE12 50            push eax
  005DEE13 8D4DCC        lea ecx,[ebp-$34]
  005DEE16 8B45FC        mov eax,[ebp-$04]
  005DEE19 8B80EC030000  mov eax,[eax+$000003ec]
  005DEE1F 8B80C0020000  mov eax,[eax+$000002c0]
  005DEE25 8B55F8        mov edx,[ebp-$08]
  005DEE28 8B18          mov ebx,[eax]
  005DEE2A FF530C        call dword ptr [ebx+$0c]
  005DEE2D 8B45CC        mov eax,[ebp-$34]
  005DEE30 E85FBAE2FF    call @UStrLen
  005DEE35 50            push eax
  005DEE36 8D4DC8        lea ecx,[ebp-$38]
  005DEE39 8B45FC        mov eax,[ebp-$04]
  005DEE3C 8B80EC030000  mov eax,[eax+$000003ec]
  005DEE42 8B80C0020000  mov eax,[eax+$000002c0]
  005DEE48 8B55F8        mov edx,[ebp-$08]
  005DEE4B 8B18          mov ebx,[eax]
  005DEE4D FF530C        call dword ptr [ebx+$0c]
  005DEE50 8B55C8        mov edx,[ebp-$38]
  005DEE53 B901000000    mov ecx,$00000001
  005DEE58 B8ACEF5D00    mov eax,$005defac
  005DEE5D E8B665E2FF    call Pos
  005DEE62 40            inc eax
  005DEE63 50            push eax
  005DEE64 8D4DC4        lea ecx,[ebp-$3c]
  005DEE67 8B45FC        mov eax,[ebp-$04]
  005DEE6A 8B80EC030000  mov eax,[eax+$000003ec]
  005DEE70 8B80C0020000  mov eax,[eax+$000002c0]
  005DEE76 8B55F8        mov edx,[ebp-$08]
  005DEE79 8B18          mov ebx,[eax]
  005DEE7B FF530C        call dword ptr [ebx+$0c]
  005DEE7E 8B45C4        mov eax,[ebp-$3c]
  005DEE81 5A            pop edx
  005DEE82 59            pop ecx
  005DEE83 E854C3E2FF    call @UStrCopy
  005DEE88 FF75D0        push dword ptr [ebp-$30]
  005DEE8B 68ACEF5D00    push $005defac
  005DEE90 8D45C0        lea eax,[ebp-$40]
  005DEE93 50            push eax
  005DEE94 8D4DBC        lea ecx,[ebp-$44]
  005DEE97 8B45FC        mov eax,[ebp-$04]
  005DEE9A 8B80EC030000  mov eax,[eax+$000003ec]
```

```
ExtractCommonExpressionMain.pas.75: s := ListBox1.Items[i];
→ 005DF02B 8D4DF4        lea ecx,[ebp-$0c]
  005DF02E 8B45FC        mov eax,[ebp-$04]
  005DF031 8B80EC030000  mov eax,[eax+$000003ec]
  005DF037 8B80C0020000  mov eax,[eax+$000002c0]
  005DF03D 8B55F8        mov edx,[ebp-$08]
  005DF040 8B18          mov ebx,[eax]
  005DF042 FF530C        call dword ptr [ebx+$0c]
ExtractCommonExpressionMain.pas.76: p := Pos(' ', s);
  005DF045 B901000000    mov ecx,$00000001
  005DF04A 8B55F4        mov edx,[ebp-$0c]
  005DF04D B85CF15D00    mov eax,$005df15c
  005DF052 E8C163E2FF    call Pos
  005DF057 8945F0        mov [ebp-$10],eax
ExtractCommonExpressionMain.pas.77: ListBox1.Item[i] := Copy(s, p + 1,
  005DF05A 8D45C8        lea eax,[ebp-$38]
  005DF05D 50            push eax
  005DF05E 8B45F4        mov eax,[ebp-$0c]
  005DF061 E82EB8E2FF    call @UStrLen
  005DF066 8BC8          mov ecx,eax
  005DF068 8B55F0        mov edx,[ebp-$10]
  005DF06B 42            inc edx
  005DF06C 8B45F4        mov eax,[ebp-$0c]
  005DF06F E86BC1E2FF    call @UStrCopy
  005DF074 FF75C8        push dword ptr [ebp-$38]
  005DF077 685CF15D00    push $005df15c
  005DF07C 8D45C4        lea eax,[ebp-$3c]
  005DF07F 50            push eax
  005DF080 8B45F0        mov eax,[ebp-$10]
  005DF083 49            dec ecx
  005DF084 BA01000000    mov edx,$00000001
  005DF089 8B45F4        mov eax,[ebp-$0c]
  005DF08C E84BC1E2FF    call @UStrCopy
  005DF091 FF75C4        push dword ptr [ebp-$3c]
  005DF094 8D45CC        lea eax,[ebp-$34]
  005DF097 BA03000000    mov edx,$00000003
  005DF09C E853C0E2FF    call @UStrCatN
  005DF0A1 8B4DCC        mov ecx,[ebp-$34]
  005DF0A4 8B45FC        mov eax,[ebp-$04]
  005DF0A7 8B80EC030000  mov eax,[eax+$000003ec]
  005DF0AD 8B80C0020000  mov eax,[eax+$000002c0]
  005DF0B3 8B55F8        mov edx,[ebp-$08]
  005DF0B6 8B18          mov ebx,[eax]
  005DF0B8 FF5320        call dword ptr [ebx+$20]
```

Figure 5.6 – A comparison between the original and optimized compiled code

> **Note**
>
> If the CPU window shows addresses (for example, `call $00325418`) instead of function names (`call Pos`), disable **Support address space layout randomization (ASLR)** in the project options (from the **Building | Delphi Compiler | Linking** configuration path).

Behind the scenes

A critical part of writing fast code is to understand what happens behind the scenes. Are you appending strings? You should know how that is implemented in a compiler. Passing a dynamic array into a function? Ditto. Wondering whether you should create 10,000 instances of a class or just create a large array of records? Knowing the implementation details will give you the answer.

In this section, I'll dig down into some frequently used data types and show how using them will bring in unexpected complexity. I will discuss memory and memory allocation, but I will treat them as very abstract entities. I'll use phrases such as, "A new string gets allocated," which means that a secret part of code, called the **memory manager**, gets memory from Windows and tells the program, "You can store your string here." We'll dig deep into the bowels of memory manager in *Chapter 6, Memory Management*.

A plethora of types

The Delphi language contains an immense number of built-in types. There's no way we can cover each of them, so let us firstly establish some classification that will simplify the rest of the discussion:

- The most basic of all built-in types are **simple types**. In this group belong the following:

  - **Integer types**, such as Byte, ShortInt, Word, SmallInt, Integer, Cardinal, Int64, UInt64, NativeInt, NativeUInt, LongInt, and LongWord

  - **Character types**, such as Char, AnsiChar, WideChar, UCS2Char, and UCS4Char

  - **Boolean types**, such as Boolean, ByteBool, WordBool, and LongBool

  - **Enumerated types**, such as `TScrollBarStyle = (ssRegular, ssFlat, ssHotTrack)`

  - **Real types**, such as Real48, Single, Real, Double, Extended, Comp, and Currency

- Similar to these are **pointer types**, such as Pointer, `PByte`, and `^TObject`. They are, after all, just memory addresses, so we can always treat a pointer as a NativeUInt and vice versa.

- Following these are **strings**, a very interesting family hiding intriguing implementation details. It contains **short strings** (such as string[42]), **ansi strings** (AnsiString, RawByteString, and UTF8String), and **unicode strings** (string, UnicodeString, and WideString).

- The last group is **structured types**. This group, which will be of great interest in this chapter, is composed of **sets** (such as `TAlignSet = set of TAlign`), **arrays** (static arrays and dynamic arrays), **records** (traditional and advanced – records with methods), **classes**, and **interfaces**. Anonymous methods can also be treated as interfaces, as they are implemented in the same way.

A different classification, which I'll also occasionally use, splits all types into two groups – **managed** and **unmanaged**. Managed types are types for which the compiler automatically generates life cycle management code, and unmanaged types are all the rest.

When a type's life cycle is managed, the compiler automatically initializes variables and fields of that type. The compiler also inserts the code that automatically releases memory associated with data of such type when it is no longer used anywhere in a program (going *out of scope* is the phrase often used).

Managed types include strings, dynamic arrays, and interfaces. Records are managed only when they contain at least one field of these types. On older Delphi versions and platforms that support ARC (iOS, Android, and Linux), classes are also managed.

> **Note**
>
> Support for **Automatic Reference Counting** (**ARC**) was removed from the Linux compiler in Delphi 10.3 and from Android and iOS compilers in Delphi 10.4.

Different types (more specifically, different classes of types) are implemented differently by the Delphi compiler and, as such, present a different challenge for a developer. In the rest of this section, I'll describe the internal implementation of the most interesting data types.

> **Tip**
>
> The best resource to really understand how all data types are implemented in Delphi is available in Embarcadero's *Docwiki*, under the title *Internal Data Formats (Delphi)*: `https://docwiki.` `embarcadero.com/RADStudio/en/Internal_Data_Formats_(Delphi)`.

Simple types

Let's start with simple types, as their implementation is indeed simple. Simple types don't allocate any memory. They are either stored as part of structured types (classes, records, and so on), on the stack (local variables of such types), or in a part of a process's memory (global variables of such types). Because of this, all operations with them are very fast.

To initialize the variable of a simple type, the compiler just has to move a few bytes around. The same goes for modifying the variable or copying one variable into another. (Fields behave just the same as variables'for all Delphi data types, so in the future, I'll just talk about variables.)

Strings

String types are probably the most used data type in Delphi. They also have a very interesting implementation, optimized for fast execution. To be exact, `AnsiString` (with all its variations) and `UnicodeString` (also known as `string`) are optimized. The `WideString` type is implemented in a different manner. As *short strings* (declarations such as `string[17]`) are only used for backward compatibility, I won't discuss them in this book.

Let's deal with the more widely used `AnsiString` and `UnicodeString` first. Data of these types is represented with a pointer to a block, allocated from the memory manager. If a string is empty, this pointer will be `nil` (which is at the CPU level represented with the number zero) and if a string is not empty, it will point to some memory.

Strings are managed types and, as such, are always initialized to the default value, `nil`.

As string data is allocated from the memory manager, changing a length of a string (for example, by executing the s := s + s statement) typically results in some memory operations. In this example, a block of memory that contains the string data would be resized, and string data would be copied to the end of the original data. Appending to a string is, therefore, a relatively expensive operation, as it reallocates memory.

> **Note**
>
> The memory manager performs some interesting optimizations to make string modification much cheaper – in terms of CPU cycles – than you would expect. You can read more about that in the next chapter, *Memory management*.

Modifying part of a string without changing its length is, however, a pretty simple operation. Code just has to move the appropriate number of bytes from one place in memory to another, and that's that.

Interesting things happen when you assign one non-empty string to another (s := GenerateString(); s1 := s;). To make such operations fast and efficient, the compiler points both variables to the same memory and increments an integer-sized counter (called the *reference count*) in that same memory. A reference count represents the number of variables that currently share the same string data.

A reference count is initialized to 1 when a string is initialized (meaning that only one variable uses this string). After s1 := s, both s and s1 point to the same memory, and the reference count is set to 2.

If one of these two variables is no longer accessible (for example, it was a local variable in a method that is no longer executing), the compiler will generate life cycle management code that decrements the reference count. If the new reference count is zero, nobody uses the string data anymore, and that memory block is released back into the system.

Therefore, after s := s1, we have two strings pointing to the same memory. However, what happens if one of these strings is modified? What happens, for example, if we do s1[1] := 'a'? It would be very bad if that also modified the original string, s.

Again, the compiler comes to the rescue. Before any modification of a string, the code will check whether the string's reference count is larger than one. If so, this string is shared with another variable. To prevent any mess, the code will at that point allocate new memory for the modified string, copy the current contents into this memory, and decrement the original reference count. After that, it will change the string so that it will point to the new memory and modify the content of that memory. This mechanism is called **copy-on-write**.

You can also force the copy-on-write behavior without actually modifying the string by calling the UniqueString function. After s1 := s; UniqueString(s1);, both variables will point to separate parts of memory and have a reference count of 1.

The situation is slightly different if a string is initialized from a constant value. If you execute a statement that initializes a string from a constant value, such as s := 'Delphi';, the compiler will simply point the s variable to the part of the program data where this string was created. To indicate that this is a special, constant string that cannot be modified, it will have a reference count set to -1. If we then execute the s1 = s; statement, both strings will have a reference count set to -1.

Any attempt to modify such a string would, again, trigger the copy-on-write mechanism, which will copy string data into a memory area that can be modified.

> **Tip**
>
> Calling the SetLength method on a string also makes the string unique.

The following code, taken from the DataTypes demo, demonstrates a part of that behavior and does the following:

1. It initializes s1 from a constant and logs information (address, reference count, and content) about the string.

2. Then, the code calls UniqueString to move the string data into writeable memory and logs information about the string again.

3. After that, the code assigns s1 to s2 so that they point to the same string and logs information about both strings.

4. Finally, the code modifies s2 and logs information about both strings again:

```
procedure TfrmDataTypes.btnCopyOnWriteClick(
  Sender: TObject);
var
  s1, s2: string;
begin
  s1 := 'Delphi';
  ListBox1.Items.Add(Format('s1 = %p [%d:%s]',
    [PPointer(@s1)^, PInteger(PNativeUInt(@s1)^-8)^, s1]));
  UniqueString(s1);
  ListBox1.Items.Add(Format('s1 = %p [%d:%s]',
    [PPointer(@s1)^, PInteger(PNativeUInt(@s1)^-8)^, s1]));
  s2 := s1;
  ListBox1.Items.Add(Format(
    's1 = %p [%d:%s], s2 = %p [%d:%s]',
    [PPointer(@s1)^, PInteger(PNativeUInt(@s1)^-8)^, s1,
     PPointer(@s2)^, PInteger(PNativeUInt(@s2)^-8)^, s2]));
  s2[1] := 'd';
  ListBox1.Items.Add(Format(
    's1 = %p [%d:%s], s2 = %p [%d:%s]',
```

```
    [PPointer(@s1)^, PInteger(PNativeUInt(@s1)^-8)^, s1,
     PPointer(@s2)^, PInteger(PNativeUInt(@s2)^-8)^, s2]));
end;
```

If you run this program and click on the **string copy-on-write** button, you'll see something similar to the following screenshot (the actual hexadecimal values will be different):

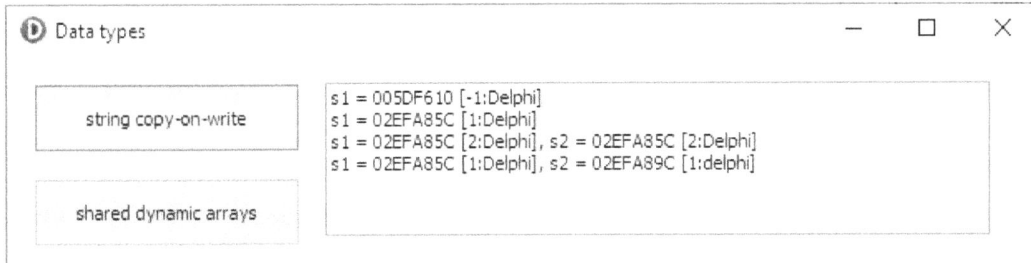

Figure 5.7 – A demonstration of a copy-on-write mechanism

If we examine the output in detail, we can reconstruct the operations behind the scenes:

1. When the string is initialized from a constant, it has a reference count of -1.

2. After calling `UniqueString`, the string is moved to a different memory area and has now a reference count of 1.

3. After `s1` is assigned to `s2`, both variables point to the same memory area and have a reference count of 2.

4. After `s2[1]` is modified, `s2` points to a new memory area, and both `s1` and `s2` have a reference count of 1.

This wraps up the implementation behind Ansi and Unicode strings. The `WideString` type is, however, implemented completely differently. It was designed to be used in OLE applications where strings can be sent from one application to another. Because of that, all `WideString` strings are allocated with Windows' OLE memory allocator, not with Delphi's memory management mechanism.

There is also no copy-on-write implemented for `WideString` strings. When you assign one `WideString` to another, new memory is allocated and data is copied. Because of that, `WideString` strings are slower than Ansi and Unicode strings and, as such, should not be used when not required.

Arrays

Delphi supports two kinds of arrays. One is of a fixed size and has settable lower and upper bounds. We call these **static arrays**. Others have a lower bound fixed at zero, while the upper bound is variable. In other words, they can grow and shrink. Because of this, they are called **dynamic arrays**.

The following code fragment shows how static and dynamic arrays are declared in Delphi:

```
var
  sarr1: array [2..22] of integer; // static array
  sarr2: array [byte] of string;   // static array
  darr1: array of TDateTime;       // dynamic array
  darr2: TArray<string>;           // dynamic array
```

Every array stores elements of some type. The compiler will treat these elements just the same as variables of that type. In the preceding second declaration, for example, the element type is `string`. This is a managed type, and all 256 elements of the array will be initialized to `nil` automatically.

Static arrays are pretty dull. They have a constant size and, as such, are created exactly the same way as simple types (except that they typically use more space). Dynamic arrays, on the other hand, are similar to strings, but with a twist.

When you declare a dynamic array, it will start as empty, and the variable will contain a `nil` pointer. Only when you change an array size (by using `SetLength` or a `dynamic array` constructor) will memory be allocated, and a pointer to that memory will be stored in the dynamic array variable. Any change to array size will reallocate this storage memory – just like with strings.

Modifying elements of an array is simple – again, just like with strings. An address of the element that is being modified is calculated, and then item data is copied from one location to another. Of course, if array elements are not of a simple type, appropriate rules for that type will apply. Setting an `array of string` element brings in the rules for the `string` type, and so on.

When you assign one dynamic array to another, Delphi, again, treats them just the same as strings. It simply copies one pointer (to the array data) into another variable. Ultimately, both variables contain the same address. A reference count in the array memory is also incremented – again, just like strings.

The similarities with string types end here. If you now modify one element in one of the arrays, the same change will apply to the second array! There is no copy-on-write mechanism for dynamic arrays, and both array variables will still point to the same array memory. We can also say that the new variable is an *alias* for the original and, sometimes, name the whole process *data aliasing*.

Aliasing is great from an efficiency viewpoint, but it may not be exactly what you wanted. Aliases make code faster, but they are also quite dangerous. If you forget that your dynamic array is just an alias, you can easily destroy data that is needed elsewhere.

> **Note**
> This is also one of the reasons why Spring's `ToArray` function returns a copy of the internal array and not just an alias.

There's, luckily, a workaround that will split both variables into two separate copies of the array data. Any time you call SetLength on a shared array, the code will make this array unique, just as UniqueString does to strings. Even if the new length is the same as the old one, a unique copy of an array will still be made.

> **Tip**
> Another way to create a true copy of a dynamic array is to call the Copy function, as in this example – arr2 := Copy(arr1).

The following code from the DataTypes demo demonstrates this. Firstly, it allocates the arr1 array by using a handy syntax introduced in Delphi XE7. Then, it assigns arr1 to arr2. This operation merely copies the pointers and increments the reference count. Both pointers and the contents of the arrays are then shown in the log.

After that, the code modifies one element of arr1 and logs again. To complete the test, the code calls SetArray to make arr2 unique and modifies arr1 again:

```
procedure TfrmDataTypes.btnSharedDynArraysClick(Sender: TObject);
var
  arr1, arr2: TArray<Integer>;
begin
  arr1 := [1, 2, 3, 4, 5];
  arr2 := arr1;
  ListBox1.Items.Add(Format(
    'arr1 = %p [%s], arr2 = %p [%s]',
    [PPointer(@arr1)^, ToStringArr(arr1), PPointer(@arr2)^,
    ToStringArr(arr2)]));

  arr1[2] := 42;
  ListBox1.Items.Add(Format(
    'arr1 = %p [%s], arr2 = %p [%s]',
    [PPointer(@arr1)^, ToStringArr(arr1), PPointer(@arr2)^,
    ToStringArr(arr2)]));

  SetLength(arr2, Length(arr2));
  arr1[2] := 17;
  ListBox1.Items.Add(Format(
    'arr1 = %p [%s], arr2 = %p [%s]',
    [PPointer(@arr1)^, ToStringArr(arr1), PPointer(@arr2)^,
    ToStringArr(arr2)]));
end;
```

The output of the program shows that, after the assignment, both arrays point to the same memory and contain the same data. This doesn't change after `arr1[2] := 42`. Both arrays still point to the same memory and contain the same (changed) data.

After `SetLength`, however, arrays point to a separate memory address, and changing `arr1` only changes that array:

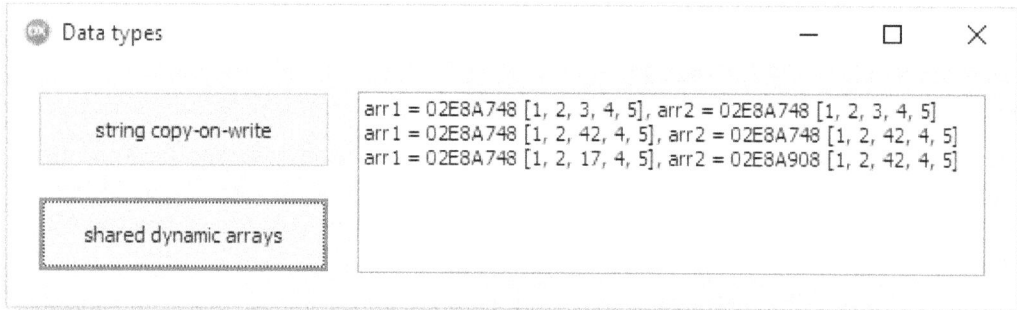

Figure 5.8 – A demonstration of dynamic array aliasing

Records

To wrap up our analysis of built-in data types, we will discuss structured types that can store elements of different types – namely, records, classes, and interfaces.

The simplest of these are records. They are actually quite similar to simple types and static arrays. As simple types, records are what we call statically allocated. Local variables of the record type are part of a thread's stack, global variables are part of the global process memory, and records that are parts of other structured types are stored as part of the owner's memory.

The important thing that you have to remember when using records is that the compiler manages the life cycle of their fields, which are themselves managed. In other words, if you declare a local variable of a record type, all its managed fields will be automatically initialized, while unmanaged types will be left at random values (at whatever value the stack at that position contains at that moment).

The following code from the `DataTypes` demo demonstrates this behavior. When you run it, it will show some random values for the a and c fields, while the b field will always be initialized to an empty string.

The code also shows the simplest way to initialize a record to default values (zero for integer and real types, an empty string for strings, `nil` for classes, and so on). The built-in (but mostly undocumented) `Default` function creates a record in which all fields are set to default values, and you can then assign it to a variable:

```
type
  TRecord = record
```

```
      a: integer;
      b: string;
      c: integer;
    end;

procedure TfrmDataTypes.ShowRecord(const rec: TRecord);
begin
  ListBox1.Items.Add(Format('a = %d, b = ''%s'', c = %d',
    [rec.a, rec.b, rec.c]));
end;

procedure TfrmDataTypes.btnRecordInitClick(Sender: TObject);
var
  rec: TRecord;
begin
  ShowRecord(rec);
  rec := Default(TRecord);
  ShowRecord(rec);
end;
```

When you run this code, you'll get some random numbers for a and c in the first log, but they will always be zero in the second log:

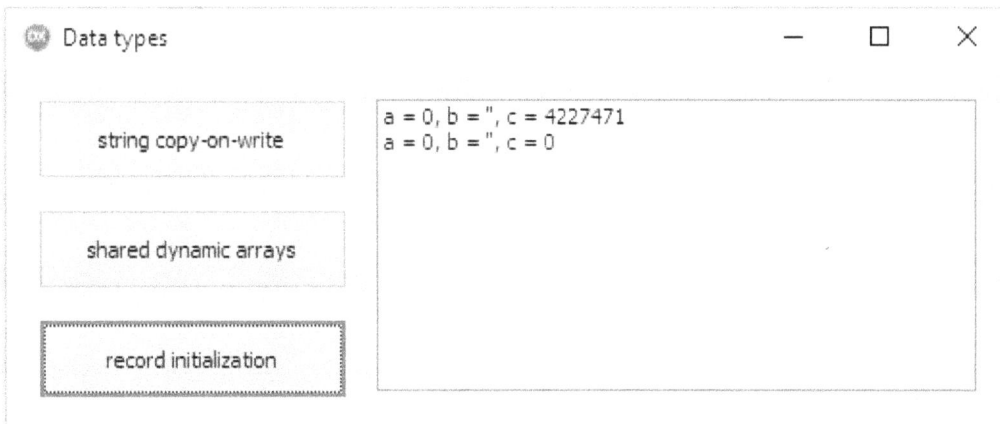

Figure 5.9 – Comparing uninitialized and initialized records

The initialization of managed fields affects the execution speed. The compiler doesn't create an optimized initialization code for each record type; instead, it does the initialization by calling general and relatively slow code.

This also happens when you assign one record to another. If all fields are unmanaged, the code can just copy data from one memory location to another. However, when at least one of the fields is of a managed type, the compiler will again call a generic copying method that is not optimized for the specific record.

The following example from the DataTypes demo shows the difference between unmanaged and managed records. It copies two different records a million times. Both records are of the same size, except that TUnmanaged contains only fields of the NativeUInt unmanaged type, and TManaged contains only fields of the IInterface managed type:

```
type
  TUnmanaged = record
    a, b, c, d: NativeUInt;
  end;

  TManaged = record
    a, b, c, d: IInterface;
  end;

procedure TfrmDataTypes.btnCopyRecClick(Sender: TObject);
var
  u1, u2: TUnmanaged;
  m1, m2: TManaged;
  i: Integer;
  sw: TStopwatch;
begin
  u1 := Default(TUnmanaged);
  sw := TStopwatch.StartNew;
  for i := 1 to 1000000 do
    u2 := u1;
  sw.Stop;
  ListBox1.Items.Add(Format('TUnmanaged: %d ms',
    [sw.ElapsedMilliseconds]));

  m1 := Default(TManaged);
  sw := TStopwatch.StartNew;
  for i := 1 to 1000000 do
    m2 := m1;
  sw.Stop;
  ListBox1.Items.Add(Format('TManaged: %d ms',
    [sw.ElapsedMilliseconds]));
end;
```

The following screenshot shows the measured difference in execution time. Granted, 31 ms to copy a million records is not a lot, but the *unmanaged* version is still five times faster. In some situations, that can mean a lot:

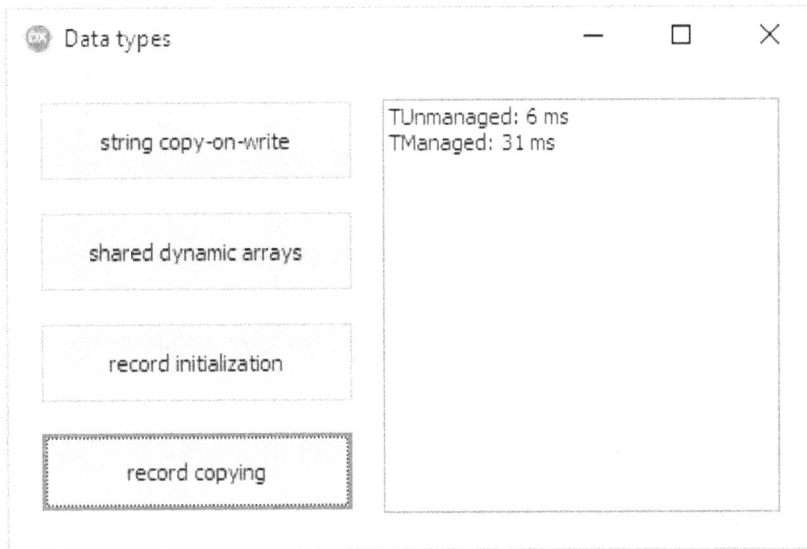

Figure 5.10 – Comparing initialization times for unmanaged and managed records

We saw how to initialize a record to empty values by using the `Default` function. As an additional option, we can use `constructor` to create a record. The syntax is just the same as if we were declaring a constructor on a class, with one small caveat – record constructors must have parameters. However, there is a workaround for that, and it is shown in the following code example.

The `TRecValue` record implements a single-parameter `constructor Create` that initializes the `Value` field. In addition to that, it also implements `class function Create`, which returns an initialized record, as shown here:

```
type
  TRecValue = record
    Value: integer;
//    constructor Create;
    constructor Create(AValue: integer); overload;
    class function Create: TRecValue; overload; static;
  end;

constructor TRecValue.Create(AValue: integer);
begin
  // inherited Create;
  Value := AValue;
```

```
end;

class function TRecValue.Create: TRecValue;
begin
  Result.Value := 42;
end;
```

Before we continue with the example, I'd like to point out a few specific parts of the code:

- We cannot create constructors without parameters. To verify, uncomment the first `Create` constructor, and the compiler will return an **E2394 Parameterless constructors not allowed on record types** error.

- The class function requires a `static` directive. This instructs the compiler to not pass a hidden `Self` parameter to the class function. Without it, the compiler would balk with **E2398 Class methods in record types must be static**. As this is a requirement, the compiler could be smart and apply the `static` directive automatically, but unfortunately, it does not.

- We cannot call the `inherited` constructor from the code, as records don't support inheritance.

When we use both `Create` overloads in code, there's no difference between the constructor and the class function, as shown in the next code fragment (assigned to the **record constructors** button in the demo program):

```
procedure TfrmDataTypes.btnRecordConstructorsClick(
  Sender: TObject);
var
  a, b: TRecValue;
begin
  a := TRecValue.Create(17);
  ListBox1.Items.Add('a = ' + IntToStr(a.Value));
  b := TRecValue.Create;
  ListBox1.Items.Add('b = ' + IntToStr(b.Value));
end;
```

Although the constructor syntax looks exactly the same as when used with classes, record constructors actually behave quite differently. Records are statically initialized, not created and destroyed, and the `constructor` syntax was implemented just as a convenience. There is also no support for record *destructors*.

Custom managed records

The support for record initialization has evolved during the initial implementation, and Delphi 10.4 Sydney brought us a true replacement for record constructors and destructors – *initializers* and *finalizers*, respectively. The new terminology makes sense, as records are not constructed/destructed but are simply declared and used. Even so, the compiler still knows when the record is used for the first time (*comes into scope*) and when it is not needed anymore (*goes out of scope*), and it can generate special code to prepare (initialize a record) and do any clean-up (finalize a record).

> **Note**
>
> Records with initializers and finalizers are called **custom managed records** in the Delphi documentation. More information about them can be found at `https://docwiki.embarcadero.com/RADStudio/en/Custom_Managed_Records`.

The following code fragment shows how to declare an `Initialize` initializer and `Finalize` finalizer. They are both implemented as *class operators* and must follow the exact syntax, as shown in the following example, where the initialized/finalized records are passed as `out` and `var` parameters, respectively:

```
type
  TCustomRecord = record
  private
    class var GNextID: integer;
  public
    Value: integer;
    Name: string;
    class constructor Create;
    constructor Create(AValue: integer;
      const AName: string);
    class operator Initialize(out Dest: TCustomRecord);
    class operator Finalize(var Dest: TCustomRecord);
    class operator Assign(var Dest: TCustomRecord;
      const [ref] Src: TCustomRecord);
  end;
```

This example also shows another new operator, `Assign`, which can be used to override the operation of copying one record to another. While `Initialize` and `Finalize` are roughly equivalent to constructors and destructors and classes, the `Assign` operator has no equivalent if we work with classes. It also uses very specific parameter-passing directives – `var` for the output (destination) record and `const [ref]` for the original (source) record. I will talk more about the `var`, `out`, and `const` parameters in the *Parameter passing* section.

The demonstration program includes the following implementation of `TCustomRecord`. The `Assign` method copies the `Value` field and changes the copy of the `Name` field, prefixing its value with `' [Copy] '`. The constructor initializes both the `Value` and `Name` fields. Initializer initializes only the Name field to 'Record N', where *N* is an incrementing number. It leaves the Value field uninitialized for demonstration purposes. All methods also extensively log the operation in progress:

```
class constructor TCustomRecord.Create;
begin
  GNextID := 1;
end;

class operator TCustomRecord.Assign(
  var Dest: TCustomRecord;
  const [ref] Src: TCustomRecord);
begin
  Dest.Value := Src.Value;
  Dest.Name := '[Copy] ' + Src.Name;
  frmDataTypes.ListBox1.Items.Add(Format(
    '...copying "%s":%d => "%s":%d',
    [Src.Name, Src.Value, Dest.Name, Dest.Value]));
end;

constructor TCustomRecord.Create(
  AValue: integer; const AName: string);
begin
  frmDataTypes.ListBox1.Items.Add(Format(
    '...creating "%s":%d (was "%s":%d)',
    [AName, AValue, Name, Value]));
  Value := AValue;
  Name := AName;
end;

class operator TCustomRecord.Finalize(
  var Dest: TCustomRecord);
begin
  frmDataTypes.ListBox1.Items.Add(Format(
    '...finalizing "%s":%d', [Dest.Name, Dest.Value]));
end;

class operator TCustomRecord.Initialize(
  out Dest: TCustomRecord);
begin
//  Dest.Value := 0;
  Dest.Name := 'Record ' + IntToStr(GNextID);
  Inc(GNextID);
```

```
  frmDataTypes.ListBox1.Items.Add(Format(
     '...initializing "%s":%d', [Dest.Name, Dest.Value]));
end;
```

The following code, attached to the **custom managed records** button, does a few simple operations on three TCustomRecord variables. Firstly, the a variable is initialized by calling the record constructor. Next, a constructor is called as a function on the b record. Finally, the c variable is initialized by assigning the a variable:

```
procedure TfrmDataTypes.btnCustomManagedRecordsClick(
  Sender: TObject);
var
  a, b, c: TCustomRecord;
begin
  Listbox1.Items.Add('Create a');
  a := TCustomRecord.Create(42, 'record A');
  ListBox1.Items.Add(Format('a = "%s":%d', [a.Name, a.Value]));
  b.Create(17, 'record B');
  ListBox1.Items.Add(Format('b = "%s":%d', [b.Name, b.Value]));
  Listbox1.Items.Add('Assign c := ' + a.Name);
  c := a;
  ListBox1.Items.Add(Format('c = "%s":%d', [c.Name, c.Value]));
  Listbox1.Items.Add('Exit');
end;
```

If you run this code and look at the output (shown in the following screenshot), you'll see that the actual behavior is quite a bit more complex than the btnCustomManagedRecordsClick method would suggest:

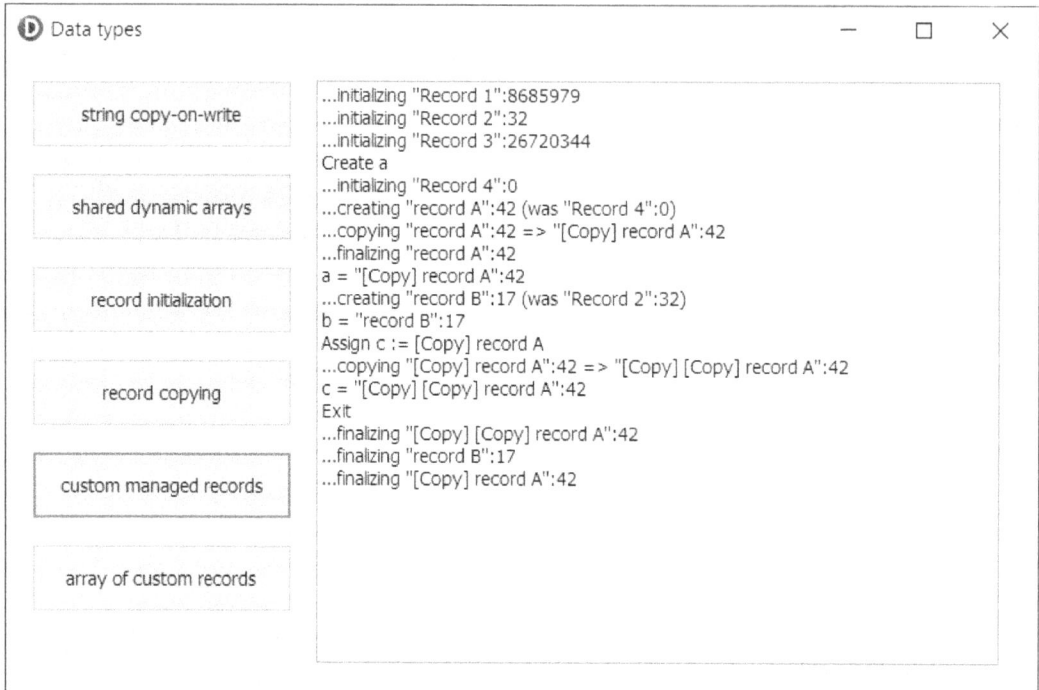

Figure 5.11 – A detailed log of a few simple operations on custom managed records

From this output, we can reconstruct the actual operations implemented by the compiler:

1. All three records (a, b, and c) are initialized even before the 'Create a' line is logged. This initialization happens inside the begin statement. This is also the place where variables of other managed types (strings, dynamic arrays, and interfaces) are initialized. You will also see that only the Name field is initialized, while the Value field contains some random value from the stack.

2. The simple a := TCustomRecord.Create line results in four record operations. The compiler first initializes a temporary record (...initializing in the log) and calls the constructor on it (...creating). Then, it assigns this record to the a variable (...copying), changing the Name during the assignment. Finally, the code finalizes the previous value of a (...finalizing). The resulting value of a.Name starts with '[Copy]', which is not something that would be obvious by looking at the code.

3. To make the code more efficient and skip all the operations on the temporary record, you should call record constructors as if they were normal functions, as in the b.Create line of code. This spares us one call each to Initialize, Assign, and Finalize.

4. Next, the code assigns `a` to `c`. This time, the resulting behavior is simple and expected – only the `Assign` operator is called.

5. When the method exits (in the `end` statement), all three local variables are finalized, and appropriate `Finalize` operators are called.

I should note that the same sequence of operation would also be generated for normal (not custom) *managed* records, except that initializers and finalizers would be automatically generated by the compiler. If the records in the example contain only unmanaged fields, the generated code would be simpler, without calling any initializers/finalizers, but the temporary record would still be generated in the `a := TCustomRecord.Create` statement.

We can compare the behavior of the previous code with the next code fragment, attached to the **array of custom records** button, which stores the `TCustomRecord` records in a dynamic array. The code firstly sets the length of the first array, `a1`, to 3. Then, it creates an alias for this array in `a2` and a true copy in `a3`. Finally, the code sets the lengths of the `a1` and `a2` arrays to zero:

```
procedure TfrmDataTypes.btnArrayOfRecordsClick(
  Sender: TObject);
var
  arr: array [1..2] of TCustomRecord;
  a1,a2,a3: TArray<TCustomRecord>;
begin
  Listbox1.Items.Add('Initialize a1');
  SetLength(a1, 3);
  ListBox1.Items.Add(Format('a1[0] = "%s":%d',
    [a1[0].Name, a1[0].Value]));
  Listbox1.Items.Add('Assign a2');
  a2 := a1;
  ListBox1.Items.Add(Format('a2[0] = "%s":%d',
    [a2[0].Name, a2[0].Value]));
  Listbox1.Items.Add('Copy a3');
  a3 := Copy(a1);
  ListBox1.Items.Add(Format('a3[0] = "%s":%d',
    [a3[0].Name, a3[0].Value]));
  Listbox1.Items.Add('Clear a1');
  SetLength(a1, 0);
  Listbox1.Items.Add('Clear a2');
  SetLength(a2, 0);
  Listbox1.Items.Add('Exit');
end;
```

The resulting log in the next figure shows some similarities and differences to the previous example:

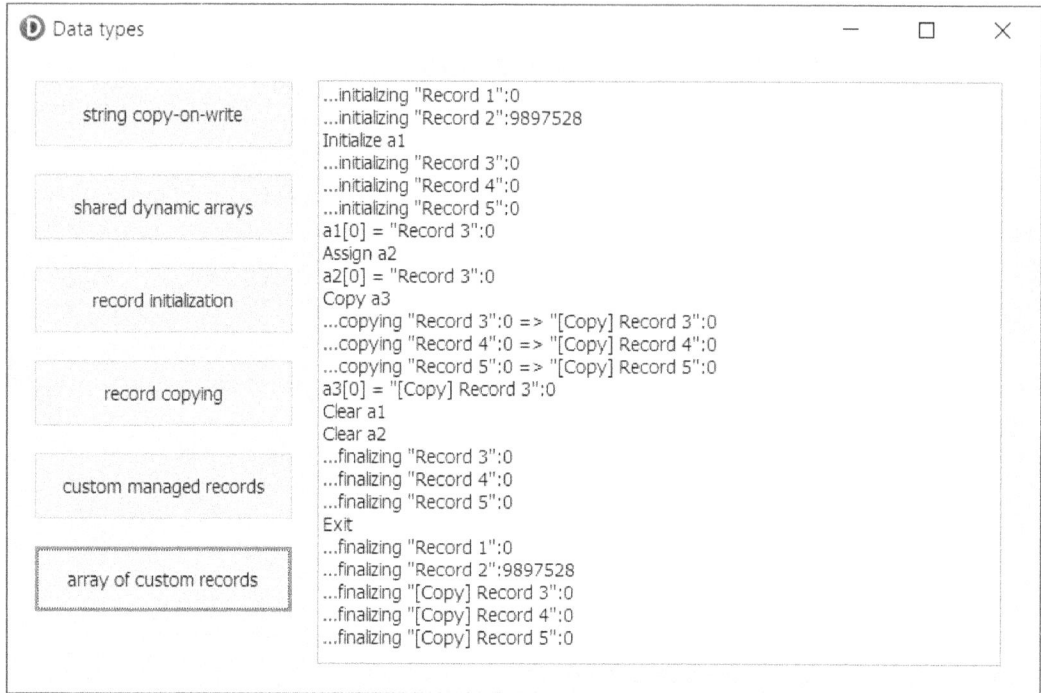

Figure 5.12 – A detailed log of manipulating custom managing records stored in a dynamic array

The operations on the records were executed in the following order:

1. In the `begin` statement, the compiler initializes the records stored in the static array, but not the records stored in the dynamic array, as dynamic arrays are initialized to a length of zero. The static array of records behaves just like normal record variables in the previous example and is just included for comparison. It is not used in the code.

2. After `SetLength` is called, the compiler initializes three records. Note that the `Value` field, which is not initialized by the initializer, is this time initialized to zero. This is because all records are part of the memory managed by the dynamic array, and this memory is initialized to zero when the array length increases.

3. The `a2 := a1` operation does nothing, as it just creates an alias to the same array data (the same records).

4. The `a3 := Copy(a1)` operation, however, creates a copy of records by calling the `Assign` operator for each array element. In this case, the compiler is intelligent enough not to call `Initialize` for all three elements first.

5. When `a1` is cleared, nothing happens, as `a2` still points to the same data.

6. When a2 is cleared, all references to the dynamic array data are gone. All three elements in the array are finalized, and array memory is released (we will only talk about the latter in the next chapter).

7. In the end statement, all the remaining records stored in a3 and the static array, arr, are also finalized.

Custom managed records are a powerful tool, but as the examples show, you should be well aware of the operations created by the compiler before you use them.

One special application of the managed records that I did not cover yet is that locally managed records are always finalized at the end of a method, even if an exception is raised during the execution. As an example, the following code fragment initializes a record, creates an object, and raises an exception. The object would not be released, but the record would be finalized:

```
procedure TfrmDataTypes.btnExceptionsClick(
  Sender: TObject);
begin
  var rec := Default(TCustomRecord);
  var obj := TCustomObject.Create;
  raise Exception.Create('Bang!');
end;
```

You can test this behavior by clicking **automatic exception handling** in the DataTypes demonstration project.

> **Tip**
>
> This example shows the powerful inline variable declaration syntax introduced in Delphi 10.3 Rio. You can read more about it at https://docwiki.embarcadero.com/RADStudio/en/Inline_Variable_Declaration.

Classes

While classes look superficially similar to records, they are, in fact, implemented in a completely different manner.

When you create an object of a class by calling the constructor for that class, the code allocates memory from the memory manager and fills it with zeroes. That, in effect, initializes all fields, managed and unmanaged, to default values. A pointer to this memory is stored in the variable that receives the result of a constructor.

When an object is destroyed, its memory is returned to the system (again, through the memory manager).

> **Note**
>
> In some Delphi versions prior to version 11, non-Windows compilers use **Automatic Reference Counting** (**ARC**) to manage objects. Objects managed by ARC behave more like interfaces, which are described in the next section. ARC was removed from the Linux compiler in Delphi 10.3 and from Android and iOS compilers in Delphi 10.4.

Copying one variable of a class type to another just copies the pointer. In effect, you then have two variables pointing to the same object. There is no reference counting associated with objects, and if you now destroy one variable (by calling the `Free` method), the other will still point to the same part of memory. Actually, both variables will point to that unused memory, and if it gets allocated later (which it will), both variables will point to memory belonging to another object. Such pointers are often referred to as **dangling pointers**. The following code fragment illustrates the problem:

```
var
  o1, o2: TObject;
begin
  o1 := TObject.Create;
  o2 := o1;
  // o1 and o2 now point to the same object
  o1.Free;
  // o1 and o2 now point to unowned memory
  o1 := nil;
  // o2 still points to unowned memory
end;
```

If you have a long-life variable or field containing an object of some class and you destroy the object, make sure that you also set the value/field to `nil`. The best way to achieve that is to use the `FreeAndNil` function instead of `Free`.

Interfaces

From the viewpoint of memory management, interfaces in Delphi are implemented as classes with added reference counting. To create an interface, you actually have to create an object of a class, which will get a reference count of 1. If you then assign this interface to another variable, both will point to the same memory, and the reference count will be incremented to 2.

> **Note**
>
> While the interfaces in Delphi inherit a COM interface model with reference counting, we can also use interfaces with customized reference counting or no reference counting. The latter can be simply implemented by deriving from the `TNoRefCountObject` class, which was added in Delphi 11 Alexandria. (In previous versions, the same functionality was wrapped in the `TSingletonObject` class.)

There is no equivalent to `SetLength` or `UniqueString` that would make a unique copy of an interface. That would require duplicating the underlying object, and Delphi has no built-in support for that.

The object implementing the interface is destroyed when its reference count falls to 0:

```
var
  i1, i2: IInterface;
begin
  i1 := TInterfacedObject.Create;
  // i1 points to an object with reference count 1
  i2 := i1;
  // both i1 and i2 point to a same object with reference count 2
  i1 := nil;
  // i2 now points to an object with reference count 1
  i2 := nil;
  // reference count dropped to 0 and object is destroyed
end;
```

Although interfaces are very similar to classes, all this reference count management has a cost. It is implemented with something called *interlocked* instructions, which are quite a bit slower than normal increment/decrement instructions. I'll discuss this in more detail in *Chapter 7, Getting Started with the Parallel World.*

This only makes a measurable difference when you assign interfaces a lot, but sometimes, this is exactly what happens. I'll show an example in the next section.

Optimizing method calls

I know you are eagerly waiting to optimize some real code, but be patient as we discuss a little bit more about the theoretical side of the process. I spent a lot of time talking about the behavior of built-in data types, but I didn't say anything about how data is passed to methods. This much shorter and more surprising section (you'll see!) will remedy this. As I'll be talking about speeding up method calls, I'll also throw in a short discussion about method inlining, just for good measure. But first, parameters!

Parameter passing

In essence, Delphi knows two ways of passing parameters to a method (or a procedure, function, or anonymous method – it's all the same). Parameters can be passed by value or by reference.

The former makes a copy of the original value and passes that copy to a method. The code inside the method can then modify its copy however it wants, without changing the original value.

The latter approach doesn't pass the value to the method but just an address of that value (a pointer to it). This can be faster than passing by value, as a pointer will generally be smaller than an array or a record. In this case, the method can then use this address to access (and modify) the original value – something that is not possible if we pass a parameter by value.

Passing by reference is indicated by prefixing a parameter name with var, out, or const. A parameter is passed by value when no such prefix is used.

The difference between var and out is mostly semantic. Prefixing a parameter by out tells the compiler that we don't pass anything into the method (it may even be an uninitialized value) and we are only interested in the result. The var prefix, on the other hand, declares that we will pass some value in and that we expect to get something out. The compiler, however, does a bad job of implementing the out prefix (as proven in https://delphisorcery.blogspot.com/2021/04/out-parameters-are-just-bad-var.html by our technical reviewer), so it is actually better to just always use the var prefix.

> **Tip**
>
> In practice, var is usually used even when a parameter is not providing any data to the method but just returning something.

The const prefix is a bit different. When this is used, the compiler will prevent us from making any changes to the parameter (with some exceptions, as we'll soon see). In essence, const is an optimization prefix used when we want to pass a larger amount of data (a record, for example). It will ensure that the parameter is passed by reference, while the compiler will make sure that the data is not changed.

> **Tip**
>
> The compiler is not *required* to pass a parameter by reference when const is used. If the size of the parameter is small enough (for example, it is of a simple type or a small record), the compiler *may* decide to pass this parameter by value. If you write some very specialized code and your method depends on always receiving the reference to the actual data, you should use the const [ref] prefix. The [ref] decorator prevents the compiler from passing the parameter by value.
>
> Earlier in the chapter, we saw const [ref] used in the Assign record operator. It is also used in the FreeAndNil method and other parts of the runtime library.

The const prefix is also useful when the parameter is a managed type (for example, a string or interface). When you pass such type by value, the code increments the reference count at the beginning of the method and decrements it at the end. If the parameter is marked as const, the reference count is not incremented/decremented, and this can make a difference in the long run. In addition to that, the compiler creates an implicit try..finally statement for non-const managed parameters, which slows down the code even more.

The `ParameterPassing` demo passes array, string, record, and interface parameters to a method multiple times and measures the time. I will skip most of the code here, as it is fairly dull and looks all the same. I'll just focus on a few interesting details first and then give an overview of measurements.

The most uninteresting of all are static arrays and records. Passing by value makes a copy, while passing by reference just passes a pointer to the data – that is all.

When you pass a string by using `const`, just an address is passed to the method, and a reference count is not touched. That makes such calls very fast.

Passing by value, however, uses the copy-on-write mechanism. When a string is passed by value, a reference count is incremented, and a pointer to string data is passed to the method. Only when the method modifies the string parameter is a real copy of the string data made.

As an example, the following code in the demo is called 10 million times when you click the **string** button:

```
procedure TfrmParamPassing.ProcString(s: string);
begin
  // Enable next line and code will suddenly become
  // much slower!
  // s[1] := 'a';
end;
```

This executes in a few hundred milliseconds. Uncomment the assignment and the code will suddenly run for 20 seconds (give or take), as a copy of quite a large string will be made each time the method is called.

Interfaces behave similarly to strings, except that there is no copy-on-write mechanism. The only difference in execution speed comes from incrementing and decrementing the reference count.

Arrays

At the beginning of this section, I promised some interesting results, so here they are! When a parameter is a dynamic array, strange things happen. However, let's start with the code.

To measure parameter passing, the code creates an array of 100,000 elements, sets the `arr[1]` element to 1, and calls `ScanDynArray` 10,000 times. `ScanDynArray` sets `arr[1]` to 42 and exits.

Another method, `btnConstDynArrayClick` (not shown), works exactly the same way, except that it calls `ScanDynArrayConst` instead of `ScanDynArray`:

```
const
  CArraySize = 100000;

procedure TfrmParamPassing.ScanDynArray(
  arr: TArray<Integer>);
begin
```

```
    arr[1] := 42;
end;

procedure TfrmParamPassing.ScanDynArrayConst(
  const arr: TArray<Integer>);
begin
  // strangely, compiler allows that
  arr[1] := 42;
end;

procedure TfrmParamPassing.btnDynArrayClick(Sender: TObject);
var
  arr: TArray<Integer>;
  sw: TStopwatch;
  i: Integer;
begin
  SetLength(arr, CArraySize);
  arr[1] := 1;
  sw := TStopwatch.StartNew;
  for i := 1 to 10000 do
    ScanDynArray(arr);
  sw.Stop;
  ListBox1.Items.Add(Format(
    'TArray<Integer>: %d ms, arr[1] = %d',
    [sw.ElapsedMilliseconds, arr[1]]));
end;
```

Can you spot the weird code in this example? It is the assignment in the ScanDynArrayConst method. I said that the const prefix will prevent such modifications of parameters, and in most cases, it does. Dynamic arrays are, however, different.

When you pass a dynamic array to a method, Delphi treats it just like a pointer. If you mark it as const, nothing changes, as only this pointer is treated as a constant, not the data it points to. That's why you can modify the original array even through the const parameter.

That's the first level of weirdness. Let's crank it up!

If you check the System unit, you'll see that TArray<T> is defined like this:

```
type
  TArray<T> = array of T;
```

So, in theory, the code will work the same if we replace `TArray<Integer>` with `array of Integer`, right? Not at all!

```
procedure TfrmParamPassing.ScanDynArray2(
  arr: array of integer);
begin
  arr[1] := 42;
end;

procedure TfrmParamPassing.ScanDynArrayConst2(
  const arr: array of integer);
begin
  // in this case following line doesn't compile
  // arr[1] := 42;
end;
```

When we use the `array of T` syntax for a method parameter, it does not represent a dynamic array but an **open array parameter**. This type of parameter data was added to the language way before dynamic arrays were introduced, and it was initially used for methods such as `Format` that accept a variable number of parameters.

For each open array parameter, the compiler actually passes two values to the method – an address of the first element in the array and the index of the highest element – the same that would be returned by calling the `High(arr)` function. The indices start at 0, as with dynamic arrays.

> **Warning**
>
> The second parameter representing an index of the highest element is always a 32-bit value. Currently, you cannot use open arrays with more than 4 GB elements, even with a 64-bit compiler.

In this case, the compiler won't allow us to modify `arr[1]` when the parameter is marked `const`. Also, in the `ScanDynArray2` method, the code makes a full copy of the array, just as if we were passing a normal static array. That makes total sense if we keep in mind that open array parameters were designed to accept static arrays of unknown size.

If we declare `type TIntArray = array of Integer` and then rewrite the code to use this array, we get the original `TArray<T>` behavior back. The array is always passed by reference, and the code can modify the original array through the `const` parameter:

```
type
  TIntArray = array of Integer;

procedure TfrmParamPassing.ScanDynArray3(arr: TIntArray);
begin
```

```
    arr[1] := 42;
  end;


procedure TfrmParamPassing.ScanDynArrayConst3(
  const arr: TIntArray);
begin
  // it compiles!
  arr[1] := 42;
end;
```

The only proper way to end this discussion about the wonderful madness of Delphi arrays is with the Slice function. It allows us to pass only a part of a *static* array into a function expecting an open array parameter, declared with the array of T syntax. Slice accepts two parameters – a static array (or an open array) and a number – and changes this into an open array containing the first *number* of elements of that array.

The following code fragment shows how this function is called in practice. You can run it by clicking the **array slice** button. The SumArr method expects an open array of integers and returns a sum of all elements in the array. Let me note again that the array of integer syntax is important – declaring the parameter as arr: TArray<integer> would not work! The btnArraySliceClick event handler creates a static array of 10 elements and initializes it to numbers from 1 to 10. Then, it adds all elements in the array together and logs the output. Finally, it uses SumArr(SliceArr(arr, 3)) to pass only the first three elements of the array:

```
function TfrmParamPassing.SumArr(
  const arr: array of integer): integer;
var
  i: Integer;
begin
  Result := arr[Low(arr)];
  for I := Low(arr) + 1 to High(arr) do
    Inc(Result, arr[i]);
end;


procedure TfrmParamPassing.btnArraySliceClick(
  Sender: tObject);
var
  arr: array [1..10] of integer;
  i: Integer;
  sum: integer;
begin
  for i := Low(arr) to High(arr) do
    arr[i] := i;
  sum := SumArr(arr);
```

```
  ListBox1.Items.Add(IntToStr(sum));
  sum := SumArr(Slice(arr, 3));
  ListBox1.Items.Add(IntToStr(sum));
end;
```

When using `Slice`, you have to keep in mind that it is not an actual function from the runtime library but a hack, implemented directly in the compiler. It works only as an intermediary function when an array is passed as an argument. Because it is implemented as an internal compiler function, it can do its magic without any copying of the array.

When the compiler passes data to a method expecting an open array, it actually passes two parameters – the address of the array data and the number of elements minus 1. In our example, where `arr` has 10 elements, `SumArr(arr)` passes the number 9 to the method. This number becomes the highest valid index of the `arr` parameter in the `SumArr` method. In other words, `High(arr)` = 9 and `Low(arr)` = 0 (as always), which gives us 10 elements with indices from 0 to 9.

When `SumArr(Slice(arr, 3))` is used, the compiler passes the number 2 to the method. In `SumArr`, `Low(arr)` is still 0 and `High(arr)` is 2, so the function accesses the first three elements of the array. The following screenshot shows the assembler code used to call the function with and without the use of the `Slice` function:

```
ParameterPassingMain.pas.271: sum := SumArr(arr);
0109012C 8D55C8                    lea edx,[ebp-$38]
0109012F B909000000                mov ecx,$00000009
01090134 8B45FC                    mov eax,[ebp-$04]
01090137 E870FFFFFF                call $010900ac
0109013C 8945F4                    mov [ebp-$0c],eax
ParameterPassingMain.pas.272: ListBox1.Items.Add(IntToStr(sum));
0109013F 8D55C4                    lea edx,[ebp-$3c]
01090142 8B45F4                    mov eax,[ebp-$0c]
01090145 E8AE4FE4FF                call $00ed50f8
0109014A 8B55C4                    mov edx,[ebp-$3c]
0109014D 8B45FC                    mov eax,[ebp-$04]
01090150 8B80F0030000              mov eax,[eax+$000003f0]
01090156 8B80C0020000              mov eax,[eax+$000002c0]
0109015C 8B08                      mov ecx,[eax]
0109015E FF513C                    call dword ptr [ecx+$3c]
ParameterPassingMain.pas.273: sum := SumArr(Slice(arr, 3));
01090161 8D55C8                    lea edx,[ebp-$38]
01090164 B902000000                mov ecx,$00000002
01090169 8B45FC                    mov eax,[ebp-$04]
0109016C E83BFFFFFF                call $010900ac
01090171 8945F4                    mov [ebp-$0c],eax
```

Figure 5.13 – The compiler implementation of the Slice function

Because `Slice` works only inside a method call, the following function will not compile:

```
var
  a: TArray<integer>;
a := Slice(arr, 3);
```

> **Note**
>
> The built-in `Slice` function can only slice an array starting at the first element. However, with some trickery, we can make it also pass a slice from the middle of the array. You can read more about this in an article written by one of the technical reviewers of this book, which can be found at `https://delphisorcery.blogspot.com/2020/09/open-array-parameters-and-subranges.html`.

Speed comparison

Let's now analyze the results of the `ParameterPassing` demo program.

Static arrays are passed as a copy (by value) or a pointer (by reference). The difference in speed clearly proves that (174 versus 0 ms). In both cases, the original value of the array element is not modified (it is 1).

The same happens when the array is declared as an open array (the lines starting with `array of Integer` and `const array of Integer`) in *Figure 5.14*.

`TArray<Integer>` and `TIntArray` behave exactly the same. An array is always passed by reference, and the original value can be modified (it shows as `42` in the log).

Records are either copied (by value) or passed as a pointer (by reference), which brings a difference of 165 versus 51 ms.

Strings are always passed as a pointer to string data. When they are passed as a normal parameter, the reference count is incremented/decremented. When they are passed as a `const` parameter, the reference count is not modified. This brings a difference of 260 versus 46 ms. A similar effect can be observed with the interface type parameters:

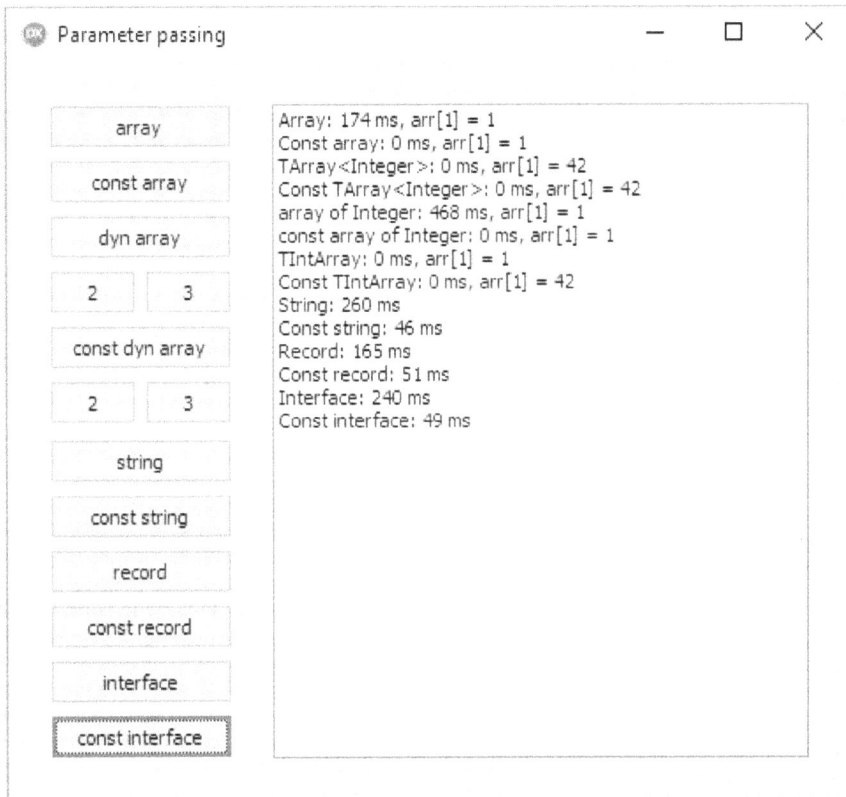

Figure 5.14 – Comparing the speed of a parameter passing for different data types

I have to point out that demos for different types cannot be compared directly with one another. For example, the loop for testing array types repeats 10,000 times, while the loop for testing strings and interfaces repeats 10,000,000 times.

Method inlining

Previously in this chapter, I spent quite some time describing the mechanics of enabling and disabling method inlining, but I never explained what that actually means. To put it simply, method inlining allows one method to be compiled as if its code is a part of another method.

When you call a method that is not marked as inlinable (i.e., doesn't have the `inline` suffix), the compiler prepares method parameters in an appropriate way (in registers and on the stack, but I won't discuss them in detail) and then executes a `CALL` assembler instruction.

If the called method is marked as inlinable, then the compiler basically just inserts the body of the inlinable method inside the code of the caller. This speeds up the code but also makes it larger, as this happens in every place that calls the inlinable method.

Let me give you an example. The following code from the `Inlining` demo increments a value 10 million times:

```
function IncrementInline(value: integer): integer; inline;
begin
  Result := value + 1;
end;

procedure TfrmInlining.Button2Click(Sender: TObject);
var
  value: Integer;
  i: Integer;
begin
  value := 0;
  for i := 1 to 10000000 do
    value := IncrementInline(value);
end;
```

As `IncrementInline` is marked as `inline` (and is, therefore, inlinable if the compiler settings don't prevent it), the code generated by the compiler doesn't actually call into that method 10 million times. The code actually looks more like what is generated in the next example:

```
procedure TfrmInlining.Button2Click(Sender: TObject);
var
  value: Integer;
  i: Integer;
begin
  value := 0;
  for i := 1 to 10000000 do
    value := value + 1;
end;
```

If you run the demo, you'll see that the inlined version executes much faster than the non-inlined code. On my test computer, the non-inlined version needed 53 ms, while the inline version executed in 26 ms.

When you call the `inline` method from the same unit, make sure that the `inline` method is implemented *before* it is called. Otherwise, the compiler will just silently use the method as if it was a normal method and won't generate even a hint.

> Tip
> A good tool for finding such problems is TMS FixInsight: `www.tmssoftware.com/site/fixinsight.asp`.

In the next example, the method is actually not inlined, although we may expect it to be:

```
type
  TfrmInlining = class(TForm)
    // some unimportant stuff ...
    procedure Button3Click(Sender: TObject);
  private
    function IncrementShouldBeInline(value: integer):
      integer; inline;
  public
  end;

procedure TfrmInlining.Button3Click(Sender: TObject);
var
  value: Integer;
  i: Integer;
begin
  value := 0;
  for i := 1 to 10000000 do
    value := IncrementShouldBeInline(value);
end;

function TfrmInlining.IncrementShouldBeInline(
  value: integer): integer;
begin
  Result := value + 1;
end;
```

Another problem that inline code can cause is that, in some situations, the result may be even slower than the non-inlined version. Sometimes, the compiler just doesn't do a good enough job. Remember, always measure!

For a long time, the compiler had another problem. When the inlined function returned an interface, it made another hidden copy of the returned value, which was only released at the end of the method that called the inlined function. This could cause the interface to not be destroyed as the programmer expected.

This problem, for example, caused threads in Delphi's TThreadPool object not to be released at the correct time. It was only fixed in the 10.2 Tokyo release, when Embarcadero introduced an improved compiler that generates better code.

And now, I'm really finished with the theory. Let's do some code!

The magic of pointers

Our friend, Mr. Smith, has progressed a lot from his first steps in Delphi. Now, he is playing with graphics. He wants to build a virtual reality app that will allow you to walk through the Antarctica forests. While he's getting some success, he has problems displaying the correct colors on the screen.

A part of his code is producing textures in **Blue-Green-Red** (**BGR**) byte order, while the graphics driver needs them in the more standard **Red-Green-Blue** (**RGB**) order. He already wrote some code to fix this problem, but his solution is a bit too slow. He'd like to push a frame or two more from the system, and so I promised that I'd help him optimize the converter. I'm glad to do it, as his problem neatly fits into the story of **pointers**.

A pointer is a variable that stores an *address* of some data (other variables, dynamically allocated memory, a specific character in a string, and so on). It is always the same size – 4 bytes on 32-bit systems and 8 bytes on 64-bit systems – just the same as `NativeInt` and `NativeUInt`, respectively. This is not a coincidence, as converting a pointer into an integer or vice versa is sometimes quite practical.

Let's say, for example, that we have a `TPoint3` record containing a three-dimensional point. We can then declare a pointer to this type with the `^TPoint3` syntax, and we can give this type a name, `PPoint3`. While the size of `TPoint3` is 24 bytes, the size of `PPoint3` is 4 or 8 bytes, depending on the compiler target (32- or 64-bit):

```
type
  TPoint3 = record
    X,Y,Z: double;
  end;
  PPoint3 = ^TPoint3;
```

If we now declare a `P3` variable of type `TPoint3` and a `PP3` variable of type `PPoint3`, we can then change the contents of `PP3` to the address of `P3` by executing `PP3 := @P3`. After that, we can use the `PP3^` notation to refer to the data that `PP3` *points to* (that is, `P3`).

Delphi also allows a shorter form to access fields via a pointer variable. We don't have to write `PP3^.X`; a simple `PP3.X` is enough:

```
var
  P3: TPoint3;
  PP3: PPoint3;
begin
  PP3 := @P3;
  PP3^.X := 1;
  // P3.X is now 1
  PP3.Y := 2;
  // P3.Y is now 2
```

```
    P3.Z := 3;
    // PP3^.Z is now 3
  end;
```

Unlike the original Pascal, Delphi allows some arithmetic operations on the pointer. For example, you can always do Inc(ptr), which will increment the address stored in ptr. It will not increment it by 1, though, but by the size of the type that ptr points to.

Some types, such as PChar, PAnsiChar, and PByte, also allow you to use them in some basic arithmetic expressions. In reality, all pointer types support this use, but only pointer types that were defined when the {$POINTERMATH ON} compiler directive was in effect will allow you to do that by default. For other types, you have to enable {$POINTERMATH ON} in the place where you do the calculation.

In the following example, the pi := pi + 1 code won't compile without an explicit {$POINTERMATH ON}, as this directive is not present in the System unit where PInteger is defined:

```
procedure TfrmPointers.btnPointerMathClick(
  Sender: TObject);
var
  pb: PByte;
  pi: PInteger;
  pa: PAnsiChar;
  pc: PChar;
begin
  pb := pointer(0);
  pb := pb + 1;
  ListBox1.Items.Add(Format('PByte increment = %d',
    [NativeUInt(pb)]));

  pi := pointer(0);
  {$POINTERMATH ON}
  pi := pi + 1;
  {$POINTERMATH OFF}
  ListBox1.Items.Add(Format('PInteger increment = %d',
    [NativeUInt(pi)]));

  pa := pointer(0);
  pa := pa + 1;
  ListBox1.Items.Add(Format('PAnsiChar increment = %d',
    [NativeUInt(pa)]));

  pc := pointer(0);
```

```
  pc := pc + 1;
  ListBox1.Items.Add(Format('PChar increment = %d',
    [NativeUInt(pc)]));
end;
```

Pointers are very useful when you create dynamic data structures, such as trees and linked lists. They are also important when you want to dynamically allocate records. We'll cover that in the next chapter. Also, they are very handy – as you'll see immediately – when we process data buffers.

Let us return to Mr. Smith's problem. He has a graphic texture, organized as an array of pixels. Each pixel is stored as a cardinal value. The highest byte of a pixel contains the **Transparency** component (also known as **Alpha**), the next byte contains the **Blue** component, the byte after that the **Green** component, and the lowest byte contains the **Red** component:

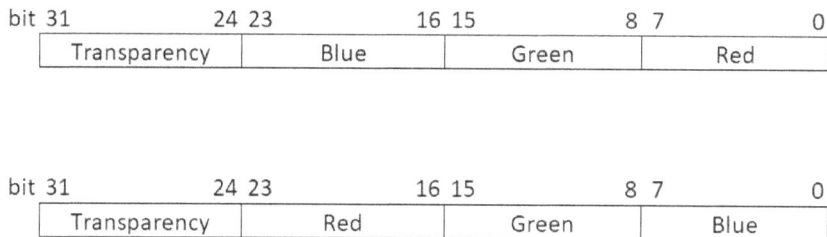

bit 31	24 23	16 15	8 7	0
Transparency	Blue	Green	Red	

bit 31	24 23	16 15	8 7	0
Transparency	Red	Green	Blue	

Figure 5.15 – Pixel storage, with the original format above and the desired format below

Mr. Smith has already created a small piece of code that you can find in the `Pointers` demo. The code creates an array with 10 million elements and fills it with the $0055AACC value. The **Blue** color is set to $55, **Green** to $AA, and **Red** to $CC for all pixels.

The code then walks over the array and processes each element. Firstly, it extracts the **Blue** component by masking only one byte (AND $00FF0000) and moving it 16 bits to the right (SHR 16). Then, it extracts the **Red** component (AND $000000FF). Finally, it clears both **Blue** and **Red** out of the original data (AND $FF00FF00) and adds the **Red** and **Blue** components back in (OR). **Red** now shifts left by 16 places (SHL 16) so that it will be stored in the second-highest byte:

```
function TfrmPointers.PrepareData: TArray<Cardinal>;
var
  i: Integer;
begin
  SetLength(Result, 100000000);
  for i := Low(Result) to High(Result) do
    Result[i] := $0055AACC;
end;
```

```
procedure TfrmPointers.btnArrayClick(Sender: TObject);
var
  rgbData: TArray<Cardinal>;
  i: Integer;
  r,b: Byte;
begin
  rgbData := PrepareData;

  for i := Low(rgbData) to High(rgbData) do
  begin
    b := rgbData[i] AND $00FF0000 SHR 16;
    r := rgbData[i] AND $000000FF;
    rgbData[i] := rgbData[i] AND $FF00FF00
                  OR (r SHL 16) OR b;
  end;
end;
```

This code runs quite fast, but all those bit operations take some time. How can we speed them up?

The best approach is just to ignore the concept that *a pixel is a Cardinal* and treat each color component as an individual byte. We can then point one pointer to the **Red** component of the first pixel and another to the **Blue** component and just swap the values. Then, we can advance each pointer to the next pixel and repeat the process.

To do that, we have to know at least a few things about how Delphi stores integer values in memory. The simple answer to that is, as the CPU expects them. Intel processors use a *little-endian* format (yes, that is a technical term) that defines that the least important byte is stored in memory first (*little end* first).

If we put three pixels into memory so that the first starts at address 0, they will look just like the following diagram:

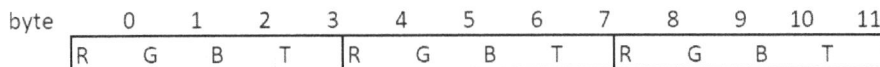

Figure 5.16 – The memory layout of three consecutive pixels

Our improved code uses that knowledge. It uses two pointers, both of type PByte (a pointer to a byte value). The pBlue pointer points to the current **Blue** byte and the pRed pointer points to the current **Red** byte. To initialize them, the code sets both to the address of the first pixel (which is the same as the address of its lowest byte) and then increments pBlue by 2 bytes.

In the loop, the code simply stores current the **Red** pixel into a temporary value (`r := pRed^`), copies the **Blue** pixel into the **Red** pixel (`pRed^ := pBlue^`), and stores the temporary **Red** pixel in the **Blue** location (`pBlue^ := r`). It then increments both pointers to the next pixel:

```
procedure TfrmPointers.btnPointerClick(Sender: TObject);
var
  rgbData: TArray<cardinal>;
  i: Integer;
  r: Byte;
  pRed: PByte;
  pBlue: PByte;
begin
  rgbData := PrepareData;

  pRed := @rgbData[0];
  pBlue := pRed;
  Inc(pBlue,2);
  for i := Low(rgbData) to High(rgbData) do
  begin
    r := pRed^;
    pRed^ := pBlue^;
    pBlue^ := r;
    Inc(pRed, SizeOf(rgbData[0]));
    Inc(pBlue, SizeOf(rgbData[0]));
  end;
end;
```

The code is a bit more convoluted but not that harder to understand, as all the pixel operations (AND, SHR, and so on) in the original code weren't easy to read either. It is also much faster. On my test computer, the original code needs 159 ms, while the new code finishes in 72 ms, which is more than twice the speed!

To cut a long story short, pointers are incredibly useful. They also make code harder to understand, so use them sparingly and document the code.

Going the assembler way

Sometimes, when you definitely have to squeeze everything from code, there is only one solution – rewrite it in an assembler. My response to any such idea is always the same – don't do it! Rewriting code in an assembler is almost always much more trouble than it is worth.

I do admit that there are legitimate reasons for writing a program in assembly language. Looking around, I quickly found five areas where assembly language is still significantly present – memory managers, graphical code, cryptography routines (encryption and hashing), compression, and interfacing with hardware.

Even in these areas, situations change quickly. I tested some small assembler routines from the graphical library *GraphicEx* and was quite surprised to find out that they are not significantly faster than equivalent Delphi code.

The biggest gain that you'll get from using an assembly language is when you want to process a large buffer of data (such as a bitmap) and then do the same operation on all elements. In such cases, you can use SSE2 instructions, which run circles around the slow 386 instruction set that the Delphi compiler uses.

As assembly language is not my game (I can read it but can't write good, optimized assembly code), my example is extremely simple. The code in the demo program, AsmCode, implements a four-dimensional vector (a record with four floating point fields) and a method that multiplies two such fields:

```
type
  TVec4 = packed record
    X, Y, Z, W: Single;
  end;

function Multiply_PAS(const A, B: TVec4): TVec4;
begin
  Result.X := A.X * B.X;
  Result.Y := A.Y * B.Y;
  Result.Z := A.Z * B.Z;
  Result.W := A.W * B.W;
end;
```

As it turns out, this is an operation that can be implemented using SSE2 instructions. In the code shown next, first, movups moves the A vector into the xmm0 register. Next, movups does the same for the other vector. Then, the magical mulps instruction multiplies four single-precision values in the xmm0 register by four single-precision values in the xmm1 register. Finally, movups copies the result of the multiplication into the function result:

```
function Multiply_ASM(const A, B: TVec4): TVec4;
asm
  movups xmm0, [A]
  movups xmm1, [B]
  mulps xmm0, xmm1
  movups [Result], xmm0
end;
```

Running the test shows a clear winner. While Multiply_PAS needs 53 ms to multiply 10 million vectors, Multiply_ASM does that in half the time – 24 ms.

As you can see in the previous example, assembly instructions are introduced with the `asm` statement and end with `end`. In the Win32 compiler, you can mix Pascal and assembly code inside one method. This is not allowed with the Win64 compiler. In 64-bit mode, a method can only be written in pure Pascal or pure assembly code.

> **Tip**
>
> The `asm` instruction is only supported by Windows and macOS compilers. In older sources, you'll also find an `assembler` instruction that is only supported for backward compatibility and does nothing.

I'll end this short excursion into the assembler world with some advice. Whenever you implement a part of your program in assembly, please also create a Pascal version. This will simplify debugging, serve as a reference implementation, and can be used on platforms that don't support inline assembly code.

> **Note**
>
> When deciding whether to compile Pascal or assembly code, you can use the following predefined conditional symbols – ASSEMBLER (defined if the `asm` syntax is supported on the current platform), CPUX86, CPUX64, CPU32BITS, CPU64BITS, CPUARM, CPUARM32, and CPUARM64. A detailed description of all predefined conditional symbols together with a list of platforms where they are defined can be found at `https://docwiki.embarcadero.com/RADStudio/en/Conditional_compilation_(Delphi)`.

With this advice in mind, we can rewrite the multiplication code as follows:

```
function Multiply(const A, B: TVec4): TVec4;
{$IFDEF ASSEMBLER}
Asm
  movups xmm0, [A]
  movups xmm1, [B]
  mulps xmm0, xmm1
  movups [Result], xmm0
end;
{$ELSE}
Begin
  Result.X := A.X * B.X;
  Result.Y := A.Y * B.Y;
  Result.Z := A.Z * B.Z;
  Result.W := A.W * B.W;
end;
{$ENDIF}
```

Returning to SlowCode

To finish this chapter, I'll return to the now well-known SlowCode example. At the end of *Chapter 3, Fixing the Algorithm*, we significantly adapted the code and ended with a version that calculates prime numbers with the Sieve of Eratosthenes (SlowCode_Sieve). That version processed 10 million numbers in 1,072 milliseconds. Let's see if we can improve on that.

The obvious target for optimization is the Reverse function, which creates a result by appending characters one at a time. We saw in this chapter that modifying a string can cause frequent memory allocations:

```
function Reverse(s: string): string;
var
  ch: char;
begin
  Result := '';
  for ch in s do
    Result := ch + Result;
end;
```

Instead of optimizing this function, let's look at how it is used. The Filter method uses it to reverse a number – reversed := StrToInt(Reverse(IntToStr(i)));.

This statement brings in another memory allocation (in the IntToStr function, which creates a new string) and executes some code that has to parse a string and return a number (StrToInt). This is quite some work for an operation that doesn't really need strings at all.

It turns out that it is very simple to reverse a decimal number without using strings. You just have to repeatedly divide it by 10 and multiply the remainders to get a new number:

```
function ReverseInt(value: Integer): Integer;
begin
  Result := 0;
  while value <> 0 do
  begin
    Result := Result * 10 + (value mod 10);
    value := value div 10;
  end;
end;
```

The new Filter method can then use this method to reverse a number:

```
function Filter(list: TList<Integer>): TArray<Integer>;
var
  i: Integer;
  reversed: Integer;
```

```
begin
  SetLength(Result, 0);
  for i in list do
  begin
    reversed := ReverseInt(i);
    if not ElementInDataDivides(list, reversed) then
    begin
      SetLength(Result, Length(Result) + 1);
      Result[High(Result)] := i;
    end;
  end;
end;
```

The new code in the demo program, SlowCode_Sieve_v2, indeed runs a bit faster. It processes 10 million elements in 851 milliseconds, which is roughly a 20% improvement.

Is there something more that we learned in this chapter that can be applied to that program? Sure – turn on optimization! Just add one line – {$OPTIMIZATION ON} – and you'll get a significantly faster version. SlowCode_Sieve_v2_opt processes 10 million elements in a mere 529 ms!

There is still room for improvement in that program. It will, however, have to wait until the next chapter.

Summary

The topic of this chapter was fine-tuning code. We started with Delphi compiler settings, which can, in some cases, significantly change code execution speed, and we learned what those situations are.

Then, I introduced a simple but effective optimization – extracting common expressions. This optimization served as an introduction to the *CPU window*, which can help us analyze compiled Delphi code.

After that, I got back to basics. Creating a fast program means knowing how Delphi works, so we looked at built-in data types. We saw what is fast and what is not.

As a logical follow-up to data types, we looked at methods – what happens when you pass parameters to a method and how to speed that up. We also reviewed a few surprising implementation details that can create problems in your code.

We ended the chapter with three practical examples. Firstly, we used pointers to speed up bitmap processing. Next, a short section on assembler code showed you how to write a fast assembler replacement for Pascal code. Finally, we revisited the SlowCode program (which isn't so slow anymore) and made it even faster.

In the next chapter, we'll go even deeper. I'll explore the topic of *memory managers* – what they are and how they work and why it is important to know how the built-in memory manager in Delphi is implemented. As usual, we'll also optimize some code. Stay tuned!

6

Memory Management

In the previous chapter, I explained a few things with a lot of hand-waving. I was talking about memory being *allocated* but I never told you what that actually means. Now is the time to fill in the missing pieces.

Memory management is part of practically every computing system. Multiple programs must coexist inside a limited memory space, and that can only be possible if the operating system is taking care of it. When a program needs some memory—for example, to create an object—it can ask the operating system and it will give it a slice of shared memory. When an object is not needed anymore, that memory can be returned to the loving care of the operating system.

Slicing and dicing memory straight from the operating system is a relatively slow operation. In lots of cases, a memory system also doesn't know how to return small chunks of memory. For example, if you call Windows' `VirtualAlloc` function to get 20 bytes of memory, it will actually reserve 4 KB (or 4,096 bytes) for you. In other words, 4,076 bytes would be wasted.

To fix these and other problems, programming languages typically implement their own internal memory management algorithms. When you request 20 bytes of memory, the request goes to that internal memory manager. It still requests memory from the operating system but then splits it internally into multiple parts.

In a hypothetical scenario, the internal memory manager would request 4,096 bytes from the operating system and give 20 bytes of that to the application. The next time the application requested some memory (30 bytes, for example), the internal memory manager would get that memory from the same 4,096-byte block.

To move from a hypothetical to a specific scenario, Delphi also includes such a memory manager. From Delphi 2006, this memory manager is called **FastMM4**. It was written as an open source memory manager by Pierre le Riche with help from other Delphi programmers and was later licensed by Borland. FastMM4 was a great improvement over the previous Delphi memory manager and, although it does not perform perfectly in the parallel programming world, it still functions very well after more than 15 years.

Delphi exposes a public interface to replace the internal memory manager, so you can easily replace FastMM4 with a different memory manager. As we'll see later in this chapter, this can sometimes be helpful.

We will cover the following topics in this chapter:

- What happens when strings and arrays are reallocated and how can we speed this up?
- Which functions can an application use to allocate memory?
- How can we use a memory manager to dynamically create a record?
- How is FastMM4 internally implemented and why does it matter to a programmer?
- How is memory allocated in a parallel world?
- How can we replace FastMM4 with a different memory manager?
- What can we do to improve `SlowCode` even more?

Technical requirements

All code in this chapter was written with Delphi 11.3 Alexandria. Most of the examples, however, could also be executed on Delphi XE and newer versions. You can find all the examples on GitHub at `https://github.com/PacktPublishing/Delphi-High-Performance---Second-Edition/tree/main/ch6`.

Optimizing strings and array allocations

When you create a string, the code allocates memory for its content, copies the content into that memory, and stores the address of this memory in the string variable. Of course, you all know that by now as this was the topic of the previous chapter.

If you append a character to this string, it must be stored somewhere in that memory. However, there is no place to store the string. The original memory block was just big enough to store the original content. The code must therefore enlarge that memory block, and only then can the appended character be stored in the newly acquired space.

As we'll see further on in the chapter, FastMM4 tries to make sure that the memory can be expanded *in place*—that is, without copying the original data into a new, larger block of memory. Still, this is not always possible, and sometimes data must be copied around.

A very similar scenario plays out when you extend a dynamic array. Memory that contains the array data can sometimes be extended in place (without moving), but often this cannot be done.

If you do a lot of appending, these constant reallocations will start to slow down the code. The `Reallocation` demo shows a few examples of such behavior and possible workarounds.

The first example, activated by the **Append String** button, simply appends the ' * ' character to a string 10 million times. The code looks simple, but the s   := s + ' * ' assignment hides a potentially slow string reallocation:

```
procedure TfrmReallocation.btnAppendStringClick(
  Sender: TObject);
var
  s: String;
  i: Integer;
begin
  s := '';
  for i := 1 to CNumChars do
    s := s + '*';
end;
```

By now, you probably know that I don't like to present problems that I don't have solutions for, and this is not an exception. In this case, the solution is called SetLength. This function sets a string to a specified size. You can make it shorter, or you can make it longer. You can even set it to the same length as before.

In case you are enlarging the string, you have to keep in mind that SetLength will allocate enough memory to store the new string, but it will not initialize it. In other words, the newly allocated string space will contain random data.

A click on the **SetLength String** button activates the optimized version of the string-appending code. As we know that the resulting string will be CNumChars long, the code can call SetLength(s, CNumChars) to preallocate all the memory in one step. After that, we should not *append* characters to the string as that would add new characters at the end of the preallocated string. Rather, we have to store characters directly in the string by writing to s[i]:

```
procedure TfrmReallocation.btnSetLengthClick(Sender: TObject);
var
  s: String;
  i: Integer;
begin
  SetLength(s, CNumChars);
  for i := 1 to CNumChars do
    s[i] := '*';
end;
```

Comparing the speed shows that the second approach is significantly faster. It runs in 33 ms instead of the original 142 ms.

> **Tip**
>
> This code still calls some string management functions internally, namely `UniqueString`. It returns fairly quickly but still represents some overhead. If you are pushing for the fastest possible execution, you can initialize the string by using a *sliding pointer*, just as in the *RGB* example from the previous chapter.

You can also remove all safety checks by casting a string into a pointer and back into `PChar`. Instead of `s[i] := '*'`, you would write `PChar(Pointer(s))[i-1] := '*'`. Be careful with indexes! String index 1 represents the first element, the one with the (pointer) offset 0. You could then rewrite the last example as follows:

```
procedure TfrmReallocation.btnSetLengthClick(
  Sender: TObject);
var
  s: String;
  i: Integer;
  p: PChar;
begin
  SetLength(s, CNumChars);
  p := Pointer(s);
  for i := 1 to CNumChars do
    p[i-1] := '*';
end;
```

A similar situation happens when you are extending a dynamic array. The code triggered by the **Append array** button shows how an array may be extended by one element at a time in a loop. Admittedly, the code looks very weird as nobody in their right mind would write a loop like this. In reality, similar code would be split into multiple longer functions and may be hard to spot:

```
procedure TfrmReallocation.btnAppendArrayClick(Sender: TObject);
var
  arr: TArray<char>;
  i: Integer;
begin
  SetLength(arr, 0);
  for i := 1 to CNumChars do begin
    SetLength(arr, Length(arr) + 1);
    arr[High(arr)] := '*';
  end;
end;
```

The solution is similar to the string case. We can preallocate the whole array by calling the `SetLength` function and then write the data into the array elements. We just have to keep in mind that the first array element always has index 0 (unlike strings, which start with index 1):

```
procedure TfrmReallocation.btnSetLengthArrayClick(Sender: TObject);
var
  arr: TArray<char>;
  i: Integer;
begin
  SetLength(arr, CNumChars);
  for i := 1 to CNumChars do
    arr[i-1] := '*';
end;
```

Improvements in speed are similar to the string demo. The original code needs 230 ms to append 10 million elements, while the improved code executes in 26 ms.

The third case when you may want to preallocate storage space is when you are appending to a list. As an example, I'll look into a `TList<T>` class. Internally, it stores the data in a `TArray<T>` class, so it again suffers from constant memory reallocation when you are adding data to the list.

The short demo code appends 10 million elements to a list. As opposed to the previous array demo, this is completely normal-looking code, found many times in many applications:

```
procedure TfrmReallocation.btnAppendTListClick(Sender: TObject);
var
  list: TList<Char>;
  i: Integer;
begin
  list := TList<Char>.Create;
  try
    for i := 1 to CNumChars do
      list.Add('*');
  finally
    FreeAndNil(list);
  end;
end;
```

To preallocate memory inside a list, you can set the `Capacity` property to an expected number of elements in the list. This doesn't prevent the list from growing at a later time; it just creates an initial estimate. You can also use `Capacity` to reduce the memory space used for the list after deleting lots of elements from it.

The difference between a list and a string or an array is that, after setting `Capacity`, you still cannot access `list[i]` elements directly. Firstly, you have to add them, just as if `Capacity` were not assigned:

```
procedure TfrmReallocation.btnSetCapacityTListClick(Sender: TObject);
var
  list: TList<Char>;
  i: Integer;
begin
  list := TList<Char>.Create;
  try
    list.Capacity := CNumChars;
    for i := 1 to CNumChars do
      list.Add('*');
  finally
    FreeAndNil(list);
  end;
end;
```

Comparing the execution speed shows only a small improvement. The original code was executed in 167 ms, while the new version needed 145 ms. The reason for that relatively small change is that `TList<T>` already manages its storage array. When it runs out of space, it will always at least double the previous size. Internal storage, therefore, grows from 1 to 2, 4, 8, 16, 32, 64, ... elements.

This can, however, waste a lot of memory. In our example, the final size of the internal array is 16,777,216 elements, which is about 60% elements too many. By setting the capacity to the exact required size, we have therefore saved 6,777,216 * `SizeOf(Char)` bytes or almost 13 MB.

> **Tip**
>
> Since Delphi 10.3, you can set a custom grow function by calling the `SetGrowCollectionFunc` function from the `SysUtils` unit.

Other data structures also support the `Capacity` property. We can find it in `TList`, `TObjectList`, `TInterfaceList`, `TStrings`, `TStringList`, `TDictionary`, `TObjectDictionary`, and others.

Memory management functions

Besides the various internal functions that the Delphi **RunTime Library** (**RTL**) uses to manage strings, arrays, and other built-in data types, RTL also implements various functions that you can use in your program to allocate and release memory blocks. In the next few paragraphs, I'll tell you a little bit about them.

Memory management functions can be best described if we split them into a few groups, each including functions that were designed to work together. Let's take a closer look at these groups:

- The first group includes `GetMem`, `AllocMem`, `ReallocMem`, and `FreeMem`.

   The `GetMem(var P: Pointer; Size: Integer)` procedure allocates a memory block of size `Size` and stores an address of this block in a pointer variable, `P`. This pointer variable is not limited to the `pointer` type but can be of any pointer type (for example, `PByte`).

   The new memory block is not initialized and will contain whatever is stored in the memory at that time. Alternatively, you can allocate a memory block with a call to the `AllocMem(Size: Integer): Pointer` function, which allocates a memory block, fills it with zeros, and then returns its address.

   To change the size of a memory block, call the `ReallocMem(var P: Pointer; Size: Integer)` procedure. The `P` variable must contain a pointer to a memory block, and `Size` can be either smaller or larger than the original block size. FastMM4 will try to resize the block in place. If that fails, it will allocate a new memory block, copy the original data into the new block, release the old block, and return an address of the new block in the `P` variable. Just as with the `GetMem` procedure, newly allocated bytes will not be initialized.

   To release memory allocated in this way, you should call the `FreeMem(var P: Pointer)` procedure.

- The second group includes `GetMemory`, `ReallocMemory`, and `FreeMemory`. These three work just the same as functions from the first group, except that they can be used from C++Builder.

- The third group contains just two functions, `New` and `Dispose`.

   These two functions can be used to dynamically create and destroy variables of any type. To allocate such a variable, call `New(var X: Pointer)`, where X is again of any pointer type. The compiler will automatically provide the correct size for the memory block, and it will also initialize all managed fields to zero. Unmanaged fields will not be initialized.

   To release such a variable, don't use `FreeMem` but `Dispose(var X: Pointer)`. In the next section, I'll give a short example of using `New` and `Dispose` to dynamically create and destroy variables of a record type.

- The fourth and last group also contains just two functions, `Initialize` and `Finalize`. Strictly speaking, they are not memory management functions. Still, they are used almost exclusively when dynamically allocating memory, and that's why they belong in this chapter.

If you create a variable containing managed fields (for example, a record) with a function other than `New` or `AllocMem`, it will not be correctly initialized. Managed fields will contain random data, and that will completely break the execution of the program. To fix that, you should call `Initialize(var V)`, passing in the variable (and not the pointer to this variable!). An example in the next section will clarify that.

Before you return such a variable to the memory manager, you should clean up all references to managed fields by calling `Finalize(var V)`. It is better to use `Dispose`, which will do that automatically, but sometimes that is not an option and you have to do it manually.

Both functions also exist in a form that accepts a number of variables to initialize. This form can be used to initialize or finalize an array of data:

```
procedure Initialize(var V; Count: NativeUInt);
procedure Finalize(var V; Count: NativeUInt);
```

In the next section, I'll dig deeper into the dynamic allocation of record variables. I'll also show how most of the memory allocation functions are used in practice.

Dynamic record allocation

While it is very simple to dynamically create new objects—you just call the `Create` constructor—dynamic allocation of records and other data types (arrays, strings ...) is a bit more complicated.

In the previous section, we saw that the preferred way of allocating such variables is with the `New` method. The `InitializeFinalize` demo shows how this is done in practice.

The code will dynamically allocate a variable of type `TRecord`. To do that, we need a pointer variable, pointing to `TRecord`. The cleanest way to do that is to declare a new `PRecord = ^TRecord` type, like so:

```
type
  TRecord = record
    s1, s2, s3, s4: string;
  end;
  PRecord = ^TRecord;
```

Now, we can just declare a variable of type `PRecord` and call `New` on that variable. After that, we can use the `rec` variable as if it were a normal record and not a pointer. Technically, we would have to always write `rec^.s1`, `rec^.s4`, and so on, but the Delphi compiler is friendly enough and allows us to drop the `^` character:

```
procedure TfrmInitFin.btnNewDispClick(Sender: TObject);
var
  rec: PRecord;
begin
  New(rec);
  try
    rec.s1 := '4';
    rec.s2 := '2';
    rec.s4 := rec.s1 + rec.s2 + rec.s4;
```

```
      ListBox1.Items.Add('New: ' + rec.s4);
    finally
      Dispose(rec);
    end;
  end;
```

> **Note**
>
> Technically, you could just use `rec: ^TRecord` instead of `rec: PRecord`, but it is customary to use explicitly declared pointer types, such as `PRecord`.

Another option is to use `GetMem` instead of `New`, and `FreeMem` instead of `Dispose`. In this case, however, we have to manually prepare allocated memory for use with a call to `Initialize`. We must also prepare it to be released with a call to `Finalize` before we call `FreeMem`.

If we use `GetMem` for initialization, we must manually provide the correct size of the allocated block. In this case, we can simply use `SizeOf(TRecord)`.

We must also be careful with parameters passed to `GetMem` and `Initialize`. You pass a pointer (`rec`) to `GetMem` and `FreeMem` and the actual record data (`rec^`) to `Initialize` and `Finalize`:

```
procedure TfrmInitFin.btnInitFinClick(Sender: TObject);
var
  rec: PRecord;
begin
  GetMem(rec, SizeOf(TRecord));
  try
    Initialize(rec^);
    rec.s1 := '4';
    rec.s2 := '2';
    rec.s4 := rec.s1 + rec.s2 + rec.s4;
    ListBox1.Items.Add('GetMem+Initialize: ' + rec.s4);
  finally
    Finalize(rec^);
    FreeMem (rec);
  end;
end;
```

This demo also shows how the code doesn't work correctly if you allocate a record with `GetMem` but then don't call `Initialize`. To test this, click the **GetMem** button. While in actual code the program may sometimes work and sometimes not, I have taken some care so that `GetMem` always returns a memory block that is not initialized to zero, and the program will certainly fail:

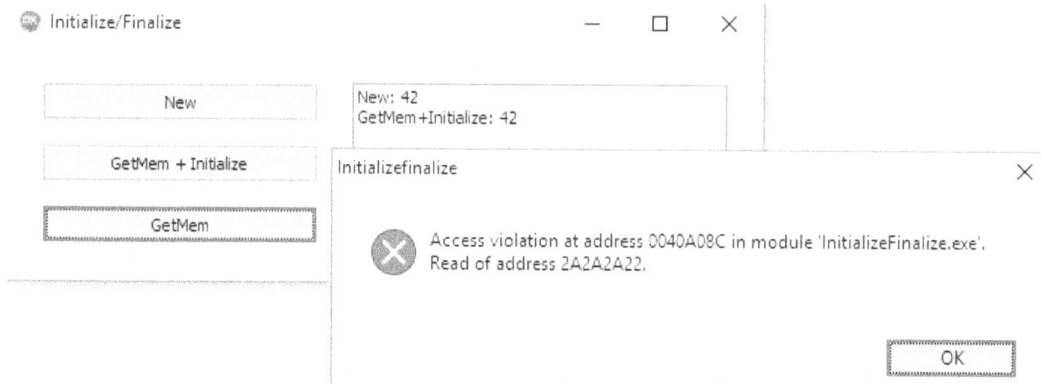

Figure 6.1 – Incorrect use of memory allocation functions can cause a program crash

It is certainly possible to create records dynamically and use them instead of classes, but one question still remains—why? Why would we want to use records instead of objects when working with objects is simpler? The answer, in one word, is *speed*.

The demo program, `Allocate`, shows the difference in execution speed. A click on the **Allocate objects** button will create 10 million objects of type `TNodeObj`, which is a typical object that you would find in an implementation of a *binary tree*. Of course, the code then cleans up after itself by destroying all those objects:

```
type
  TNodeObj = class
    Left, Right: TNodeObj;
    Data: NativeUInt;
  end;


procedure TfrmAllocate.btnAllocClassClick(Sender: TObject);
var
  i: Integer;
  nodes: TArray<TNodeObj>;
begin
  SetLength(nodes, CNumNodes);
  for i := 0 to CNumNodes-1 do
    nodes[i] := TNodeObj.Create;
  for i := 0 to CNumNodes-1 do
    nodes[i].Free;
end;
```

Similar code, activated by the **Allocate records** button, creates 10 million records of type TNodeRec, which contains the same fields as TNodeObj:

```
type
  PNodeRec = ^TNodeRec;
  TNodeRec = record
    Left, Right: PNodeRec;
    Data: NativeUInt;
  end;

procedure TfrmAllocate.btnAllocRecordClick(
  Sender: TObject);
var
  i: Integer;
  nodes: TArray<PNodeRec>;
begin
  SetLength(nodes, CNumNodes);
  for i := 0 to CNumNodes-1 do
    New(nodes[i]);
  for i := 0 to CNumNodes-1 do
    Dispose(nodes[i]);
end;
```

Running both methods shows a big difference. While the class-based approach needs 366 ms to initialize objects and 76 ms to free them, the record-based approach needs only 76 ms to initialize records and 56 ms to free them. Where does that big difference come from?

When you create an object of a class, lots of things happen. Firstly, TObject.NewInstance is called to allocate an object. That method uses TObject.InstanceSize (which is compiled inline) to get the size of the object, then GetMem to allocate the memory, and, in the end, InitInstance, which fills the allocated memory with zeroes. It then walks the inheritance chain, checks if any interface is implemented, and sets one up if not. If a custom-managed record is used in the fields of the class, it also calls its custom initializer (if present).

Secondly, a chain of constructors is called. After all that, a chain of AfterConstruction methods is called (if such methods exist). All in all, that is quite a process and takes some time.

Much less is going on when you create a record. If it contains only unmanaged fields, as in our example, a GetMem procedure is called, and that's all. If the record contains managed fields, this GetMem procedure is followed by a call to the _Initialize method in the *System* unit, which initializes managed fields. If the record is a custom-managed record with a custom initializer, this initializer also gets called.

The problem with records is that we cannot declare generic pointers. When we are building trees, for example, we would like to store some data of type T in each node. The initial attempt at that, however, fails. The following code does not compile with the current Delphi compiler:

```
type
  PNodeRec<T> = ^TNodeRec<T>;
  TNodeRec<T> = record
    Left, Right: PNodeRec<T>;
    Data: T;
  end;
```

We can circumvent this by moving the TNodeRec<T> declaration inside the generic class that implements a tree. The following code from the Allocate demo shows how we could declare such an internal type as a generic object and as a generic record:

```
type
  TTree<T> = class
  strict private type
    TNodeObj<T1> = class
      Left, Right: TNodeObj<T1>;
      Data: T1;
    end;

    PNodeRec = ^TNodeRec;
    TNodeRec<T1> = record
      Left, Right: PNodeRec;
      Data: T1;
    end;
    TNodeRec = TNodeRec<T>;
  end;
```

If you click the **Allocate node<string>** button, the code will create a TTree<string> object and then create 10 million class-based nodes and the same number of record-based nodes. This time, New must initialize the Data: string managed field, but the difference in speed is still big. The code needs 669 ms to create and destroy class-based nodes and 133 ms to create and destroy record-based nodes.

Another big difference between classes and records is that each object contains two hidden pointer-sized fields. Because of that, each object is 8 bytes larger than you would expect (16 bytes in 64-bit mode). That amounts to 8 * 10,000,000 bytes or a bit over 76 MB. Records are therefore not only faster but also save space! You can see an illustration of this here:

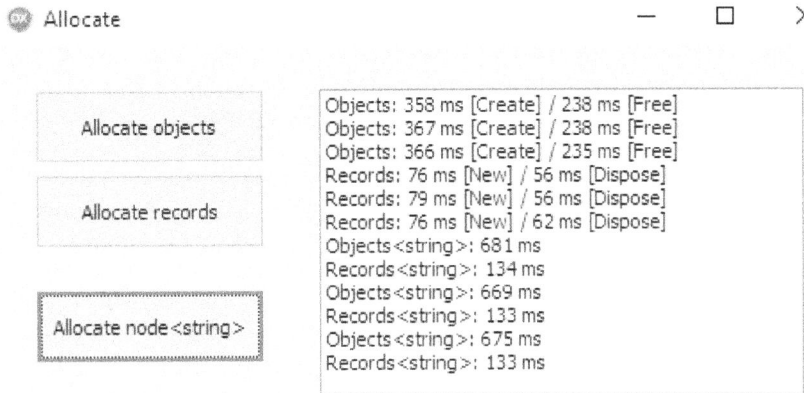

Figure 6.2 – Comparing record and object allocation speed

Before we can tackle more complex data types, such as strings and dynamic arrays, we have to take a closer look at the internals of Delphi's memory manager, FastMM4.

FastMM4 internals

To get full speed out of anything, you have to understand how it works, and memory managers are no exception to this rule. To write very fast Delphi applications, you should therefore understand how Delphi's default memory manager works.

FastMM4 is not just a memory manager—it is three memory managers in one! It contains three significantly different subsystems— a *small block allocator*, a *medium block allocator*, and a *large block allocator*.

The first one, the allocator for *small blocks*, handles all memory blocks smaller than 2.5 KB. This boundary was determined by observing existing applications. As it turned out, in most Delphi applications, this covers 99% of all memory allocations. This is not surprising, as in most Delphi applications most memory is allocated when an application creates and destroys objects and works with arrays and strings, and those are rarely larger than a few hundred characters.

Next comes the allocator for *medium blocks*, which are memory blocks with a size between 2.5 KB and 160 KB. The last one, the allocator for *large blocks*, handles all other requests.

The difference between allocators lies not just in the size of memory that they serve, but in the strategy they use to manage memory.

The *large* block allocator implements the simplest strategy. Whenever it needs some memory, it gets it directly from Windows by calling `VirtualAlloc`. This function allocates memory in 4 KB blocks, so this allocator could waste up to 4,095 bytes per request. As it is used only for blocks larger than 160 KB, this wasted memory doesn't significantly affect the program, though.

The *medium* block allocator gets its memory from the large block allocator. It then carves this larger block into smaller blocks, as they are requested by the application. It also keeps all unused parts of the memory in a linked list so that it can quickly find a memory block that is still free.

The *small* block allocator is where the real smarts of FastMM4 lies. There are actually 56 small memory allocators, each serving only one size of memory block. The first one serves 8-byte blocks, the next one 16-byte blocks, followed by the allocator for 24-, 32-, 40-, ... 256-, 272-, 288-, ... 960-, 1056-, ... 2384-, and 2608-byte blocks. They all get memory from the medium block allocator.

> **Tip**
>
> If you want to see block sizes for all 56 allocators, open `FastMM4.pas` (or `GetMem.inc` when using the built-in memory manager) and search for `SmallBlockTypes`.

What that actually means is that each memory allocation request will waste some memory. If you allocate 28 bytes, they'll be allocated from the 32-byte allocator, so 4 bytes will be wasted. If you allocate 250 bytes, they'll come from the 256-byte allocator, and so on. The sizes of memory allocators were carefully chosen so that the amount of wasted memory is typically below 10%, so this doesn't represent a big problem in most applications.

Each allocator is basically just an array of equally sized elements (memory blocks). When you allocate a small amount of memory, you'll get back one element of an array. All unused elements are connected into a linked list so that the memory manager can quickly find a free element of an array when it needs one.

The following diagram shows a very simplified representation of FastMM4 allocators. Only two small block allocators are shown. The boxes with thick borders represent allocated memory. The boxes with thin borders represent unused (free) memory. Free memory blocks are connected into linked lists. The block sizes in different allocators are not to scale.

Small block allocators

8-byte allocator

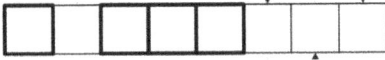

16-byte allocator

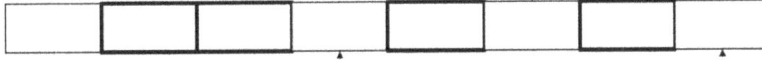

other allocators

Medium block allocator

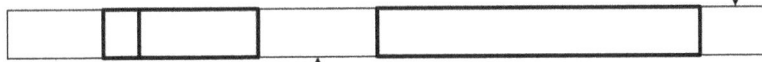

Large block allocator

Figure 6.3 – Simplified representation of FastMM4 memory manager

FastMM4 implements a neat trick that helps a lot when you resize strings or arrays by a small amount. At the beginning of this chapter, I talked about that (*Optimizing strings and array allocations*), and showed a small program that demonstrated how appending characters one by one works much more slowly than preallocating a whole string. There was a 4x speed difference (142 versus 33 ms) between these approaches.

Well, truth be told, I had to append lots and lots of characters—10 million of them—for this difference to show. If I were appending only a few characters, both versions would run at nearly the same speed. If you can, on the other hand, get your hands on a pre-2006 Delphi and run the demo program there, you'll see that the *one-by-one* approach runs terribly slow. The difference in speed will be a few more orders of magnitude larger than in my example.

The trick I'm talking about assumes that if you resized memory once, you'll probably want to do it again, soon. If you are enlarging the memory, it will limit the smallest size of the new memory block to be at least twice the size of the original block plus 32 bytes. Next time you want to resize, FastMM4 will (hopefully) just update the internal information about the allocated memory and return the same block, knowing that there's enough space at the end.

All that trickery is hard to understand without an example, so here's one. Let's say we have a string of 5 characters that neatly fits into a 24-byte block. Sorry—what am I hearing? *"What? Why!? 5 Unicode characters need only 10 bytes!"* Oh, yes—strings are more complicated than I told you before.

In reality, each Delphi `UnicodeString` and `AnsiString` variable contains some additional data besides the actual characters that make up the string. Parts of the string are also made up of the following:

- 4-byte length of string
- 4-byte reference count
- 2-byte field storing the size of each string character (either 1 for `AnsiString` or 2 for `UnicodeString`)
- 2-byte field storing the character code page

In addition to that, each string includes a terminating `Chr(0)` character. For a 5-character string, this gives us *4 (length) + 4 (reference count) + 2 (character size) + 2 (codepage) + 5 (characters) * 2 (size of a character) + 2 (terminating Chr(0)) = 24 bytes.*

When you add one character to this string, the code will ask the memory manager to enlarge a 24-byte block to 26 bytes. Instead of returning a 26-byte block, FastMM4 will round that up to *2 * 24 + 32 = 80* bytes. Then, it will look for an appropriate allocator, find one that serves 80-byte blocks (great—no memory loss!), and return a block from that allocator. It will, of course, also have to copy data from the original block to the new block.

> **Note**
>
> This formula, *2 * size + 32*, is used only in small block allocators. A medium block allocator only overallocates by 25%, and a large block allocator doesn't implement this behavior at all.

Next time you add one character to this string, FastMM4 will just look at the memory block, determine that there's still enough space inside this 80-byte memory block, and return the same memory. This will continue for quite some time while the block grows to 80 bytes in 2-byte increments. After that, the block will be resized to *2 * 80 + 32 = 192* bytes (yes, there *is* an allocator for this size), data will be copied, and the game will continue.

This behavior indeed wastes some memory but, in most circumstances, significantly boosts the speed of code that was not written with speed in mind.

Memory allocation in a parallel world

We've seen how FastMM4 boosts reallocation speed. Let's take a look at another optimization that helps a lot when you write multithreaded code—as we will in the next three chapters.

The life of a memory manager is simple when there is only one thread of execution inside a program. When the memory manager is dealing out the memory, it can be perfectly safe in the knowledge that nothing can interrupt it in this work.

When we deal with parallel processing, however, multiple paths of execution simultaneously execute the same program and work on the same data. (We call them *threads*, and I'll explain them in the next chapter.) Because of that, life from the memory manager's perspective suddenly becomes very dangerous.

For example, let's assume that one thread wants some memory. The memory manager finds a free memory block on a free list and prepares to return it. At that moment, however, another thread also needs some memory from the same allocator. This second execution thread (running in parallel with the first one) would also find a free memory block on the free list. If the first thread didn't yet update the free list, that may even be the same memory block! That can only result in one thing—complete confusion and crashing programs.

It is extremely hard to write code that manipulates some data structures (such as a free list) in a manner that functions correctly in a multithreaded world. So hard that FastMM4 doesn't even try it. Instead of that, it regulates access to each allocator with a lock. Each of the 56 small block allocators gets its own lock, as do medium and large block allocators.

When a program needs some memory from, say, a 16-byte allocator, FastMM4 locks this allocator until the memory is returned to the program. If, during this time, another thread requests memory from the same 16-byte allocator, it will have to wait until the first thread finishes.

This indeed fixes concurrency problems but introduces a *bottleneck*—a part of the code where threads must wait to be processed in a serial fashion. If threads do lots of memory allocation, this serialization will completely negate the speedup that we expected to get from the parallel approach. Such a memory manager would be useless in a parallel world.

To fix that, FastMM4 introduces memory allocation optimization, which only affects small blocks.

When accessing a small block allocator, FastMM4 will try to lock it. If that fails, it will not wait for the allocator to become unlocked but will try to lock the allocator for the next block size. If that succeeds, it will return memory from the second allocator. That will indeed waste more memory but will help with the execution speed. If the second allocator also cannot be locked, FastMM4 will try to lock the allocator for yet the next block size. If the third allocator can be locked, you'll get back memory from it. Otherwise, FastMM4 will repeat the process from the beginning.

This process can be somehow described with the following pseudo-code:

```
allocIdx := find best allocator for the memory block
repeat
  if can lock allocIdx then
    break;
  Inc(allocIdx);
  if can lock allocIdx then
```

```
      break;
   Inc(allocIdx);
   if can lock allocIdx then
      break;
   Dec(allocIdx, 2)
until false


allocate memory from allocIdx allocator


unlock allocIdx
```

A careful reader would notice that this code fails when the first line finds the last allocator in the table, or the one before that. Instead of adding some conditional code to work around the problem, FastMM4 rather repeats the last allocator in the list three times. The table of small allocators actually ends with the following sizes: 1,984; 2,176; 2,384; 2,608; 2,608; 2,608. When requesting a block size above 2,384, the first line in the preceding pseudo-code will always find the first 2,608 allocator, so there will always be two more after it.

This approach works great when memory is allocated, but hides another problem. And how could I better explain a problem than with a demonstration ...?

An example of this problem can be found in the `ParallelAllocation` program. If you run it and click the **Run** button, the code will compare the serial version of some algorithm with a parallel one. I'm aware that I did not explain parallel programming at all, but the code is so simple that even somebody without any understanding of the topic will guess what it does.

The core of a test runs a loop with the `Execute` method on all objects in a list. If a `parallelTest` flag is set, the loop is executed in parallel; otherwise, it is executed serially. The only mystery part in the code, `TParallel.For`, does exactly what it says—it executes a `for` loop in parallel:

```
if parallelTest then
  TParallel.For(0, fList.Count - 1,
    procedure(i: integer)
    begin
      fList[i].Execute;
    end)
else
  for i := 0 to fList.Count - 1 do
    fList[i].Execute;
```

If you'll be running the program, make sure that you execute it without the debugger (*Ctrl* + *Shift* + *F9* will do that). Running with the debugger slows down parallel execution and can skew the measurements.

On my test machine, I got the following results:

Figure 6.4 – Comparing serialized and parallel memory allocation speed

In essence, parallelizing the program made it almost four times faster. Great result!

Well, no. Not a great result. You see, the machine I was testing on has 8 hyperthreaded cores (that is, 8 physical cores, each being able to execute some instructions in parallel). If all would be running in parallel, I would expect an almost 8x speedup, not a mere 3-4 times improvement!

If you take a look at the code, you'll see that each `Execute` method allocates a ton of objects. It is obvious (even more, given the topic of the current chapter) that a problem lies in the memory manager. The question remains, though, where exactly lies this problem and how can we find it?

I ran into exactly the same problem a few years ago. A highly parallel application that processes gigabytes and gigabytes of data was not running fast enough. There were no obvious problematic points, and I suspected that the culprit was FastMM4. I tried swapping the memory manager for a more multithreading-friendly one and, indeed, the problem was somehow reduced, but I still wanted to know where the original sin lay in my code. I also wanted to continue using FastMM4 as it offers great debugging tools.

In the end, I found no other solution than to dig into the FastMM4 internals, find out how it works, and add some logging there. More specifically, I wanted to know when a thread is waiting for a memory manager to become unlocked. I also wanted to know at which locations in my program this happens the most.

To cut a (very) long story short, I extended FastMM4 with support for this kind of logging. This extension was later integrated into the main FastMM4 branch. As these changes are not included in Delphi, you have to take some steps to use this code. Before we do that, however, we have to learn how to replace Delphi's memory manager with a different one.

Replacing the default memory manager

While writing a new memory manager is a hard job, installing it in Delphi—once it is completed—is very simple. The *System* unit implements `GetMemoryManager` and `SetMemoryManager` functions that help with that, as illustrated here:

```
type
  TMemoryManagerEx = record
```

```
    {The basic (required) memory manager functionality}
    GetMem: function(Size: NativeInt): Pointer;
    FreeMem: function(P: Pointer): Integer;
    ReallocMem: function(P: Pointer; Size: NativeInt): Pointer;
    {Extended (optional) functionality.}
    AllocMem: function(Size: NativeInt): Pointer;
    RegisterExpectedMemoryLeak: function(P: Pointer): Boolean;
    UnregisterExpectedMemoryLeak: function(P: Pointer): Boolean;
  end;

procedure GetMemoryManager(
  var MemMgrEx: TMemoryManagerEx); overload;
procedure SetMemoryManager(
  const MemMgrEx: TMemoryManagerEx); overload;
```

A proper way to install a new memory manager is to call GetMemoryManager and store the result in some global variable. Then, the code should populate the new TMemoryManagerEx record with pointers to its own replacement methods and call SetMemoryManager. Typically, you would do this in the initialization block of the unit implementing the memory manager.

The new memory manager *must* implement the GetMem, FreeMem, and ReallocMem functions. It *may* implement the other three functions (or only some of them). Delphi is smart enough to implement AllocMem internally if it is not implemented by the memory manager, and the other two will just be ignored if you call them from the code.

When the memory manager is uninstalled (usually from the finalization block of the memory manager unit), it has to call SetMemoryManager and pass to it the original memory manager configuration stored in the global variable.

The nice thing with memory managers is that you can call functions from the previous memory manager in your code. That allows you to just add some small piece of functionality that may be useful when you are debugging some hard problem or researching how Delphi works.

Logging memory manager

To demonstrate this approach, I have written a small logging memory manager. After it is installed, it will log all memory calls (except the two related to memory leaks) to a file. It has no idea how to handle memory allocation, though, so it forwards all calls to the existing memory manager.

The demonstration program, LoggingMemoryManager, shows how to use this logging functionality. When you click the **Test** button, the code installs it by calling the InstallMM function. In the current implementation, it logs information into the <projectName>_memory.log file, which is saved to the <projectName>.exe folder.

Then, the code creates a TList<integer> object, writes 1,024 integers into it, and destroys the list.

In the end, the memory manager is uninstalled by calling `UninstallMM`, and the contents of the log file are loaded into the listbox:

```
procedure TfrmLoggingMM.Button1Click(Sender: TObject);
var
  list: TList<integer>;
  i: Integer;
  mmLog: String;
begin
  mmLog := ChangeFileExt(ParamStr(0), '_memory.log');
  if not InstallMM(mmLog) then
    ListBox1.Items.Add('Failed to install memory manager');

  list := TList<integer>.Create;
  for i := 1 to 1024 do
    list.Add(i);
  FreeAndNil(list);

  if not UninstallMM then
    ListBox1.Items.Add(
      'Failed to uninstall memory manager');

  LoadLog(mmLog);
end;
```

The memory manager itself is implemented in the `LoggingMM` unit. It uses three global variables. `MMIsInstalled` contains the current installed/not installed status, `OldMM` stores the configuration of the existing state, and `LoggingFile` stores a handle of the logging file:

```
var
  MMIsInstalled: boolean;
  OldMM: TMemoryManagerEx;
  LoggingFile: THandle;
```

The installation function first opens a logging file with a call to the `CreateFile` Windows API. After that, it retrieves the existing memory manager state, sets a new configuration with pointers to the logging code, and exits. Memory leak functions are not used, so corresponding pointers are set to `nil`:

```
function InstallMM(const fileName: string): boolean;
var
  myMM: TMemoryManagerEx;
begin
  if MMIsInstalled then
    Exit(False);
```

```
  LoggingFile := CreateFile(PChar(fileName), GENERIC_WRITE,
    0, nil, CREATE_ALWAYS, FILE_ATTRIBUTE_NORMAL, 0);
  if LoggingFile = INVALID_HANDLE_VALUE then
    Exit(False);

  GetMemoryManager(OldMM);

  myMM.GetMem := @LoggingGetMem;
  myMM.FreeMem := @LoggingFreeMem;
  myMM.ReallocMem := @LoggingReallocMem;
  myMM.AllocMem := @LoggingAllocMem;
  myMM.RegisterExpectedMemoryLeak := nil;
  myMM.UnregisterExpectedMemoryLeak := nil;
  SetMemoryManager(myMM);

  MMIsInstalled := True;
  Result := True;
end;
```

Using the Windows API makes my logging memory manager bound to the operating system, so why don't I simply use a TStream class for logging?

Well, you have to remember that the purpose of this exercise is to log all memory requests. Using any built-in Delphi functionality may cause hidden memory manager operations. If I'm not careful, an object may get created somewhere or a string could get allocated. That's why the code stays on the safe side and uses the Windows API to work with the logging file.

Uninstalling the memory manager is simpler. We just have to close the logging file and restore the original configuration, like so:

```
function UninstallMM: boolean;
begin
  if not MMIsInstalled then
    Exit(False);

  SetMemoryManager(OldMM);

  if LoggingFile <> INVALID_HANDLE_VALUE then begin
    CloseHandle(LoggingFile);
    LoggingFile := INVALID_HANDLE_VALUE;
  end;

  MMIsInstalled := False;
```

```
    Result := True;
  end;
```

Logging itself is again tricky. The logging code is called from the memory manager itself, so it can't use any functionality that would use the memory manager. Most important of all, we cannot use any `UnicodeString` or `AnsiString` variable. The logging code, therefore, creates the log output together from small parts, as follows:

```
function LoggingGetMem(Size: NativeInt): Pointer;
begin
  Result := OldMM.GetMem(Size);
  Write('GetMem(');
  Write(Size);
  Write(') = ');
  Write(NativeUInt(Result));
  Writeln;
end;
```

Logging a string is actually pretty easy as Delphi already provides a terminating `Chr(0)` character at the end of a string. A `Write` method just passes the correct parameters to the `WriteFile` Windows API:

```
procedure Write(const s: PAnsiChar); overload;
var
  written: DWORD;
begin
  WriteFile(LoggingFile, s^, StrLen(s), written, nil);
end;

procedure Writeln;
begin
  Write(#13#10);
end;
```

Logging a number is tricky, as we can't call `IntToStr` or `Format`. Both use dynamic strings, which means that the memory manager would be used to manage strings, but as we are already inside the memory manager, we cannot use memory management functions.

The logging function for numbers, therefore, implements its own conversion from `NativeUInt` to a buffer containing a hexadecimal representation of that number. It uses the knowledge that `NativeUInt` is never more than 8 bytes long, which generates, at max, 16 hexadecimal numbers:

```
procedure Write(n: NativeUInt); overload;
var
  buf: array [1..18] of AnsiChar;
  i: Integer;
  digit: Integer;
begin
  buf[18] := #0;
  for i := 17 downto 2 do
  begin
    digit := n mod 16;
    n := n div 16;
    if digit < 10 then
      buf[i] := AnsiChar(digit + Ord('0'))
    else
      buf[i] := AnsiChar(digit - 10 + Ord('A'));
  end;
  buf[1] := '$';
  Write(@buf);
end;
```

Even with these complications, the code is far from perfect. The big problem with the current implementation is that it won't work correctly in multithreaded code. A real-life implementation would need to add locking around all the `Logging*` methods.

Another problem with the code is that logging is really slow because of the frequent `WriteFile` calls. A better implementation would collect log data in a larger buffer and only write it out when the buffer becomes full. An improvement in that direction is left as an exercise for the reader.

The following screenshot shows the demonstration program, `LoggingMemoryManager`, in action. The first `GetMem` call creates a `TList<Integer>` object. The second creates a `TArray<Integer>` object used internally inside the `TList<Integer>` object to store the list data. After that, `ReallocMem` is called from time to time to enlarge this `TArray` object. We can see that it is not called for every element that the code adds to the list, but in larger and larger steps. Memory is firstly enlarged to $10 bytes, then to $18, $28, $48, and so on, to $1,008 bytes. This is a result of the optimization inside the `TList<T>` class that I mentioned at the beginning of this chapter.

Furthermore, we can see that the built-in memory manager doesn't always move the memory around. When memory is enlarged to $28 bytes, the memory manager returns a new pointer (the original value was $35FFBA0; the new value is $35EA850). After the next allocation, this pointer stays the same.

Only when the memory is enlarged to $78 bytes does the memory manager have to allocate a new block and move the data:

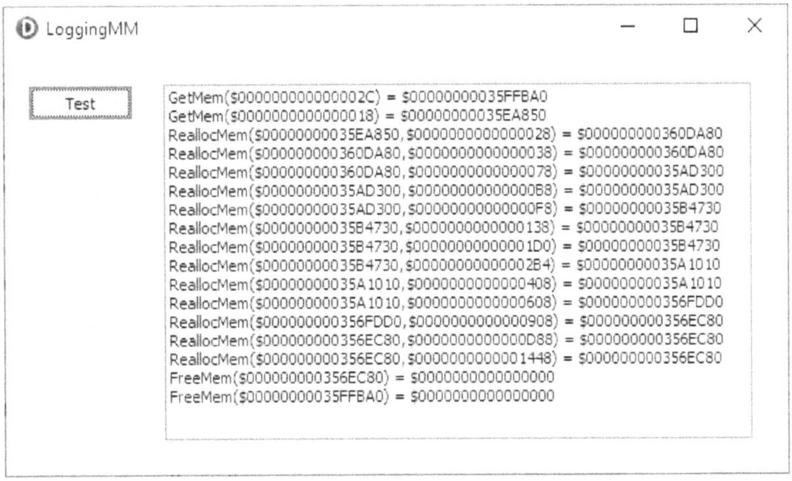

Figure 6.5 – A logging memory manager in action

I'm pretty sure you will not be writing a new memory manager just to improve the speed of multithreaded programs. I'm sure I won't do it. Luckily, there are crazier and smarter people out there who put their work out as open source, and we can build on their work. In most cases, we can simply download a new manager, use it (as a first unit!) in our project, and press *F9*.

Tip

After changing the memory manager, rebuild your program and test whether it is still working well.

Let's take a look at a few different memory managers that can be used to speed up our programs.

FastMM4 with release stack

For starters, let's return to FastMM4. While it is included in Delphi's RTL, FastMM4 still exists as an external open source project with a dual license. It can be used under a **Mozilla Public License 1.1 (MPL 1.1)** or with a **GNU Lesser General Public License 2.1 (LGPL 2.1)**. Both licenses are compatible with commercial programs.

To use it in your code, you have to clone the FastMM4 repository at `https://github.com/pleriche/FastMM4` (or download the ZIP file and unpack it somewhere on the disk). Then, you have to add `FastMM4` as the first unit in the project file (`.dpr`) and, if necessary, update search paths so that Delphi will find the new unit. For example, the `ParallelAllocation` program starts like this:

```
program ParallelAllocation;
uses
  FastMM4 in ,FastMM\FastMM4.pas',
  Vcl.Forms,
  ParallelAllocationMain in 'ParallelAllocationMain.pas'
    {frmParallelAllocation};
```

To enable the memory manager logging, you have to define a `LogLockContention` conditional symbol, rebuild (as `FastMM4` has to be recompiled), and, of course, run the program without the debugger.

If you do that, you'll see that the program runs quite a bit slower than before. On my test machine, the parallel version was only 1.6x faster than the serial one. The logging takes its toll, but that is not important. The important part will appear when you close the program.

At that point, the logger will collect all results and sort them by frequency. The 10 most frequent sources of locking in the program will be saved to a file called `<programname>_MemoryManager_EventLog.txt`. You will find it in the folder with the `<programname>.exe` file. The three most frequent sources of locking will also be displayed on the screen.

The following screenshot shows a cropped version of this log. Some important parts are marked out:

Figure 6.6 – Example of a lock contention log

For starters, we can see that at this location the program waited **19,020** times for a memory manager to become unlocked. Next, we can see that the memory function that caused the problem was `FreeMem`. Furthermore, we can see that somebody tried to delete from a list (`InternalDoDelete`) and that this deletion was called from `TSpeedTest.Execute`, line 130. `FreeMem` was called because the list in question is actually a `TObjectList` type, and deleting elements from the list caused it to be destroyed.

The most important part here is the memory function causing the problem—`FreeMem`. Of course! Allocations are optimized. If an allocator is locked, the next one will be used, and so on. Releasing memory, however, is not optimized! When we release a memory block, it *must* be returned to the same allocator that it came from. If two threads want to release memory to the same allocator at the same time, one will have to wait.

I had an idea of how to improve this situation by adding a small stack (called a **release stack**) to each allocator. When `FreeMem` is called and it cannot lock the allocator, the address of the memory block that is to be released will be stored on that stack. `FreeMem` will then quickly exit.

When a `FreeMem` function successfully locks an allocator, it first releases its own memory block. Then, it checks if anything is waiting on the release stack for that allocator and releases these memory blocks too (if there are any).

This change is also included in the main FastMM4 branch, but it is not activated by default as it increases the overall memory consumption of the program. However, in some situations, it can do miracles, and if you are developing multithreaded programs you certainly should test it out.

To enable release stacks, open the project settings for the program, remove the `LogLockContention` conditional define (as that slows the program down), and add the `UseReleaseStack` conditional define. Rebuild, as `FastMM4.pas` has to be recompiled.

On my test machine, I got much better results with this option enabled. Instead of a 3.5x speedup, the parallel version was 5.4x faster than the serial one. The factor is still not close to 8x, as the threads do too much fighting for the memory, but the improvement is still significant:

Figure 6.7 – Comparing serialized and parallel memory allocation speed with enabled release stack

That is as far as FastMM4 will take us. For faster execution, we need a more multithreading-friendly memory manager.

FastMM5

Years after writing FastMM4, Pierre le Riche started work on another memory manager—FastMM5. It advertises better multithreading performance and is configurable at runtime. It can also be configured into three different modes—speed, memory usage efficiency, and balanced modes.

FastMM5 can be downloaded from `https://github.com/pleriche/FastMM5`. Unlike the previous version, it cannot be freely used in commercial software. You can use it under **GNU General Public License v3** (**GPL v3**) or purchase a commercial license. More information is available on the GitHub page.

When running the `ParallelAllocation` test, FastMM5 works almost the same as FastMM4 with a release stack. The results are shown in the next screenshot:

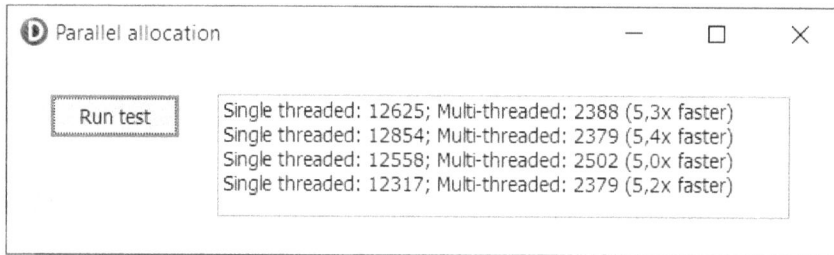

Figure 6.8 – Comparing serialized and parallel memory allocation speed in FastMM5

This simple test doesn't really show us how well FastMM5 will perform in a real-life situation. The only way to determine that is to test it with your code.

TBBMalloc

Another good alternative for a memory manager is Intel's **TBBMalloc**, part of their **oneAPI Threading Building Blocks** (**oneTBB**) library, formerly known as **Threading Building Blocks**. It is released under the Apache 2.0 license, which allows usage in open source and commercial applications. You can also buy a commercial license if the Apache 2.0 license doesn't suit you.

Intel's library was designed for C and C++ users. While you can go and download it from `https://github.com/oneapi-src/oneTBB`, you will still need a Delphi unit that will connect Delphi's memory management interface with the functions exported from Intel's `tbbmalloc` DLL. Alternatively, you can use the version that is included in the GitHub repository for the book (in the `tbbmalloc` subfolder).

In order to be able to run the application, you'll have to make sure that it will find `tbbmalloc.dll`. The simplest way to do that is to put the DLL in the `exe` folder. For 32-bit applications, you should use the DLL from the `tbbmalloc32` subfolder, and for 64-bit applications, you should use the DLL from the `tbbmalloc64` subfolder.

Intel's TBBMalloc Interface actually implements three different interface units, all named cmem. A version in the cmem_delphi subfolder can be used with Delphis up to 2010, the cmem_xe subfolder contains a version designed for Delphi XE and newer, and there's also a version for Free Pascal in the cmem_fps subfolder.

Creating an interface unit to a DLL is actually pretty simple. The *cmem* unit firstly imports functions from the tbbmalloc.dll file:

```
function scalable_getmem(Size: nativeUInt): Pointer; cdecl;
  external 'tbbmalloc' name 'scalable_malloc';

procedure scalable_freemem(P: Pointer); cdecl;
  external 'tbbmalloc' name 'scalable_free';

function scalable_realloc(P: Pointer; Size: nativeUInt):
  Pointer; cdecl;
  external 'tbbmalloc' name 'scalable_realloc';
```

After that, writing memory management functions is a breeze. You just have to redirect calls to the DLL functions:

```
function CGetMem(Size: NativeInt): Pointer;
begin
  Result := scalable_getmem(Size);
end;

function CFreeMem(P: Pointer): integer;
begin
  scalable_freemem(P);
  Result := 0;
end;

function CReAllocMem(P: Pointer; Size: NativeInt): Pointer;
begin
  Result := scalable_realloc(P, Size);
end;
```

The AllocMem function is implemented by calling the scalable_getmem DLL function and filling the memory with zeroes afterward:

```
function CAllocMem(Size : NativeInt) : Pointer;
begin
  Result := scalable_getmem(Size);
```

```
   if Assigned(Result) then
     FillChar(Result^, Size, 0);
 end;
```

The `ParallelAllocation` demo is all set for testing with TBBMalloc. You only have to copy `tbbmalloc.dll` from the `tbbmalloc32` folder to `Win32\Debug` and change the `ParallelAllocation.dpr` file so that it will load *cmem* instead of FastMM4:

```
program ParallelAllocation;

uses
  cmem in 'tbbmalloc\cmem_xe\cmem.pas',
  Vcl.Forms,
  ParallelAllocationMain in 'ParallelAllocationMain.pas'
    {frmParallelAllocation};
```

What about the speed? Unbeatable! Running the `ParallelAllocation` demo with TBBMalloc shows parallel code being up to 6 or 7 times faster than the serial code:

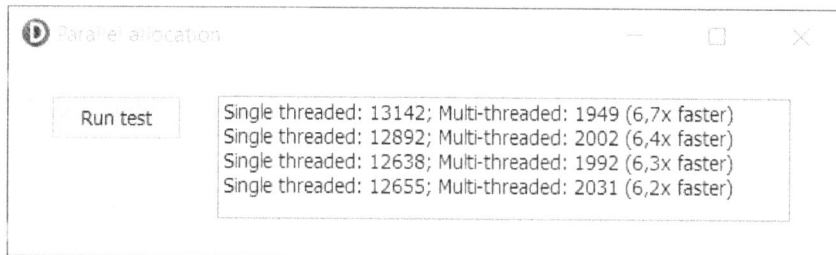

Figure 6.9 – Comparing serialized and parallel memory allocation speed in TBBMalloc

Comparing memory managers

Instead of comparing the single-threaded and multithreaded versions of code in one memory manager, we can also compare execution times under all different memory managers. The following figure shows this comparison for the four memory managers covered and tested in this chapter:

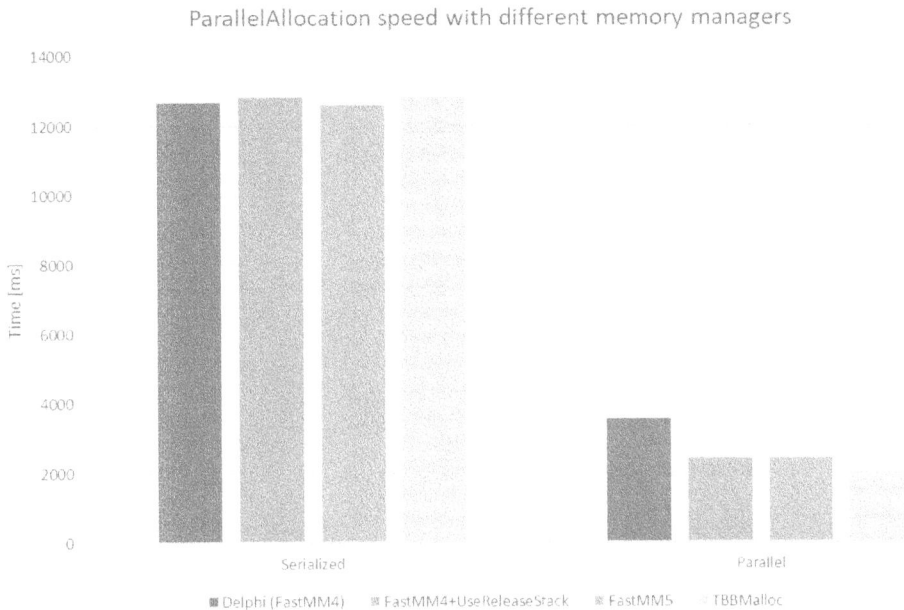

Figure 6.10 – Comparing ParallelAllocation program execution time with different memory managers

We can see that all four behave about the same when the code runs in a single thread. The parallel versions differ more. The default Delphi memory manager is the slowest, Intel's implementation is almost twice as fast, and both external FastMM implementations (FastMM4 with release stack and FastMM5) are only slightly slower than Intel's memory manager.

Still, this is a very limited test, designed to test only one stress point of memory managers. It does not tell us much about memory manager behavior in a complex system, so it is always best to test your own application with different memory managers and measure execution speed and memory consumption by yourself.

There is no silver bullet

While preparing material for the second edition of the book, I encountered something quite surprising. Compared to the first edition, which used Delphi 10.2 Tokyo as the newest Delphi release at the time, the second edition uses Delphi 11 Alexandria. And that, somehow, made the `ParallelAllocation` test much faster.

In the following screenshot, you can see execution times in ms from the same program (built with FastMM4 and release stacks), compiled with Delphi 10.2 Tokyo and Delphi 11 Alexandria:

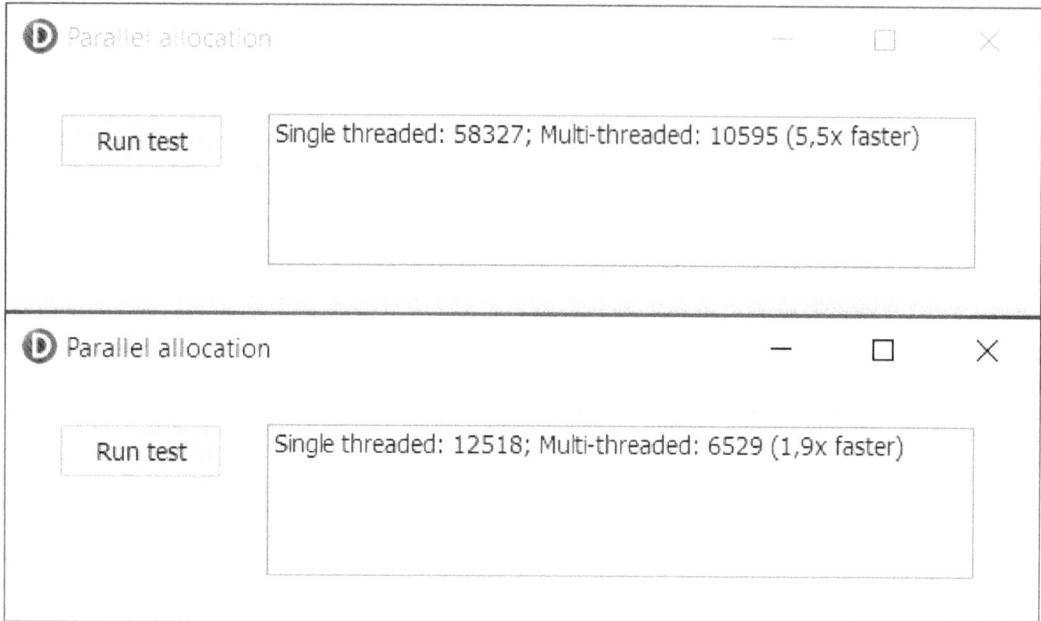

Figure 6.11 – Comparing program speed in Delphi 10.2 Tokyo (top) and Delphi 11 Alexandria (bottom)

We can see that the single-threaded allocations are 4,6x faster in Alexandria than in Tokyo, while multithreaded allocations got a lesser boost (but they are still 1,9x faster). All this with the same FastMM4 memory manager downloaded from GitHub, so the change must be in the compiler or in the runtime library.

It took me some time to get to the result, but in the end, we found (thank you, Stefan!) that the `ParallelAllocation` demo tests the speed of the `System.Move` function more than it tests the allocation/deallocation speed, and this RTL function has been much improved in Delphi 11 Alexandria. The main culprit is the following code fragment, part of the `TFooList.Execute` function:

```
while FList.Count > 0 do
begin
  FValue := FList[0].Value;
  Result := FList[0].Value;
  FList.Delete(0);
end;
```

As it deletes elements from the beginning of the list, `TList` has to move all remaining elements to a lower index each time `Delete` is called. This code should actually be written as follows:

```
for i := FList.Count - 1 downto 0 do
begin
  FValue := FList[i].Value;
```

```
    result := FList[i].Value;
    FList.Delete(i);
  end;
```

With this change, the single-threaded part of the test runs in 631 ms (instead of 12,518 ms), which is almost 20x faster! The multithreaded part is now actually slower than the single-threaded part because of all the overhead that multithreading brings. This shows us yet again that *fixing the algorithm* is much preferred to any other approach!

Fine-tuning SlowCode

We are near the end of the chapter. Let's step back a little and take a breath.

This chapter is packed with information, but if I had to pick the most important lesson, it would be: don't enlarge strings (arrays, lists) element by element. If you know how much space some data will take, allocate space at the beginning.

As all recommendations go, this too must be taken with a grain of salt. Most of the time, such optimizations will not matter. You should always have the following checklist in mind:

1. Measure. Find where the problem actually is.
2. See if you can improve the algorithm. This will give the best results.
3. Fine-tune the code.

In the previous chapter, we optimized the heart out of our old friend SlowCode. Just to remind you, the latest version, SlowCode_Sieve_v2_opt, processes 10 million elements in a mere 529 ms. Still, there is a little room for improvement that I've neglected so far.

The problem lies in the Filter function, which expands the Result array, element by element:

```
function Filter(list: TList<Integer>): TArray<Integer>;
var
  i: Integer;
  reversed: Integer;
begin
  SetLength(Result, 0);
  for i in list do
  begin
    reversed := ReverseInt(i);
    if not ElementInDataDivides(list, reversed) then
    begin
      SetLength(Result, Length(Result) + 1);
      Result[High(Result)] := i;
    end;
```

```
      end;
  end;
```

A much better approach would be to create a "just right" result array at the beginning. But how big is "just right"? We don't know how many elements the resulting array will have to store.

That's true, but we know one thing—it will not be larger than the list. All elements that are inserted into Result come from this list, so Result can only contain the same amount or fewer elements, not more.

If we don't know how big the resulting array will be, we'll do the next best thing. At the beginning, we'll set it to the maximum possible value. Then, we'll add elements to it. At the end, we'll truncate it to the correct size.

This method is implemented in the Filter function in the SlowCode_Sieve_v3 program:

```
function Filter(list: TList<Integer>): TArray<Integer>;
var
  i: Integer;
  reversed: Integer;
  outIdx: Integer;
begin
  SetLength(Result, list.Count);
  outIdx := 0;
  for i in list do
  begin
    reversed := ReverseInt(i);
    if not ElementInDataDivides(list, reversed) then
    begin
      Result[outIdx] := i;
      Inc(outIdx);
    end;
  end;
  SetLength(Result, outIdx);
end;
```

How much faster is this "improved and optimized" version? Sadly, not very much. The time to process 10 million numbers went from 529 ms to 512 ms.

So, what does this prove? Exactly what I was telling you before. Most of the time, you don't gain much by fiddling with small details. That, and—oh my, is FastMM4 fast!

Don't forget, though, that in a multithreaded world, the overhead from the memory manager is automatically higher (as I showed earlier). When you write multithreaded programs, such small optimizations can give you a bigger improvement than in a single-threaded world.

Just to be sure I'm not missing something, I've enlarged the test set from 10 million to 100 million numbers. `SlowCode_Sieve_v2_opt` needed 7,524 ms, and `SlowCode_Sieve_v3` needed 7,512 ms. The difference is minuscule. It is quite obvious that the biggest problem is not memory reallocations but other processing. Maybe a parallel approach will be able to help? We'll see.

Summary

This chapter dove deep into the murky waters of memory managers. We started on the light side with a discussion of strings, arrays, and how memory for them is managed.

After that, I enumerated all memory management functions exposed by Delphi. We saw how we can allocate memory blocks, change their size, and free them again. We also learned about memory management functions designed to allocate and release managed data—strings, interfaces, dynamic arrays, and so on.

The next part of the chapter was dedicated to records. We saw how to allocate them, how to initialize them, and even how to make dynamically allocated records function together with generic types. We also saw that records are not just faster to create than objects but also use less memory.

Then, we went really deep into the complicated world of memory manager implementation. I explained—from a very distant viewpoint—how FastMM4 is organized internally and what tricks it uses to speed up frequently encountered situations.

After that, I proved that even FastMM4 is not perfect. It offers great performance in single-threaded applications but can cause slowdowns in a multithreaded scenario. I showed a way to detect such bottlenecks in a multithreaded program and also how to improve FastMM4's speed in such situations.

Sometimes, this is still not enough, but Delphi offers a solution. You can swap the memory manager for another. To show this approach, I developed a small memory manager wrapper that logs all memory manager calls to a file. After that, I showed how you can replace FastMM4 with two alternative memory managers, FastMM5 and TBBMalloc.

The question may arise whether to always replace FastMM4 with an alternative. After all, alternatives are faster. The answer to that question is a resounding *no*! I believe that the program should be—if at all possible—deployed with the same memory manager that it was tested with. FastMM4 offers such great capabilities for debugging (that I sadly didn't have space to discuss here) that I simply refuse to use anything else during development. I would create a release version of a program with a non-default memory manager only in very specific circumstances and only after extensive testing.

At the end of the chapter, we returned to the `SlowCode` program. Although it contained excessive memory reallocations, it turned out that they didn't slow down the program in a significant way. Still, fixing that was an interesting exercise.

In the next chapter, I'll return to the bigger picture. We'll forget all about the optimizations at the level of memory allocations and object creation, and we'll again start fiddling with the algorithms. This time, we'll look in a completely new direction. We'll force our programs to do multiple things at once! In other words, we'll learn to multithread.

7
Getting Started with the Parallel World

If you are reading this book from start to finish, without skipping chapters, you've been through quite a lot. I've discussed algorithms, optimization techniques, memory management, and more, but I've quite pointedly stayed away from **parallel programming** (or **multithreading**, as it is also called).

I had a very good reason for that. Parallel programming is hard. It doesn't matter if you are an excellent programmer. It doesn't matter how good the supporting tools are. Parallel programming gives you plenty of opportunities to introduce weird errors into the program: errors that are hard to repeat and even harder to find. That's why I wanted you to explore other options first.

If you can make your program fast enough without going the parallel way, then make it so! Classical non-parallel (*single-threaded*) code will always contain fewer bugs and hidden traps than parallel code.

Sometimes this is not possible, and you have to introduce parallel programming techniques into the code. To do it successfully, it's not enough to know how to *do* parallel programming. Much more important is that you know what you should absolutely *never* do in parallel code.

To set you in the right mood, this chapter will cover the following topics:

- What are parallel programming and multithreading?
- What are the most common causes of program errors in parallel code?
- How should we handle the user interface in a parallel world?
- What is synchronization, and why is it both good and bad?
- What is interlocking, and when is it better than synchronization?
- How can we remove synchronization from the picture?
- What communication tools does Delphi offer to a programmer?
- What other libraries are there to help you?

Technical requirements

All code in this chapter was written with Delphi 11.3 Alexandria. Most of the examples, however, could also be executed on Delphi XE and newer versions. You can find all the examples on GitHub: `https://github.com/PacktPublishing/Delphi-High-Performance---Second-Edition/tree/main/ch7`.

Processes and threads

As a programmer, you probably already have some understanding of what a process is. As operating systems look at it, a process is a rough equivalent of an application. When a user starts an application, an operating system creates and starts a new process. The process owns the application code and all the resources that the code uses—memory, file handles, device handles, sockets, windows, and so on.

When the program is executing, the system must also keep track of the current execution address, the state of the CPU registers, and the state of the program's stack. This information, however, is not part of the process, but of a **thread** belonging to this process. Even the simplest program uses one thread.

In other words, the process represents the program's *static* data while the thread represents the *dynamic* part. During the program's lifetime, the thread describes its line of execution. If we know the state of the thread at every moment, we can fully reconstruct the execution in all its details.

Multithreading

All operating systems support one thread per process (the main thread) but some go further and support multiple threads in one process. Actually, most modern operating systems support **multithreading**, as this approach is called. With multithreading, the operating system manages multiple execution paths through the same code. Those paths may execute at the same time (and then again, they may not—but more on that later).

> **Note**
> The default thread created when the program starts is called the **main thread**. Other threads that come afterward are called **worker** or **background** threads.

In most operating systems (including Windows, OS X, iOS, and Android), processes are **heavy**. It takes a long time (at least at the operating system level, where everything is measured in microseconds) to create and load a new process. In contrast to that, threads are **light**. New threads can be created almost immediately—all that the operating system has to do is allocate some memory for the stack and set up some control structures used by the kernel.

Another important point is that processes are **isolated**. The operating system does its best to separate one process from another so that buggy (or malicious) code in one process cannot crash another process (or read private data from it). Threads, however, don't benefit from this protection.

If you're old enough to remember Windows 3, where this was not the case, you can surely appreciate the stability this isolation brings to the user. In contrast to that, multiple threads inside a process **share** all process resources—memory, file handles, and so on. Because of that, threading is inherently fragile—it is very simple to bring down one thread with a bug in another.

Multitasking

In the beginning, operating systems were single-tasking. In other words, only one task (that is, a process) could be executed at a time, and only when it completed the job (when the task terminated) could a new task be scheduled (started).

As soon as the hardware was fast enough, **multitasking** was invented. Most computers still had only one processor, but through operating system magic, it looked like this processor was executing multiple programs at the same time.

Each program was given a small amount of time to do its job. After that, it was paused and another program took its place. After some indeterminate time (depending on the system load, the number of higher priority tasks, and so on), the program could execute again and the operating system would run it from the position in which it was paused, again only for a small amount of time.

There are two very different approaches to multitasking. In **cooperative** multitasking, the process itself tells the operating system when it is ready to be paused. This simplifies the operating system but gives a badly written program an opportunity to bring down the whole computer. Remember Windows 3? That was cooperative multitasking at its worst.

A better approach is **pre-emptive** multitasking, where each process is given its allotted time (typically about a few tens of milliseconds in Windows) and is then *pre-empted*; that is, the hardware timer fires, takes control from the process, and gives it back to the operating system, which can then schedule the next process.

This approach is used in current Windows, macOS, and all other modern desktop and mobile operating systems. That way, a multitasking system can appear to execute multiple processes at once even if it has only one processor core. Things get even better if there are multiple cores inside the computer, as multiple processes can then really execute at the same time.

The same goes for threads. Single-tasking systems were limited to one thread per process by default. Some multitasking systems were single-threaded (that is, they could only execute one thread per process), but all modern operating systems support multithreading—they can execute multiple threads inside one process. Everything I said about multitasking applies to threads too. Actually, it is the threads that are scheduled, not the processes.

When to parallelize code

Before you start parallelizing code, you should understand whether the particular code is a good candidate for parallelization or not. There are some typical examples where parallelization is particularly simple, and there are some where it is really hard to implement.

One of the most common examples is executing long parts of code in the *main thread*. In Delphi, the main thread is the only one responsible for managing the user interface. If it is running a long task and not processing user interface events, then the user interface is blocked. We can solve this problem by moving the long task into a background thread, which will allow the main thread to manage the user interface. A responsive program makes for a happy user, as I like to say.

> **Note**
> Android applications have a separate thread dedicated to running the user interface – the **UI thread**. This thread is similar to the main thread on Windows as you should not run any long-running operations in it.

Another problem that is usually relatively simple to solve is enabling a server application to handle multiple clients at once. Instead of processing client requests one by one, we can create a separate thread for each client (up to some reasonable limit) and process requests from that client in that thread.

As requests from different clients really should not interact directly with one another, we don't have to take care of data sharing or inter-thread communications, which are, as you'll see, the biggest source of problems in multithreaded programming. The biggest problem we usually have is determining the upper limit for concurrent running threads, as that is affected both by the problem we are solving and the number of processors in the user's computer.

The last class of problems, which is also the hardest to implement, is speeding up an algorithm. We must find a way to split an algorithm into parts, and that way it will be different in each application. Sometimes we can just process part of the input data in each thread and aggregate partial results later, but that will not always be possible. We'll return to this advanced topic in *Chapter 9, Exploring Parallel Practices*.

The most common problems

Before we start writing multithreaded code, I'd like to point out some typical situations that represent the most common sources of problems in multithreaded programs. After that, I'll look into possible ways of solving such situations.

The biggest problem with the situations I'm about to describe is that they are all completely valid programming approaches if you are writing single-threaded code. Because of that, they sometimes even slip into (multithreaded) code written by the best programmers.

As we'll see later in the chapter, the best way to work around them is just to stay away from problematic situations. Instead of data sharing, for example, we can use data duplication and communication channels. But I'm getting ahead of myself ...

All of the situations I'm going to describe have something in common. They are a source of problems that can stay well hidden. Often, parallel programs seem to be working during testing but then randomly fail for only some customers.

The examples in this chapter are carefully chosen so they *always* cause problems in multithreaded code. In real life, sadly, the situation is quite different.

The only way to detect such problems in multithreaded code is to put a lot of effort into testing. Only automated unit tests running for a long time have any chance of finding them.

You'll notice that I'm skipping ahead a bit here. I did not tell you how to write multithreaded code at all, but I'm already using it in examples. For now, you can just believe me that the code in the examples does what I'm saying. In the next chapter, you'll learn all you need to know to really understand what I'm doing here, and then you can return to the examples from this chapter and re-examine them.

Never access the UI from a background thread

Let's start with the biggest source of hidden problems—manipulating a user interface from a background thread. This is, surprisingly, quite a common problem—even more so as all Delphi resources on multithreaded programming will simply say to never do that. Still, it doesn't seem to reach some programmers, and they will always try to find an excuse to manipulate the user interface from a background thread.

Indeed, there *may* be a situation where VCL or FireMonkey *may* be manipulated from a background thread, but you'll be treading on thin ice if you do that. Even if your code works with the current Delphi, nobody can guarantee that changes in graphical frameworks introduced in future Delphis won't break your code. It is *always* best to cleanly decouple background processing from a user interface.

> **Note**
> On all Delphi-supported operating systems except Android, you should only access the user interface functionality from the main thread. On Android, however, user interface functions should only be used from the UI thread.

Let's look at an example that nicely demonstrates the problem. The `ParallelPaint` demo has a simple form, with eight `TPaintBox` components and eight threads. Each thread runs the same drawing code and draws a pattern into its own `TPaintBox`. As every thread accesses only its own `Canvas`, and no other user interface components, a naive programmer would therefore assume that drawing into paintboxes directly from background threads would not cause problems. A naive programmer would be very much mistaken.

If you run the program, you will notice that, although the code paints constantly into some of the paint boxes, others stop being updated after some time. You may even get a **Canvas does not allow drawing** exception. It is impossible to tell in advance which threads will continue painting and which will not.

The following screenshot shows an example of output. The first two paintboxes in the first row and the last one in the last row were not being updated anymore when I grabbed the screenshot:

Figure 7.1 – A screenshot of a partially broken user interface

Lines are drawn in the DrawLine method. It does nothing special, just sets the color for the next line and draws it. Still, that is enough to break the user interface when this is called from multiple threads at once, even though each thread uses its own Canvas:

```
procedure TfrmParallelPaint.DrawLine(canvas: TCanvas;
  p1, p2: TPoint; color: TColor);
begin
  Canvas.Pen.Color := color;
  Canvas.MoveTo(p1.X, p1.Y);
  Canvas.LineTo(p2.X, p2.Y);
end;
```

Is there a way around this problem? Indeed there is. Delphi's TThread class implements a method, Queue, that executes some code in the main thread. (Actually, TThread has multiple methods that can do that; I'll return to this topic later in the chapter.)

Queue takes a procedure or anonymous method as a parameter and sends it to the main thread. After a short amount of time, the code is then executed *in the main thread*. It is impossible to tell how much time will pass before the code is executed, but that delay will typically be very short, in the order of milliseconds. As it accepts an anonymous method, we can use the magic of *variable capturing* and write the corrected code as shown here:

```
procedure TfrmParallelPaint.QueueDrawLine(canvas: TCanvas;
  p1, p2: TPoint; color: TColor);
begin
  TThread.Queue(nil,
    procedure
    begin
      Canvas.Pen.Color := color;
      Canvas.MoveTo(p1.X, p1.Y);
      Canvas.LineTo(p2.X, p2.Y);
    end);
end;
```

> **Note**
> I will talk more about variable capturing in *Chapter 9, Exploring Parallel Practices*. Another good source of information is the Delphi documentation on anonymous methods, which can be found at https://docwiki.embarcadero.com/RADStudio/en/Anonymous_Methods_in_Delphi.

If you run the corrected program, the final result should always be similar to the following screenshot, with all eight TPaintBox components showing a nicely animated image:

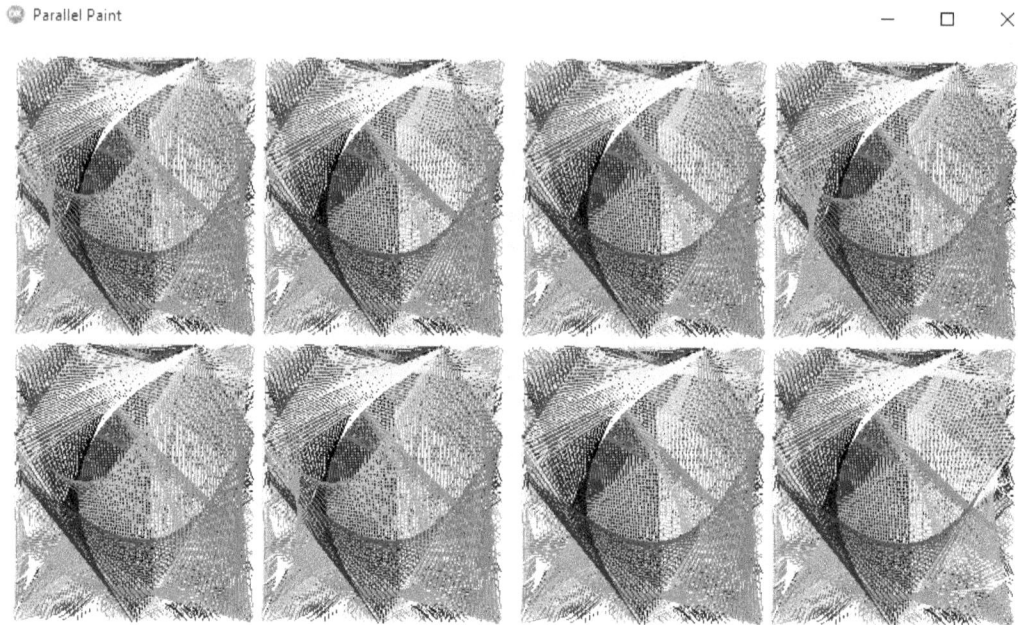

Figure 7.2 – User interface of a corrected Parallel Paint program

As accessing the UI from a background thread is a frequent problem, Delphi 11 introduced a new tool that helps find the reasons for some such problems. It cannot yet detect the incorrect usage from the previous example, but it can detect if a UI control (such as an edit box, list box, memo, etc.) is created from a background thread. This functionality exists only for the VCL framework and is not enabled by default. To turn it on, execute the following statement in your code (for example, from the main form's OnCreate event):

```
TControl.RaiseOnNonMainThreadUsage := true;
```

This is a welcome addition to the framework but it will not detect many problems. The only way to fight such problems is to stay vigorous and careful when coding.

Simultaneous reading and writing

The next situation that I regularly see when looking at badly written parallel code is simultaneous reading and writing from/to a shared data structure, such as a list. The SharedList program demonstrates how things can go wrong when you share a data structure between threads. Actually, scrap that – it shows how things *will* go wrong if you do that.

This program creates a shared list, FList: TList<Integer>. Then, it creates one background thread that runs the ListWriter method and multiple background threads, each running the

`ListReader` method. Indeed, you can run the same code in multiple threads. This is perfectly normal behavior and is sometimes extremely useful.

The `ListReader` method is incredibly simple. It just reads all the elements in a list and does that over and over again. As I've mentioned before, the code in my examples makes sure that problems in multithreaded code really *do* occur, but because of that, my demo code, most of the time, also looks terribly stupid. In that case, the reader just reads and rereads the data because that's the best way to expose the problem:

```
procedure TfrmSharedList.ListReader;
var
  i, j, a: Integer;
begin
  for i := 1 to CNumReads do
    for j := 0 to FList.Count - 1 do
      a := FList[j];
end;
```

The `ListWriter` method is a bit different. It also loops around, and in each loop, it adds or removes an element from the shared list. This is designed so that the problem is quick to appear:

```
procedure TfrmSharedList.ListWriter;
var
  i: Integer;
begin
  for i := 1 to CNumWrites do
  begin
    if FList.Count > 10 then
      FList.Delete(Random(10))
    else
      FList.Add(Random(100));
  end;
end;
```

If you start the program in a debugger, and click on the **Shared lists** button, there's a high possibility that you'll get an `EArgumentOutOfRangeException` exception. (In reality, you'll maybe have to try multiple times. Parallel programming errors tend to go away if you try to repeat them.) A look at the stack trace will show that it appears in the line `a := FList[j];`.

In retrospect, this is quite obvious. The code in `ListReader` starts the inner `for` loop and reads the `FListCount`. At that time, `FList` has 11 elements so `Count` is 11. At the end of the loop, the code tries to read `FList[10]`, but in the meantime, `ListWriter` has deleted one element and the list now only has 10 elements. Accessing element `[10]` therefore raises an exception.

We'll return to this topic later, in the *Locking* section. For now, you should just keep in mind that sharing data structures between threads causes problems.

Sharing a variable

OK, so rule number two is "Shared structure is *bad*." What about sharing a simple variable? Nothing can go wrong there, right? Wrong! There are actually multiple ways something can go wrong.

The `IncDec` program demonstrates one of the bad things that can happen. The code contains two methods: `IncValue` and `DecValue`. The former increments a shared `FValue: integer;` a number of times, and the latter decrements it the same number of times:

```
procedure TfrmIncDec.IncValue;
var
  i: integer;
  value: integer;
begin
  for i := 1 to CNumRepeat do begin
    value := FValue;
    FValue := value + 1;
  end;
end;


procedure TfrmIncDec.DecValue;
var
  i: integer;
  value: integer;
begin
  for i := 1 to CNumRepeat do begin
    value := FValue;
    FValue := value - 1;
  end;
end;
```

A click on the **Inc/Dec** button sets the shared value to *0*, runs `IncValue`, then runs `DecValue`, and logs the result:

```
procedure TfrmIncDec.btnIncDec1Click(Sender: TObject);
begin
  FValue := 0;
  IncValue;
  DecValue;
  LogValue;
end;
```

I know you can all tell what FValue will hold at the end of this program. Zero, of course. But what will happen if we run IncValue and DecValue in parallel? That is, actually, hard to predict!

A click on the **Multithreaded** button does almost the same, except that it runs IncValue and DecValue in parallel. How exactly that is done is not important at the moment (but feel free to peek into the code if you're interested):

```
procedure TfrmIncDec.btnIncDec2Click(Sender: TObject);
begin
  FValue := 0;
  RunInParallel(IncValue, DecValue);
  LogValue;
end;
```

Running this version of the code may still sometimes put zero in FValue, but that will be extremely rare. You most probably won't be able to see that result unless you are very lucky. Most of the time, you'll just get a seemingly random number from the range -10,000,000 to 10,000,000 (which is the value of the CnumRepeat constant).

In the following screenshot, the first number is a result of the single-threaded code, while all the rest were calculated by the parallel version of the algorithm:

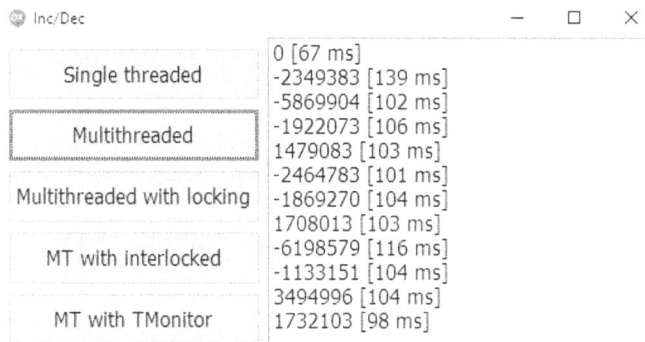

Figure 7.3 – Changing a shared variable from two threads can generate the wrong result

To understand what's going on, you should know that Windows (and all other operating systems) does many things at once. At any given time, there are hundreds of threads running in different programs and they are all fighting for the limited number of CPU cores. As our program is the active one (has focus), its threads will get most of the CPU time, but still, they'll sometimes be paused for some amount of time so that other threads can run.

Because of that, it can easily happen that IncValue reads the current value of FValue into value (let's say that the value is *100*) and is then paused. DecValue reads the same value and then runs for some time, decrementing FValue. Let's say that it gets it down to *-20,000*. (That is just a number without any special meaning.)

After that, the IncValue thread is awakened. It should increment the value to *-19,999*, but instead of that it adds *1* to *100* (stored in value), gets *101*, and stores *that* in FValue. Ka-boom! In each repetition of the program, this will happen at different times and will cause a different result to be calculated.

You may complain that the problem is caused by the two-stage increment and decrement, but you'd be wrong. I dare you—go ahead, and change the code so that it will modify FValue with Inc(FValue) and Dec(FValue) and it still won't work correctly.

Well, I hear you say, "so I shouldn't even modify one variable from two threads at the same time?" I can live with that. But surely, it is OK to write into a variable from one thread and read from another?

The answer, as you can probably guess given the general tendency of this section, is again—no, you may not. There are some situations where this is OK (for example, when a variable is only one byte long) but, in general, even simultaneous reading and writing can be a source of weird problems.

The ReadWrite program demonstrates this problem. It has a shared buffer, FBuf: Int64, and a pointer variable used to read and modify the data, FPValue: PInt64. At the beginning, the buffer is initialized to an easily recognized number and a pointer variable is set to point to the buffer:

```
FPValue := @FBuf;
FPValue^ := $7777777700000000;
```

The program runs two threads. One just reads from the location and stores all the read values in a list. This value is created with Sorted and Duplicates properties, set in a way that prevents it from storing duplicate values:

```
procedure TfrmReadWrite.Reader;
var
  i: integer;
begin
  for i := 1 to CNumRepeat do
    FValueList.Add(FPValue^);
end;
```

The second thread repeatedly writes two values into the shared location:

```
procedure TfrmReadWrite.Writer;
var
  i: integer;
begin
```

```
  for i := 1 to CNumRepeat do begin
    FPValue^ := $7777777700000000;
    FPValue^ := $0000000077777777;
  end;
end;
```

In the end, the contents of the FValueList list are logged on the screen. We would expect to see only two values—*$7777777700000000* and *$0000000077777777*. In reality, we see four, as the following screenshot demonstrates:

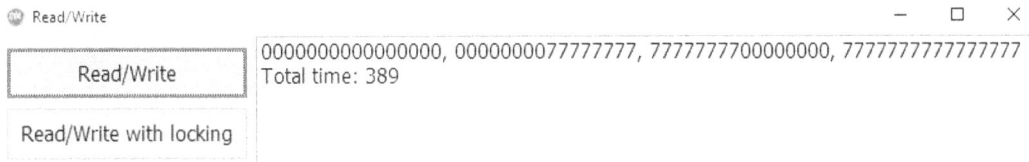

Figure 7..4 – Reading and writing shared variables may result in the wrong data

The reason for that strange result is that Intel processors in 32-bit mode can't write a 64-bit number (as int64 is) in one step. In other words, reading and writing 64-bit numbers in 32-bit code is not *atomic*.

> **Note**
>
> To demonstrate the issue, you should compile the program as a 32-bit application. It will work correctly if compiled for a 64-bit platform.

When multithreading programmers talk about something being *atomic*, they want to say that an operation executes in one indivisible step. Any other thread will either see a state before the operation or a state after the operation, but never some undefined intermediate state.

How do values *$7777777777777777* and *$0000000000000000* appear in the test application? Let's say that FValue^ contains *$7777777700000000*. The code then starts writing *$0000000077777777* into FValue by firstly storing *$77777777* in the bottom four bytes. After that, it starts writing *$00000000* in the upper four bytes of FValue^, but in the meantime, Reader reads the value and gets *$7777777777777777*.

In a similar way, Reader will sometimes see *$0000000000000000* in FValue^.

We'll look into a way to solve this situation in the following section, *Synchronization*, but in the meantime, you may wonder—when is it okay to read/write from/to a variable at the same time? Sadly, the answer is… it depends. Not even just on the CPU family (Intel and ARM processors behave completely differently), but also on a specific architecture used in a processor. For example, older and newer Intel processors may not behave the same in that respect.

You can always depend on access to byte-sized data being atomic, but that is that. Access (reads and writes) to larger quantities of data (words, integers) is atomic only if the data is correctly *aligned*. You can access word-sized data atomically if it is word-aligned, and integer data if it is double-word-aligned. If the code was compiled in 64-bit mode, you can also atomically access int64 data if it is quad-word-aligned.

When you are not using data packing (such as packed records), the compiler will take care of alignment, and data access should automatically be atomic. You should, however, still check the alignment in code – if nothing else, to prevent stupid programming errors.

If you want to write and read larger amounts of data, modify the data, or if you want to work on shared data structures, correct alignment will not be enough. You will need to introduce *synchronization* into your program. Before we tackle that, however, I would like to present another problem when even simultaneous *reading* from multiple threads can fail.

Hidden behavior

One would expect that there would be no problems if two threads are just *reading* from the same data buffer. In theory, that is correct, but in practice, the act of reading may be implemented by a complex process that may introduce bugs into the code. This has actually happened to me and the whole experience was quite sobering—so much so that I was quite careful not to fall into any of the traps mentioned in this chapter while coding!

The problem is hard to repeat in a simple program with the tools we have at our disposal at this moment, so I'll try to describe it in words. The application in question had one thread that generated some data and multiple threads that processed this data. Each data-processing thread needed a full copy of the generated data (and there was lots of data) so I decided to cheat and send to each thread the same TMemoryStream. After all, threads will only read from the data and, if I am careful, what could go wrong? A lot, as it turned out.

I was, of course, careful. I knew that I must not use any TStream reading mechanism inside the data-processing thread as that would certainly break. The code instead only accessed the memory block, managed by the TMemoryStream, and then used pointer magic to access the data, somewhat like this:

```
procedure ConsumeData(memStr: TMemoryStream);
var
  localSize: integer;
  p: PByte;
  i: integer;
begin
  localSize := memStr.Size;
  p := memStr.Memory;
  for i := 0 to localSize - 1 do begin
    Process(p^);
```

```
    Inc(p);
  end;
end;
```

Looks safe enough? I sure thought so! The application, however, crashed regularly. It took us some time to find the real source of the problem—reading the `Size` property.

As well-seasoned Delphi programmers know (and as I should definitely remember from writing that code), `TStream.Size` doesn't access an internal property but is implemented in a very convoluted manner, as shown here:

```
function TStream.GetSize: Int64;
var
  Pos: Int64;
begin
  Pos := Seek(0, soCurrent);
  Result := Seek(0, soEnd);
  Seek(Pos, soBeginning);
end;
```

This code does the following sequence of operations:

1. It starts by jumping 0 bytes from the current position. This returns the current position in the stream.

2. Then the code jumps 0 bytes from the end of the stream. This returns the number of bytes in the stream.

3. In the end, the code jumps to the initial position in the stream.

This already doesn't look very safe. What happens if two threads execute the `Seek` function on the same object? What does `Seek` do? Let's see:

```
function TCustomMemoryStream.Seek(const Offset: Int64;
  Origin: TSeekOrigin): Int64;
begin
  case Origin of
    soBeginning: FPosition := Offset;
    soCurrent: Inc(FPosition, Offset);
    soEnd: FPosition := FSize + Offset;
  end;
  Result := FPosition;
end;
```

Here it is! The code modifies an `FPosition` field. When two threads read from the `Size`, this field becomes a shared variable, modified in both threads. And as we know, this fails sooner or later.

In this case, I decided not to change the algorithm but to fix the underlying data structure. Instead of `TMemoryStream`, the code now uses my own reimplementation of `TMemoryStream`, which stores stream size in a field and returns it with a simple operation:

```
function TGpMemoryStream.GetSize: int64;
begin
  Result := FSize;
end;
```

You can also use this stream (or a very useful wrapper, `IGpBuffer`) in your application. It is available at `https://github.com/gabr42/GpDelphiUnits/blob/master/src/GpStuff.pas`.

And the moral of this story? Don't make assumptions. I was so sure that reading `Size` is thread-safe that I didn't check the implementation.

But enough of introducing problems—let's start solving them! To prevent multiple threads from modifying shared data at the same time, we'll introduce *synchronization* into the code.

Synchronization

Whenever you need to access the same data from multiple threads, and at least one thread is modifying the data, you have to *synchronize* access to the data. As we've just seen, this holds for shared data structures and for simple variables.

Synchronization will make sure that one thread cannot see invalid intermediate states that another thread creates temporarily while updating the shared data. In a way, this is similar to database transactions at the *read committed* level, when other users cannot see changes applied to the database while a transaction is in progress.

The simplest way to synchronize two (or more) threads is to use *locking*. With locking, you can protect a part of the program so that only one thread will be able to access it at any time. If one thread has successfully *acquired* but not yet *released* the lock (we also say that the thread now *owns* the lock), no other threads will be able to acquire that same lock. If any thread tries to acquire a lock, it will be paused until the lock is available (after the original lock owner releases the lock).

We could compare such a lock with a real-world lock. When somebody wants to access a critical resource that must not be shared (for example, a toilet), they will lock the door behind them. Any other potential users of the critical resource will queue in a line. When the first user is finished using the critical resource, they will unlock it and leave, allowing the next user access.

Synchronization mechanisms are almost always implemented directly by the operating system. It is possible to implement them directly in the code, but that should only be used to solve very specific circumstances. As a general rule, you should not attempt to write your own synchronization mechanism.

If you are writing a multi-platform application, accessing operating system synchronization mechanisms can be a pain. Luckily, Delphi's runtime library provides a very nice platform-independent way to work with them.

The simplest way to implement locking is to use a *critical section*. In Delphi, you should use the `TCriticalSection` wrapper implemented in the `System.SyncObjs` unit instead of accessing the operating system directly.

Critical sections

The easiest way to explain the use of critical sections is with an example. As the `ReadWrite` example is still fresh in your mind, I'll return to it. This program implements—in addition to the unsafe reading/writing approach—code that reads and writes data with additional protection from a critical section lock.

This critical section is created when a form is created and destroyed when the form is destroyed. There is only one critical section object shared by both threads:

```
procedure TfrmReadWrite.FormCreate(Sender: TObject);
begin
  FLock := TCriticalSection.Create;
end;

procedure TfrmReadWrite.FormDestroy(Sender: TObject);
begin
  FreeAndNil(FLock);
end;
```

When a reader wants to read from `FPValue^`, it will firstly acquire the critical section by calling `FLock.Acquire`. At that point, the thread will either successfully acquire the ownership of the lock and continue execution or it will block until the lock becomes unowned.

After some finite time, the thread will manage to acquire the lock and continue with the next line, `value := FPValue^`. This will safely read the value into a temporary variable, knowing that nobody can write to the value at the same time. After that, the critical section is immediately released by calling `FLock.Release`, which also makes the lock unowned:

```
procedure TfrmReadWrite.LockedReader;
var
  i: integer;
  value: int64;
begin
  for i := 1 to CNumRepeat do begin
    FLock.Acquire;
    value := FPValue^;
    FLock.Release;
```

```
        FValueList.Add(value);
    end;
  end;
```

The writer does something similar. Before an FPValue^ is written to, the lock is acquired, and after that, the lock is released:

```
procedure TfrmReadWrite.LockedWriter;
var
  i: integer;
begin
  for i := 1 to CNumRepeat do begin
    FLock.Acquire;
    FPValue^ := $7777777700000000;
    FLock.Release;
    FLock.Acquire;
    FPValue^ := $0000000077777777;
    FLock.Release;
  end;
end;
```

Before we look into the result of that change, I'd like to point out something very important. Introducing a lock doesn't automatically solve problems. You also have to use it in the right place (in this example, when reading and writing the shared value), and you have to use it in *all* places where shared data is accessed.

For example, if LockedReader will be using a critical section, as it is now, but LockedWriter will not, access to the shared value will still be unsynchronized and we will get wrong results from the program.

Let's see how introducing a critical section affects our program. The following screenshot shows the result of both versions of the code—unprotected and synchronized:

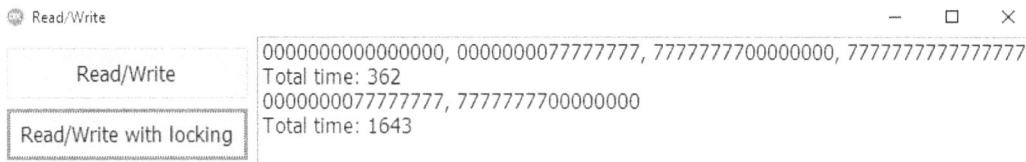

Figure 7.5 – Introducing synchronization fixes problems with bad data but slows down the program

We can see two important consequences of locking here. One, the output from the code is now correct. Two, the program runs much longer with locking.

This is a typical result of synchronizing access with locks. They are slow—at least if you are trying to acquire them many thousands of times per second, as my test program does. If you are using them sparingly, and only lock a short part of the code, they will not affect the program that much.

An interesting thing about critical sections is that they will happily let through a thread that already owns a lock. A thread that already owns a lock can therefore call `Acquire` a second time and be sure that it will not be blocked. It must then also call `Release` twice to release the critical section.

For example, the following code is perfectly valid:

```
FLock.Acquire;
FLock.Acquire;
value := FPValue^;
FLock.Release;
FLock.Release;
```

This situation sometimes comes up when procedure A locks a critical section and calls procedure B, which also locks a critical section. Instead of refactoring the code so that the critical section is only the acquired one, we can rely on this fact and acquire the critical section in both places. Acquiring an already acquired critical section is very fast.

> **Note**
> This behavior is called **re-entrancy**, and a locking mechanism that allows such coding is deemed **re-entrant**.

There are two big problems that critical sections can introduce into your code. One we've already seen—the program runs slower with locking. The second is even worse—if such synchronization is not implemented correctly, it may cause the program to endlessly wait for a lock that will never become free. We call that **deadlocking**.

Deadlocking happens when thread 1 acquires critical section A and then tries to acquire critical section B while, in the meantime, thread 2 acquires critical section B and then tries to acquire critical section A. Both threads will be blocked as they will wait for a critical section that will never be released.

Deadlocking is not specific to critical sections. It can appear regardless of the locking mechanism used in the code.

This situation is demonstrated in program *deadlocking*. It contains two threads, both accessing the shared variable, Counter, stored in the shared object, Shared. To synchronize access to this shared variable, they both acquire and release the critical section, shared.LockCounter.

To demonstrate the problem, both threads also acquire the additional critical section, shared. LockOther. In this demo, the second critical section is meaningless, but, in badly designed real-life code, such situations occur a lot. Code may acquire a critical section, then do some processing, and in the middle of the processing access different data, protected with a different critical section.

When multiple critical sections are always acquired in the same order, everything works just fine. But if—because of a programming error or a weird system design—one thread acquires critical sections in a different order than another thread, a deadlock may occur.

The same situation is programmed into the Deadlocking program. The TaskProc1 method acquires LockCounter first and LockOther second, while the TaskProc2 method (running in a different thread) acquires LockOther first and LockCounter second. Both methods are shown here:

```
procedure TaskProc1(const task: ITask;
  const shared: TSharedData);
begin
  while frmDeadlock.FTask1.Status <> TTaskStatus.Canceled
  do begin
    shared.LockCounter.Acquire;
    shared.LockOther.Acquire;
    shared.Counter := shared.Counter + 1;
    shared.LockOther.Release;
    shared.LockCounter.Release;
  end;
end;

procedure TaskProc2(const task: ITask;
  const shared: TSharedData);
begin
  while frmDeadlock.FTask1.Status <> TTaskStatus.Canceled
  do begin
    shared.LockOther.Acquire;
    shared.LockCounter.Acquire;
    shared.Counter := shared.Counter + 1;
    shared.LockCounter.Release;
```

```
        shared.LockOther.Release;
    end;
  end;
```

If you start only one thread, everything will work just fine. The thread will continue to acquire both critical sections, increment the shared resource, and release both critical sections. If you start both threads, however, the program will quickly deadlock. `TaskProc1` will acquire `LockCounter` and block on waiting for `LockOther`, while at the same time, `TaskProc2` will acquire `LockOther` and block on waiting for `LockCounter`.

Luckily, sources of deadlocking are easy to find if you can repeat the problem in a debugger. When the program blocks, just click the *Pause* icon and go through the threads in the **Thread status** window. Double-click each thread and check if the *Call stack* for the thread is currently inside an *Acquire* call. Any such thread should then be carefully investigated to see if it is part of the problem.

The following screenshot shows the `TaskProc1` method paused during a deadlock:

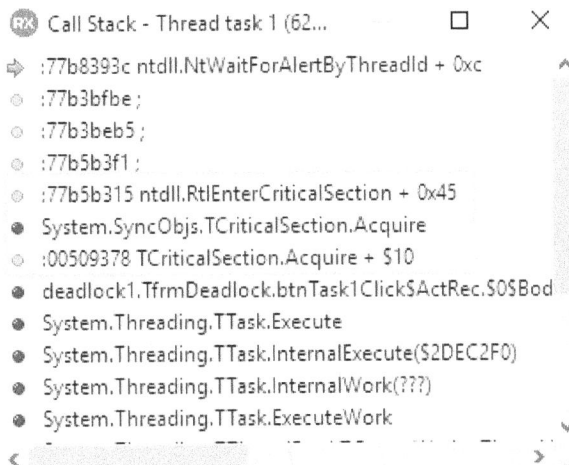

Figure 7.6 – Call stack during a deadlock

The best solution for such problems is good program design. The part of the code where the critical section is acquired should always be very short and should—if at all possible—never call other code that acquires critical sections. If that cannot be prevented, you should design the code so that it always acquires all the critical sections first (and always in the same order), and only then calls the protected code.

If that is also not possible, the last solution is to not use `Acquire` (or its namesake, `Enter`), but the `TryEnter` function. That one tries to acquire the critical section, returns `False` if the critical section is already locked, or acquires the critical section and returns `True`. It will never block. An adapted version of the code is shown here:

```
procedure TryTaskProc1(const task: ITask;
  const shared: TSharedData);
begin
  while frmDeadlock.FTask1.Status <> TTaskStatus.Canceled
  do begin
    shared.LockCounter.Acquire;
    if shared.LockOther.TryEnter then
    begin
      shared.Counter := shared.Counter + 1;
      shared.LockOther.Leave;
    end;
    shared.LockCounter.Release;
  end;
end;

procedure TryTaskProc2(const task: ITask;
  const shared: TSharedData);
begin
  while frmDeadlock.FTask1.Status <> TTaskStatus.Canceled
  do begin
    shared.LockOther.Acquire;
    if shared.LockCounter.TryEnter then
    begin
      shared.Counter := shared.Counter + 1;
      shared.LockCounter.Leave;
    end;
    shared.LockOther.Release;
  end;
end;
```

If you enter a critical section with `TryEnter`, you should leave it with a call to `Leave`, just for readability reasons.

If you run the demo and use the **TryTask1** and **TryTask2** buttons to start the `TryTaskProc1` and `TryTaskProc2` methods, you'll notice that the program continues to work but also that it increments the counter much more slowly than with only one thread running. This happens because both threads are constantly fighting for the second critical section. Most of the time, each thread acquires the first critical section, then tries to acquire the second one, fails, and does nothing.

To learn more about deadlocks and ways to fix them, go to the beautiful *The Deadlock Empire* website (`http://deadlockempire.4delphi.com/delphi/`), where you can explore this topic through a game.

Other locking mechanisms

Critical sections are not the only mechanisms that operating systems expose to applications. There are also other approaches that may be preferable in different situations. In Delphi, most of them are nicely wrapped in system-independent wrappers, implemented in the `System.SyncObjs` unit.

Mutex

The first such object is a *mutex*. It is very similar to a critical section, as you can use it to protect access to a critical resource. The big difference between a mutex and a critical section is that a critical section can only be shared between the threads of one program. A mutex, on the other hand, can have a *name*. Two (or more) programs can create mutexes with the same name and use them to access some shared resource (for example, a file) that may only safely be used by one program at a time.

A mutex is implemented in a `TMutex` class that exposes the same API as a `TCriticalSection`. After you `Create` a mutex, you can call `Acquire` to access and lock the mutex and `Release` to unlock it.

If you create a mutex without a name, it can only be used inside one program and functions exactly as a critical section. Locking with mutexes, however, is *significantly* slower than locking with critical sections. As the `IncDec` demo shows, locking with mutexes (activated with the **MT with mutex** button) can be more than 50 times slower than locking with critical sections (the **MT with locking** button). A figure with a comparison of the timing for different locking approaches is shown later in this chapter.

Mutexes do provide one advantage over a critical section. You can acquire a mutex by calling a `WaitFor` function, which accepts a timeout value in milliseconds. If the code fails to lock a mutex in that amount of time, it will return `wrTimeout` and continue. This works just the same as `TCriticalSection.TryEnter`, except with an additional timeout. Nevertheless, if you need such functionality, you'll probably be better off using the `TSpinlock` mechanism described later in this chapter.

Semaphore

The next standard synchronization mechanism is a *semaphore*. A **semaphore** is used to synchronize access to resources that can be used by more than one user (more than one code path) at the same time, but have an upper limit on the number of concurrent users.

For example, if you want to protect a resource that supports up to three concurrent users, you can create a semaphore with a count of *three*. When you *acquire* such a semaphore, you decrement the number of available resources (the semaphore's internal counter) by one. If you try to acquire a semaphore when its internal counter is at zero, you'll be blocked. `Release`, on the other hand, increments the internal counter and allows another thread to enter the critical path.

Semaphores, like mutexes, can have names and can be used to synchronize multiple programs. They are also equally slow, so I won't peruse them in this book. Still, semaphores are an interesting topic as they can be used to solve many difficult problems. For more information on the topic, I'd like to recommend a beautiful (free) book, *The Little Book of Semaphores*, which you can read or download at `http://greenteapress.com/wp/semaphores/`.

TMonitor

One of the problems with critical sections is that you have to create and manage them in parallel to the data being protected. To solve that, in Delphi 2009, the developers extended the `TObject` class with an additional pointer field used by the new `TMonitor` record, which was added to the *System* unit.

As the `TObject` is the base parent of each and every class type in Delphi programs, this causes the size of every object created in your program to increment by 4 bytes (8 if you're compiling for a 64-bit processor). This and the abysmally bad `TMonitor` implementation have caused some programmers to call this the worst abuse of space in the entire history of Delphi.

The idea behind `TMonitor` was completely solid. It was designed to provide a per-object locking capability together with some additional functionality that allowed for the creation of a fast in-process event and fast in-process semaphore (more on that in a moment). In reality, the implementation was so flawed that the only part that you could safely use was the basic locking, implemented with the `Enter`, `TryEnter`, and `Exit` functions. All other implementation was broken for many Delphi versions and was only fixed a few years later.

As far as I know, in recent releases (Tokyo and a few before), `TMonitor` doesn't have any known bugs. The implementation is now mature and I can definitely recommend using it along with other locking mechanisms.

`TMonitor` can safely be used to function as a critical section. It is simple to use (you don't have to create a separate object) and comparable in speed (most of the time, it is faster). It is also sometimes inconvenient, as it can only be used to lock an existing object and an appropriate object may not exist in the code.

The `IncDec` demo works around that by locking the form object itself. This doesn't affect the normal VCL operations in any way, as VCL doesn't use `TMonitor` to lock `TForm` objects. In a larger program, however, that would be frowned upon, as the meaning of the code is hard to grasp. You should use `TMonitor` only when you really work with a shared **object** so you can put a lock directly on it.

Another small problem with `TMonitor` is a naming conflict. Delphi VCL already defines a `TMonitor` class in the `Vcl.Forms` unit. If your unit includes `Vcl.Forms` (and every form unit does that), you'll have to type `System.TMonitor` instead of `TMonitor` to make the code compile. Alternatively, you can call the `MonitorEnter` and `MonitorExit` procedures, which do the same.

Let's look at the code that protects access to a shared counter with TMonitor. It was taken from the IncDec demo. To synchronize access to the counter, it first locks the form object by calling System.TMonitor.Enter(Self). After that, it can safely operate on the counter. In the end, it unlocks the form object by calling System.TMonitor.Exit(Self). The other part of the demo, MonitorLockedIncValue (not shown here), does the same:

```
procedure TfrmIncDec.MonitorLockedDecValue;
var
  value: integer;
  i: Integer;
begin
  for i := 1 to CNumRepeat do begin
    System.TMonitor.Enter(Self);
    value := FValue;
    FValue := value + 1;
    System.TMonitor.Exit(Self);
  end;
end;
```

> **Note**
>
> To acquire and release a TMonitor, you can also call the MonitorEnter and MonitorExit helper methods.

If you run the demo, you'll notice that the TMonitor approach is much faster than working with TCriticalSection. On my test computer, it is about three times faster. This is a result of the internal implementation of TMonitor, which doesn't use a critical section, but an improved idea called a **spinlock**.

Spinlock

The difference between a critical section and a spinlock lies in the implementation of the Acquire call. When a thread cannot access a critical section because it is already locked, Windows puts this thread to sleep. (Other operating systems behave mostly the same.) When a critical section becomes available, Windows selects one of the threads waiting for it and wakes it up.

A spinlock, on the other hand, assumes that the code protected with it is very short and that the spinlock will be released quickly. If a spinlock is already acquired from another thread, the code first tries to *actively wait* or *spin*. Instead of going to sleep, the code runs in a tight loop and constantly checks whether the spinlock has become available. Only if that doesn't happen, after some time, the thread goes to sleep.

Delphi implements a spinlock object that you can use directly in your code. It is called `TSpinLock` and can be found in the `System.SyncObjs` unit.

A `TSpinLock` is a record, not an object, so there's no need to free it. You still have to create it though, as some data is initialized in the constructor. The code in the `IncDec` demo does that by calling `FSpinlock := TSpinLock.Create(false)`.

The code then uses `FSpinlock.Enter` to acquire a spinlock and `FSpinlock.Exit` to release it. `TSpinLock` also implements a `TryEnter` method, which accepts a *timeout* parameter. If the spinlock cannot be acquired in *timeout* milliseconds, `TryEnter` returns `False`:

```
procedure TfrmIncDec.SpinlockDecValue;
var
  i: integer;
  value: integer;
begin
  for i := 1 to CNumRepeat do begin
    FSpinlock.Enter;
    value := FValue;
    FValue := value - 1;
    FSpinlock.Exit;
  end;
end;
```

The only problem with `TSpinLock` is that it is not *re-entrant*. If a thread that has already acquired a spinlock calls `Enter` for a second time, the code will either raise an exception (if you have passed `True` to the constructor) or block. The implementation, however, provides the `IsLocked` and `IsLockedByCurrentThread` functions, which you can use to write a re-entrant spinlock, using `TSpinLock` as a base.

The `System.SyncObjs` unit also implements a faster implementation of two synchronization primitives—an event, `TLightweightEvent`, and a semaphore, `TLightweightSemaphore`. They are both based on the idea of actively spinning in a loop (for a short time) if the object cannot be immediately acquired.

The following screenshot shows the execution time comparison for all the locking mechanisms that I've discussed so far. Well, all and one more. The last approach, *Interlocking*, will be discussed in the *Interlocked operations* section later in this chapter.

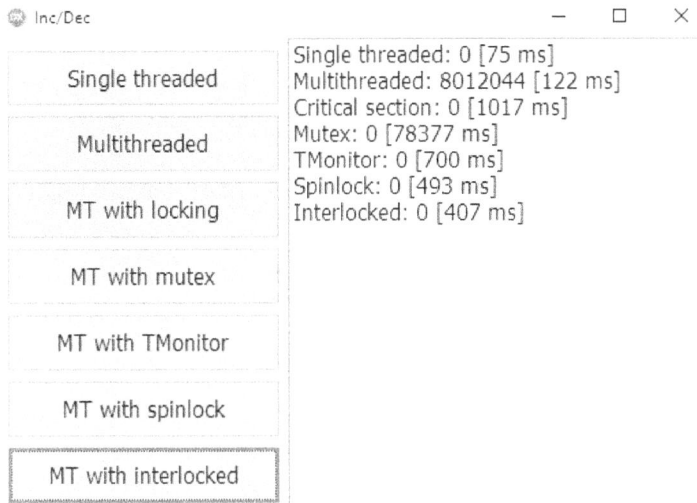

Figure 7.7 – Comparing different locking mechanisms

All the mechanisms I've discussed so far are completely symmetrical. They treat all the threads the same. Sometimes, though, this is not what we would like.

Readers-writer locks

Lots of the time, threads in a program have to share access to data that is mostly read from and almost never changed. In this case, it is inefficient if the reader threads block one another. The only time the access must be protected is when a thread has to modify the shared data.

In such cases, we can use a synchronization mechanism that allows multiple simultaneous readers to access a shared resource, but at the same time allows only one writer. In addition, a writer must block out all the readers, as they should not read from a shared resource while it is being modified. We've seen before, in the SharedList example, that this is not going to work.

This is a common situation and, to help, modern operating systems (including all those supported by Delphi) implement a locking mechanism called **readers-writer lock**, also known as a single-writer lock, a multi-reader lock, an MRSW (multiple readers, single writer) lock, an SWMR (single writer, multiple readers) lock, or an MREW (multiple readers, exclusive writer) lock. In Delphi, such a mechanism is called TLightweightMREW and is defined in the System.SyncObjs unit.

> **Note**
>
> TLightweightMREW is a recent addition to the Delphi environment. It was added in version 10.4.1. Older versions of Delphi came with a different implementation—TMREWSync—which is still included for backward compatibility. I do not recommend using it as it is very slow and works only on Windows. On all other platforms, it just delegates the work to TMonitor, which basically turns it into a normal critical section.

With `LightweightMREW` and similar mechanisms that provide multiple levels of access, we cannot just call `Acquire` or `Enter` to lock the resource. We also have to state our intentions in advance. With `TLightweightMREW`, we have to call `BeginRead` if we want to read from a resource and `BeginWrite` if we will write to it. Similarly, we have to call `EndRead` or `EndWrite` to release a resource.

> **Warning**
>
> Trying to end `BeginRead` with `EndWrite` or `BeginWrite` with `EndRead` will crash the program or end in a deadlock (depending on the operating system). `TLightweightMREW` should be used with care!

Keep in mind that the compiler cannot tell what you are doing—reading or writing—with a shared resource. Actually, the compiler cannot even tell what the shared resource is. It is entirely your responsibility as a programmer to use the correct locking methods in appropriate places in the code.

To compare readers-writer lock with a critical section, I have written the `ReadersWriterLock` program. It runs one writer thread and six reader threads. Readers read from a shared list and run a short calculation for each entry. The writer tries to update the list every 100 ms (for details, see the code). After running all threads for three seconds, the program logs how many times the writer was able to update the list and how many times each reader was able to process the whole list. The results are shown in the following screenshot:

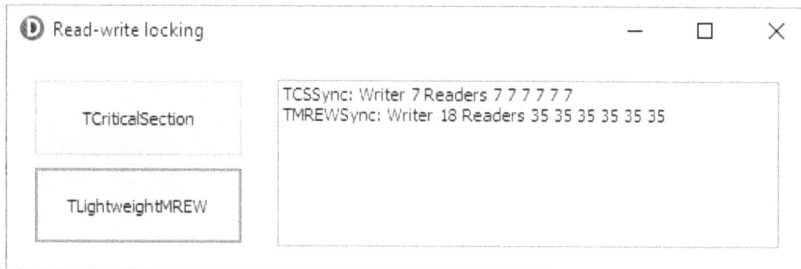

Figure 7.8 – Comparing a critical section with a readers-writer lock

We can see that when a critical section was used, the list was only updated seven times and each reader also managed to process the list seven times. With a readers-writer lock, the reader was able to update the list 18 times and each reader was able to process the list 35 times, which is a big improvement.

Although the usage of the TLightweightMREW is simple to deduce from the demo program, I have to point out one implementation detail. While most synchronization objects in Delphi are implemented as objects, the readers-writer lock is implemented as a **record**. This has two important consequences:

- You don't create and destroy a TLightweightMREW lock; instead, you just declare a field of that type

- The TLightweightMREW lock must be part of a bigger object (typically the object that we want to protect) as we can't share the lock itself between threads.

Before we move on, I have to warn you about one TLightweightMREW peculiarity. While read locks are re-entrant (a thread that already owns a read lock can call BeginRead again), write locks are not. If a thread gains write access to the lock and calls BeginWrite again, the program will either deadlock (on Windows) or raise an exception (on all other platforms).

To fix this, I have written a wrapper record, TLightweightMREWEx, which implements reentrant write locking. More details are available on my blog at https://www.thedelphigeek.com/2021/02/readers-writ-47358-48721-45511-46172.html.

A short note on coding style

In all the examples so far, I used a simple coding style:

```
// acquire access
// work with shared resource
// release access
```

For example, the IncDec demo from this chapter contains the following code:

```
FSpinlock.Enter;
value := FValue;
FValue := value - 1;
FSpinlock.Exit;
```

In reality, I don't code resource locking in that way. I'm a strong proponent of the thought that all resource management pairs, such as create/destroy, acquire/release, beginupdate/endupdate, and so on should, if at all possible, be written in a try..finally block. I would therefore write the previous example as follows:

```
FSpinlock.Enter;
try
  value := FValue;
  FValue := value - 1;
finally
  FSpinlock.Exit;
end;
```

This approach has two big advantages. One is that acquiring and releasing a resource are now visibly connected with a `try..finally` construct. When you browse through the code, this pair of operations will immediately jump out at you.

The other advantage is that the resource will be released if your code crashes while the resource is allocated. This is important when the code that reacts to the exception also tries to access the shared resource. Or, maybe the program tries to shut down cleanly and during the cleanup, accesses the shared resource. Without a `try..finally`, the resource would stay acquired forever and that second part of the code would just deadlock.

Saying all that, it may seem weird that I didn't insert `try..finally` blocks everywhere in the test code. Well, I had a reason for that – two, actually.

Firstly, I wanted to keep code examples short. I don't like to read books that have long code examples spanning multiple pages and so I kept all code examples in this book very short.

Secondly, I wanted to write this note on coding style and that was a good excuse for it.

Shared data with built-in locking

Sometimes you don't need to implement an additional locking mechanism because Delphi already does that for you. The runtime library contains three such data structures with terrible names—`TThreadList`, `TThreadList<T>`, and `TThreadedQueue<T>`. The first one can be found in the `System.Classes` unit, while the other two live in `System.Generics.Collections`.

Both variants of `TThreadList` work the same. They don't expose the normal list-management functions (such as `Add`, `Remove`, `Items[]`, and so on). Instead of that, they implement the `LockList` function, which uses `TMonitor` to acquire access to an internal `TList` (or `TList<T>`) object and returns that object. To release access to this list, the program has to call the `UnlockList` function.

The following example from the `SharedList` demo shows how to use `TThreadList<Integer>` to represent the shared storage. To access the shared list, the code calls `FThreadList.LockList`. This function returns a `TList<Integer>` object, which can then be used in the code. To release the shared list, the code calls `FThreadList.UnlockList`:

```
procedure TfrmSharedList.ThreadListReader;
var
  i, j, a: Integer;
  list: TList<Integer>;
begin
  for i := 1 to CNumReads do
  begin
    list := FThreadList.LockList;
    for j := 0 to FList.Count - 1 do
      a := FList[j];
```

```
    FThreadList.UnlockList;
  end;
end;
```

> **Warning**
>
> Using the list returned from the `LockList` after `UnlockList` was called is a programming error that cannot be caught by the compiler. Be careful!

Comparing the various approaches in the `SharedList` demo gives interesting results. As expected, the critical section-based locking is the slowest. More interestingly, the `TMonitor` locking used in the `TThreadList` is faster than the `TLightweightMREW` implementation. As it turns out, the spinlock approach used in `TMonitor` really pays off in this scenario. The following screenshot shows a comparison of execution times measured on the test computer:

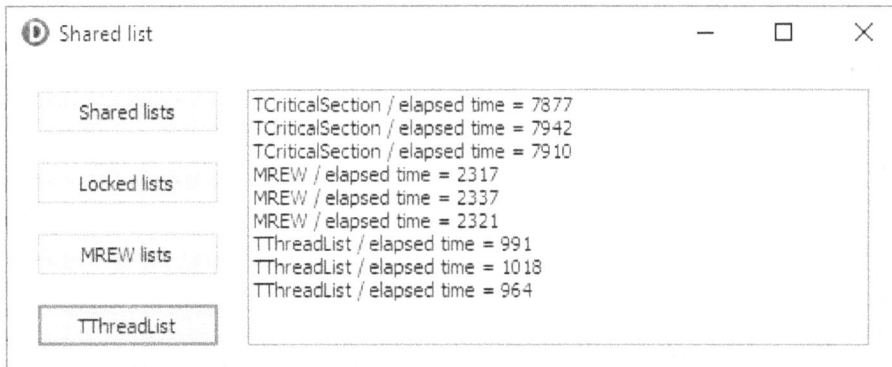

Figure 7.9 – Comparing different ways of locking a list

The third data structure, `TThreadedQueue<T>`, implements a standard first-in, first-out fixed-size queue. All operations on the queue are protected with a `TMonitor`. I will use this data structure in practical examples near the end of this chapter and all through *Chapter 8, Working with Parallel Tools*.

Interlocked operations

When shared data is small enough and you only need to increment it or swap two values, there's an option to do that without locking. All modern processors implement instructions that can do simple operations on memory locations in such a way that another processor cannot interrupt the operation in progress.

Such instructions are sometimes called **interlocked operations**, while the whole idea of programming with them is called **lock-free** programming or, sometimes, **microlocking**. The latter term is actually more appropriate, as the CPU indeed does some locking. This locking happens inside the CPU and is therefore much faster than operating system-based locking, such as critical sections.

As these operations are implemented on an assembler level, you have to adapt the code to the specific CPU target. Or, and this is very much advised, you just use the `TInterlocked` class from the `System.SyncObjs` unit.

Functions in the `TInterlocked` class are mostly just simple wrappers for the `Atomic*` family of the CPU-dependent function from the *System* unit. For example, `TInterlocked.Add` is implemented as a simple call to `AtomicIncrement`:

```
class function TInterlocked.Add(var Target: Integer;
  Increment: Integer): Integer;
begin
  Result := AtomicIncrement(Target, Increment);
end;
```

Interlocked functions are declared `inline`, so you won't lose any cycles if you call `TInterlocked.Add` instead of `AtomicIncrement` directly.

Windows implements more interlocked functions than are exposed in the `TInterlocked` class. If you want to call them directly, you can find them in the `Winapi.Windows` unit. All their names start with `Interlocked`.

We can split interlocked functions into two big families. One set of functions modifies a shared value. You can use them to increment or decrement a shared value by some amount. Typical representatives of this family are `Increment`, `Decrement`, and `Add`.

The other set of functions is designed to exchange two values. It contains only two functions with multiple overloads that support different types of data, `Exchange` and `CompareExchange`.

The functions from the first family are easy to use. They take a value, increment or decrement it by some amount, and return the new value. All that is done *atomically*, so any other thread in the system will always see the old value or the new value but not some intermediate value.

The only exceptions to that pattern are the `BitTestAndSet` and `BitTestAndClear` functions. They both test whether some bit in a value is set and return that as a `boolean` function result. After that, they either set or clear the same bit, depending on the function. Of course, all that is again done atomically.

We can use `TInterlocked.Increment` and `TInterlocked.Decrement` to manipulate the shared value in the `IncDec` demo. Although `TInterlocked` is a class, it is composed purely of class functions. As such, we never create an instance of the `TInterlocked` class, but just use its class methods directly on our code, as shown in the following example:

```
procedure TfrmIncDec.InterlockedIncValue;
var
  i: integer;
begin
```

```
  for i := 1 to CNumRepeat do
    TInterlocked.Increment(FValue);
end;

procedure TfrmIncDec.InterlockedDecValue;
var
  i: integer;
begin
  for i := 1 to CNumRepeat do
    TInterlocked.Decrement(FValue);
end;
```

As we've seen in the previous section, this approach beats even spinlocks, which were the fastest locking mechanism so far.

Functions in the second family are a bit harder to use. Exchange takes two parameters: shared data and a new value. It returns the original value of the shared data and sets it to the new value.

In essence, the Exchange function implements the steps outlined in the following function, except that they are done in a safe, atomic way:

```
class function TInterlocked.Exchange(var Target: Integer;
  Value: Integer): Integer;
begin
  Result := Target;
  Target := Value;
end;
```

The second set of data-exchanging functions, CompareExchange, is a bit more complicated. CompareExchange takes three parameters—shared data, new value, and test value. It compares shared data to the test value and, if the values are equal, sets the shared data to the new value. In all cases, the function returns the new value of the shared data.

The following code exactly represents the CompareExchange behavior, except that the real function implements all the steps as one atomic operation:

```
class function CompareExchange(var Target: Integer;
  Value: Integer; Comparand: Integer): Integer;
begin
  Result := Target;
  if Target = Comparand then
    Target := Value;
end;
```

How can we know whether the shared data was replaced with the new value? We can check the function result. If it is equal to the test value, `Comparand`, then the value was replaced, otherwise it was not.

> **Note**
>
> There is also an overloaded `CompareExchange`, which returns the success status in a boolean `Succeeded` parameter.

A helper function, `TInterlocked.Read`, can be used to atomically read from the shared data. It is implemented as follows:

```
class function TInterlocked.Read(var Target: Int64): Int64;
begin
  Result := CompareExchange(Target, 0, 0);
end;
```

If the original value is equal to 0, it will be replaced with 0. In other words, it will not be changed. If it is not equal to 0, it also won't be changed because `CompareExchange` guarantees that. In both cases, `CompareExchange` will return the original value, which is exactly what the function needs.

We can use `TInterlocked.Read` and `TInterlocked.Exchange` to re-implement the `ReadWrite` demo without locking:

```
procedure TfrmReadWrite.InterlockedReader;
var
  i: integer;
  value: int64;
begin
  for i := 1 to CNumRepeat do begin
    value := TInterlocked.Read(FPValue^);
    FValueList.Add(value);
  end;
end;

procedure TfrmReadWrite.InterlockedWriter;
var
  i: integer;
begin
  for i := 1 to CNumRepeat do begin
    TInterlocked.Exchange(FPValue^, $7777777700000000);
    TInterlocked.Exchange(FPValue^, $0000000077777777);
  end;
end;
```

If you run the code, you'll see that it is indeed faster than the locking approach. On the test computer, it executes about 20% faster, as you can verify in the following screenshot:

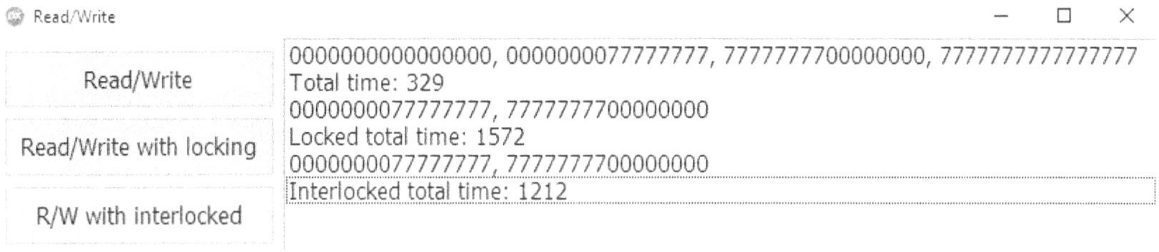

Figure 7.10 – Comparing interlocked operation speed with the locking approach

Similar to locking, interlocking is not a magic bullet. You have to use it correctly and you have to use it in all the relevant places. If you modify shared data with interlocked instructions but read that same data with a normal read, the code will not work correctly.

You can verify that by changing the `value := TInterlocked.Read(FPValue^);` line in the `InterlockedReader` method to `value := FPValue^;`. If you rerun the test, it will give the same (wrong) results as the original, fully unsafe version.

Object life cycle

Special care should be given to the way objects are created and destroyed in a multithreaded environment. Not every approach is safe, and it is quite simple to make a mistake that would cause occasional crashes.

The simplest way is to lock all access to an object. If you create it in a critical section, use it from a critical section, and destroy it in a critical section, everything will work. This way, however, is very slow and error-prone, as it is quite simple to forget to lock something.

If you create the object before the background tasks are created, and destroy it after the background tasks are destroyed, everything will work, too. If this is a normal object on a non-ARC platform, you are just sharing a pointer and that is never a problem. If you are sharing an interface-implementing object or if you are compiling for an ARC platform, the reference counting mechanism kicks into action. As the reference counting is threadsafe, that is again not a problem. Of course, all methods of that object that are called from the parallel code must work correctly if they are called from two or more threads at the same time.

> **Note**
> Delphi does not support ARC anymore. I nevertheless discuss ARC in this section because not every reader will be using the latest Delphi version.

Problems appear if you would like to create or destroy the shared object when the parallel tasks using that object are already running. Creation is possible—with some care. Destruction, you should stay away from.

Let's discuss the latter first. If you have a normal object in a non-ARC environment, then it is obvious that you cannot destroy it in one thread while another thread is using it. With interfaces or ARC, that is not very obvious. You may think that it is safe to assign nil to a shared interface in one thread (`FShared := nil;`) while another thread copies it into a different interface and uses that copy (`tmp := FShared; tmp.DoSomething;`). Trust me, that will not work. One thread will try to clear the object by calling `_IntfClear`, while another will try to copy the object by calling `_IntfCopy`. Depending on luck, this will either work or the second thread will own an invalid reference to a destroyed object and will crash when trying to use it.

Contrary to that, creating an object from a thread is indeed possible. (As you dig deeper and deeper into parallel code, you'll start to notice that the biggest problem in the parallel world is not how to set up parallel computation, but how to clean it up when the task is done.) Let's assume that we have a shared object of type `TSharedObject`, which for some reason cannot be initialized before the worker threads are started. At some time, a thread will need this object and will try to create it. But, hey, we're living in a parallel world, so it's entirely possible that some other thread will try to do that at the same time too!

One way to approach the problem is with locking. We have to establish a shared lock that will be used just for the purpose of initializing an object and use it to synchronize access to the object creation. The following code fragment shows an idea of how that could be implemented:

```
//In main thread
FInitializationLock := TCriticalSection.Create;

//In a thread
FInitializationLock.Acquire;
if not assigned(FSharedObject) then
  FSharedObject := TSharedObject.Create;
FInitializationLock.Release;
```

The important part of this approach is that we need to test whether the shared object is assigned from within the acquired critical section (or a spinlock or `TMonitor`). If we do the test before acquiring the critical section, the code will sometimes fail. It would be entirely possible for two threads to test whether the object is assigned, then for both to acquire the critical section (one after another, of course), and for both to store a newly created `TSharedObject` in the shared variable. After that, one instance of the shared object would be lost forever.

This approach wastes lots of time because it locks the critical section each time, even when the object has already been created. A better way to do it is with *double-checked locking*. This pattern follows the following algorithm:

- Check if the condition is met

- Lock access to the critical path

- Check if the condition is met again

Doing the second check (which gives double-checked locking its name) is crucial. Without it, the condition could change after it was tested in the first step and before exclusive access is granted.

By using this pattern, we would rewrite object creation as follows:

```
//In main thread
FInitializationLock := TCriticalSection.Create;

//In a thread
if not assigned(FSharedObject) then
begin
  FInitializationLock.Acquire;
  if not assigned(FSharedObject) then
    FSharedObject := TSharedObject.Create;
  FInitializationLock.Release;
end;
```

This approach is almost preferable to the simple lock-and-test as most of the time, the object will already be created when the thread reaches this point in code and locking will simply be skipped.

We can even remove the locking altogether and use interlocked operations to create an object. The following example code shows how:

```
procedure CreateSharedObject;
var
  value: TSharedObject;
begin
  if not assigned(FSharedObject) then
  begin
    value := TSharedObject.Create;
    if TInterlocked.CompareExchange(pointer(FSharedObject),
        pointer(value), nil) <> nil
    then
      value.Free;
  end;
end;
```

The code first checks whether the `FSharedObject` is already created. If not, it will create a new instance of `TSharedObject` and store it in a temporary value. After that, it will use `CompareExchange` to store the temporary value in the shared variable, but only if `FSharedObject` still contains `nil`.

`CompareExchange` can return a value that is not `nil`. That will happen if another thread has tested whether `FSharedObject` is assigned at the same time as the first thread. Both threads will then create a new `TSharedObject` and execute `CompareExchange`. One of them will succeed and return the old state, `nil`, while another will be executed a bit later and will return the new state of the `FSharedObject` (the object that was put there by the first thread). The second thread will then throw the `TSharedObject` stored in a temporary variable away.

This approach is also called *optimistic* initialization because the thread optimistically creates an object and expects that it will be able to store it in the shared value. As, in most cases, a conflict between threads rarely occurs, this is, in general, a fast and useful way to create shared objects. It also doesn't require a separate synchronization primitive, which may, on some occasions, be a big advantage.

> **Warning**
>
> This approach is only useful if creating an object is a fast operation without side effects that would cause problems if executed more than once. It is, for example, not suitable for creating true singletons that really must be created only once.

Communication

Seeing all that, you may now agree with me when I say that data sharing is hard. It is hard to do it safely as there are so many opportunities to make a mistake in the code. It is also hard to do it fast because locking approaches don't *scale* well. In other words, locking will prevent the code from working two times faster when you run the code on twice the CPU cores.

Luckily, there's a better way. Better, but way harder. Instead of sharing, we can do what I've been advocating since the beginning of this book and change the algorithm. And to do that, we'll need some *communication* techniques. In my parallel projects, I always prefer communication to synchronization, and I strongly suggest that you try to do the same.

You may find it strange that I've dedicated that much space to data sharing and synchronization techniques if I don't recommend using them. Well, sometimes you just have to bite the bullet and share data because nothing else makes sense. In such cases, you also have to intimately know everything that can go wrong and ways to fix it.

The best way to remove data sharing from the equation is to replace it with *data duplication* and *aggregation*. Before creating the threads, we make a copy of the original data, one for each thread. A thread can then do its work in peace as it is not sharing data with anybody. In the end, the thread sends its result back to the main thread (here, the *communication* comes into play) so that the main thread can then combine (aggregate) partial results together.

Sending data from one thread to another is simpler if you don't have to care about object ownership. It is best to always use just simple data types (such as integers) and managed data types (strings and interfaces). As this data is never simultaneously used in two threads, the rules from the *Object life cycle* section don't apply. We can create and destroy such objects in completely normal ways.

Sending data from the main thread to the worker thread and sending data in the opposite direction are two completely different problems, and we have to approach them in different ways. The way we choose to send data always depends on the implementation of the receiving thread. Before we learn how to send data to worker threads, we have to learn how to create and manage them. I'll therefore cover this in later chapters.

Sending data from worker threads to the main thread depends only on the implementation of the main thread, and that is something we are familiar with. In the rest of the chapter, we'll look into four techniques for communicating with the main thread—sending Windows messages, Synchronize, Queue, and polling.

To demonstrate these techniques, I have re-implemented the code from IncDemo in a new program, IncDecComm. All background tasks in that program are started in the same way. A click on a button calls the RunInParallel method and passes, as parameters, the names of two methods, one to increment and one to decrement the initial value.

RunInParallel creates a new background task for each worker method and passes to it the initial value, 0. In this program, passing the same value to both worker methods represents the most basic implementation of *data copying*:

```
procedure TfrmIncDecComm.RunInParallel(
  task1, task2: TProc<integer>);
begin
  FNumDone := 0;
  FTasks[0] := TTask.Run(procedure begin task1(0); end);
  FTasks[1] := TTask.Run(procedure begin task2(0); end);
end;
```

Result *aggregation* depends on the communication method and will be evaluated separately for each approach.

Windows messages

The first method works by sending Windows messages to the main thread. As the name suggests, it is limited to the Windows operating system. I'll show two ways to achieve the same result. The first works only in VCL applications, while the second also works with FireMonkey.

Before the tasks are started, the code initializes the result FValue to zero so that partial results can be aggregated correctly. It also sets the FNumDone variable to zero. This variable counts the tasks that have completed their work:

```
procedure TfrmIncDecComm.btnMessageClick(Sender: TObject);
begin
  FValue := 0;
  FNumDone := 0;
  RunInParallel(IncMessage, DecMessage);
end;
```

Both tasks are implemented in the same way, so I'll only show one method here. IncMessage increments the starting value CNumRepeat times. In the end, it posts a MSG_TASK_DONE message back to the main form and passes the resulting value as a parameter. DecMessage looks the same except that the value is decremented inside the for loop:

```
procedure TfrmIncDecComm.IncMessage(startValue: integer);
var
  i: integer;
  value: integer;
begin
  value := startValue;
  for i := 1 to CNumRepeat do
    value := value + 1;
  PostMessage(Handle, MSG_TASK_DONE, value, 0);
end;
```

The message parameters wParam and lParam can be used to pass data alongside the message. They have the same size as a pointer, which allows us to use an object or an interface as data storage. I'll show an example of such code near the end of this section.

To process this message in the main form, we have to declare the MSG_TASK_DONE variable (WM_USER indicates the start of the message range reserved for custom use) and add a message-handling method, MsgTaskDone, to the form:

```
const
  MSG_TASK_DONE = WM_USER;

procedure MsgTaskDone(var msg: TMessage);
  message MSG_TASK_DONE;
```

This method is called every time the main form receives the MSG_TASK_DONE message. Inside, we first increment the FValue value by the partial result, passed in the msg.WParam field (*aggregation* step). After that, we increment the FNumDone counter. We know that we will receive exactly two messages, so we can call the cleanup method, Done, when FNumDone is equal to 2:

```
procedure TfrmIncDecComm.MsgTaskDone(var msg: TMessage);
begin
  Inc(FValue, msg.WParam);
  Inc(FNumDone);
  if FNumDone = 2 then
    Done('Windows message');
end;
```

In the Done method, we only have to clean the worker task interfaces:

```
procedure TfrmIncDecComm.Done(const name: string);
begin
  FTasks[0] := nil;
  FTasks[1] := nil;
end;
```

This seems like a lot of work, but it is worth it. I'll show the actual results after I finish discussing all four methods, but I can already tell you that this approach is finally faster than the single-threaded code.

This method of declaring a message handler is specific to VCL. On FireMonkey, you have to invest a bit more work and create a hidden window that will receive and process messages.

When you click on the **Message + AllocateHwnd** button, the following code executes. It initializes FValue and FNumDone, as in the previous example, and then creates a hidden window by calling AllocateHwnd:

```
procedure TfrmIncDecComm.btnAllocateHwndClick(
  Sender: TObject);
begin
  FValue := 0;
  FNumDone := 0;
  FMsgWnd := AllocateHwnd(MsgWndProc);
  Assert(FMsgWnd <> 0);
  RunInParallel(IncMsgHwnd, DecMsgHwnd);
end;
```

The task method, IncMsgHwnd, looks the same as IncMethod, except the last line, which must now post the message to the FMsgWnd handle:

```
PostMessage(FMsgWnd, MSG_TASK_DONE, value, 0);
```

The `MsgWndProc` method processes all messages sent to this hidden window. We are only interested in `MSG_TASK_DONE`. Any other messages are passed to the default Windows message handler, `DefWindowProc`. The code that handles `MSG_TASK_DONE` is identical to the code in the `MsgTaskDone` method:

```
procedure TfrmIncDecComm.MsgWndProc(var msg: TMessage);
begin
  if Msg.Msg = MSG_TASK_DONE then
  begin
    Inc(FValue, msg.WParam);
    Inc(FNumDone);
    if FNumDone = 2 then
      Done('Windows message');
  end
  else
    DefWindowProc(FMsgWnd, msg.Msg, msg.wParam,
      msg.lParam);
end;
```

To clean up this hidden window, we have to call the `DeallocateHwnd` method:

```
if FMsgWnd <> 0 then
begin
  DeallocateHwnd(FMsgWnd);
  FMsgWnd := 0;
end;
```

I promised before to show how you can send an object or an interface as message parameters, so here's a short example. To send an object, we simply have to cast it to the appropriate type. In Delphi 11.3 Alexandria, that is `NativeUInt` for wParam and `NativeInt` for lParam.

If we did the same with an interface, it would be destroyed on exit from `btnObjIntClick`. To prevent that, we have to increase its reference count by calling `_AddRef`:

```
procedure TfrmIncDecComm.btnObjIntClick(Sender: TObject);
var
  tobj: TObject;
  iint: IInterface;
begin
  tobj := TObject.Create;
  iint := TInterfacedObject.Create;
  iint._AddRef;
  PostMessage(Handle, MSG_OBJ_INT, NativeUInt(tobj),
    NativeInt(iint));
end;
```

On the receiving side, we can simply cast the object back to the appropriate object type (`TObject` in this example). We also have to remember to destroy it when we're done with it.

An interface can also be cast to the appropriate interface type (`IInterface` in this example) but then we also have to remove the reference added in `btnObjIntClick` by calling `_Release`. We can then use this interface as we need, and it will be destroyed once it is no longer in use and its reference count drops to zero. The following code demonstrates this:

```
procedure TfrmIncDecComm.MsgObjInt(var msg: TMessage);
var
  tobj: TObject;
  iint: IInterface;
begin
  tobj := TObject(msg.WParam);
  tobj.Free;
  iint := IInterface(msg.LParam);
  iint._Release;
end;
```

Synchronize and Queue

The second and third approaches are so similar that I've condensed them into one section.

The setup for this approach is the same as with the first shown in the Windows message method—we only have to initialize `FValue` and `FNumTasks`. The code in *increment* and *decrement* tasks also starts the same, with a `for` loop that increments or decrements the value but then uses a different approach to pass the value back to the main thread.

The following code uses `TThread.Synchronize` to execute an anonymous method in the target thread. `Synchronize` accepts a `TThread` object as the first parameter and an anonymous method in the second. It will execute the second parameter in the main thread. In the first parameter, we can pass either the current `TThread` object or, as the following code does, a `nil` value. In the next chapter, I will discuss this parameter in more detail:

```
procedure TfrmIncDecComm.IncSynchronize(startValue: integer);
var
  i: integer;
  value: integer;
begin
  value := startValue;
  for i := 1 to CNumRepeat do
    value := value + 1;

  TThread.Synchronize(nil,
    procedure
```

```
      begin
        PartialResult(value);
      end);
  end;
```

Synchronize **pauses** the execution in the worker thread, then somehow (the implementation depends on the graphical framework) passes the second parameter to the main thread, waits for the main thread to execute that code, and only then resumes the worker thread. Let me explain this again as it is important. While the TThread.Synchronize call executes in the **worker** thread, the anonymous method passed as the second parameter—and by extension the PartialResult method—executes in the **main** thread.

The fact that Synchronize pauses the execution of a worker thread is not important in our case, as the worker had already finished the job. However, sometimes this slows down the background task. A typical example would be sending notifications about the progress to the main thread. Each time Synchronize is called, the worker thread pauses for some time, which makes the CPUs underutilized.

In such cases, it is better to use TThread.Queue instead of Synchronize. This method accepts the same parameters as Synchronize and works almost completely the same, with one exception. It does not wait for the main thread to execute the anonymous method. Instead of that, it returns immediately, which allows the background task to continue processing data.

Result processing in the PartialResult method is completely the same as in MsgTaskDone. I'm just using different methods in different approaches so that each implementation is separated from the others. It also allows me to log appropriate method names in the log:

```
  procedure TfrmIncDecComm.PartialResult(value: integer);
  begin
    Inc(FValue, value);
    Inc(FNumDone);
    if FNumDone = 2 then
      Done('Synchronize');
  end;
```

Polling

Instead of reacting to notifications from the thread, we can also periodically check the status of background calculations from the main thread. In other words, we *poll* their status.

If we want to send some data from a background worker, we can simply insert it into a TThreadedList<T> or, for a faster operation, into TThreadedQueue<T>. In the main thread, we must then periodically check if something has appeared in this queue and process the data.

As TThreadedQueue<T> has quite an unwieldy interface, which doesn't allow us to check whether the queue is empty, we have to take care while creating it. The following code initializes the queue with two parameters (queue size—we know that we'll only put two values in the queue), *0* (push timeout), and *0* (pop timeout). The last parameter is the most important because it allows our polling code to work correctly. I'll return to that shortly.

The polling is done from a `TimerCheckQueue` timer, which is set to trigger every *10* milliseconds but is disabled when the program starts. The following code also starts this timer:

```
procedure TfrmIncDecComm.btnThQueueAndTImerClick(
  Sender: TObject);
begin
  FValue := 0;
  FNumDone := 0;
  FResultQueue := TThreadedQueue<integer>.Create(2, 0, 0);
  RunInParallel(IncThQueue, DecThQueue);
  TimerCheckQueue.Enabled := true;
end;
```

The worker task starts with a now familiar `for` loop. After that, it stores the partial result in the queue by calling `PushItem`:

```
procedure TfrmIncDecComm.IncThQueue(startValue: integer);
var
  i: integer;
  value: integer;
begin
  value := startValue;
  for i := 1 to CNumRepeat do
    value := value + 1;
  Assert(FResultQueue.PushItem(value) = wrSignaled);
end;
```

If the queue is full, `PushItem` waits for some user-configurable timeout before it returns. If a free space appears in the queue in that time, the item is pushed to the queue and the call returns `wrSignaled`. If, after the timeout, the queue is still full, the call returns `wrTimeout`. It would be great if we could specify the timeout as a parameter to `PushItem` but, sadly, this is not implemented. We have to pass it to the constructor and it defaults to infinite waiting (constant `INFINITE`).

The timer method triggers every *10* milliseconds and tries to fetch a value from the queue. If the queue is empty, `PopItem` waits up to a configurable timeout before it returns wrTimeout. If there's data in the queue, the method returns `wrSignaled`. As with `PushItem`, we can only configure the timeout in the constructor. The default value for timeout is again `INFINITE`.

When the timer manages to retrieve partial data from the queue, it aggregates it to the result, increments the number of tasks, and terminates the operation when both tasks are completed. After that, it breaks out of the while loop because the Done method destroys the queue, so we can't read from it anymore:

```
procedure TfrmIncDecComm.TimerCheckQueueTimer(Sender: TObject);
var
  qsize: integer;
  value: integer;
begin
  while FResultQueue.PopItem(qsize, value) = wrSignaled do
  begin
    FValue := FValue + value;
    Inc(FNumDone);
    if FNumDone = 2 then begin
      Done('TThreadedQueue + TTimer');
      break; //while
    end;
  end;
end;
```

In the Done method, we have to stop the timer and destroy the thread in addition to stopping the worker tasks:

```
TimerCheckQueue.Enabled := false;
FreeAndNil(FResultQueue);
```

The TThreadedQueue<T> is also useful in combination with *push* notifications. We can, for example, store some complex data type in a queue and then send a simple Windows message to notify the main thread that it should read from the queue.

Performance

Let's complete this section by examining the execution times of different communication approaches. Even more importantly, let's compare them with the original single-threaded implementation and with locking methods.

The original single-threaded code needed *75* ms to increment and decrement values. Critical section-based synchronization needed *1,017* ms, TMonitor implementation took *700* ms, spinlock needed *493* ms, and the fastest parallel implementation so far, TInterlocked, used *407* ms.

In comparison, communication-based approaches are blindingly fast. Both message-based mechanisms needed only *16-23* milliseconds, Synchronize and Queue are even faster with *13-14* milliseconds, and even the slowest timer-based code needed *24* milliseconds at worst:

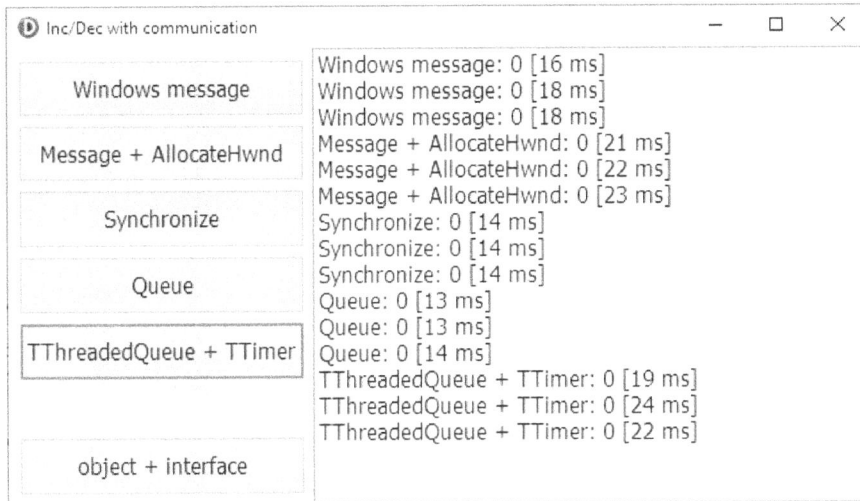

Figure 7.11 – Using communication instead of locking

It is obvious why the timer-based code is the slowest in the pack. After the last value is calculated, we may have to wait up to *10* ms for the timer to trigger.

What is interesting is that other approaches are more than twice as fast as the single-threaded code. The main reason for that is the way to increase and decrease a value, which is in the new code done in one line (`value := value + 1`), while in the old code it needed two assignments:

```
value := FValue;
FValue := value + 1;
```

If we correct the original code, then it needs around *34* ms to do the job. In other words, the `Synchronize` and `Queue` based approaches are exactly twice as fast as the original code, which is the best possible result we can get from running parallel code on two cores.

Third-party libraries

While this book focuses almost exclusively on the out-of-the-box Delphi experience, sometimes I do point to external libraries that can simplify your programming experience. In the context of this chapter, such libraries are Spring4D (`www.spring4d.org`) and OmniThreadLibrary (`www.omnithreadlibrary.com`).

Spring4D is a multipurpose library with very limited support for parallel programming. It is, however, widely used in the Delphi community. We have talked extensively about its collection support, in *Chapter 5, Fine-Tuning the Code*. It can also help with thread synchronization issues.

I've said before that creating and destroying TCriticalSection objects is a pain. To fix that, Spring4D introduces a Lock record, which doesn't require initialization. You just declare a variable of that type and then use its Enter and Leave methods.

Another useful Spring4D addition is the optimistic initializer, TLazyInitializer. In the *Object life cycle*, section, I showed how to safely initialize the FSharedObject: TSharedObject variable and it was not very simple. With TLazyInitializer we can do it in one line:

```
TLazyInitializer.EnsureInitialized<TSharedObject>(FSharedObject);
```

As opposed to Spring4D, **OmniThreadLibrary** has everything to do with parallel programming. Similar to Spring4D, it contains an easy-to-use critical section, TOmniCS, with the Acquire and Release methods. It also implements an optimistic initializer, Atomic<T>, which can be used as shown here:

```
Atomic<TSharedObject>.Initialize(FSharedObject);
```

OmniThreadLibrary also implements very fast multiple readers, and an exclusive writer synchronization mechanism called TOmniMREW. It should only be used to lock very short parts of code because access methods never sleep when access is locked. They always use the spinning mechanism, which can use lots of CPU time when waiting for a longer time. This mechanism is also not re-entrant and should be used with care. If you are using an older Delphi version (pre-10.4.1), however, you can use it instead of the extremely slow TMREWSync.

Finally, I would like to mention OmniThreadLibrary lock-free data structures. There's a fixed-size stack, TOmniBoundedStack, fixed-size queue, TOmniBoundedQueue, and a very fast dynamically-allocated queue, TOmniBaseQueue. All of them can easily be used with standard Delphi multithreading code.

We will explore OmniThreadLibrary's support for parallel programming patterns in *Chapter 10, More Parallel Patterns*.

Summary

In this chapter, we finally started learning how to write parallel code. I started with a short introduction about processes and threads, single and multithreading, single-tasking, and multitasking. I also explained the most important differences between processes and threads.

After that, we started learning what *not* to do when writing multithreaded code. Firstly, I brought up the most important dogma—never access the user interface from a background thread. Such strong words deserve proof and I gave you some.

In the next, largest part of the chapter, I slowly explained why you should be extremely careful if you want to access shared data from multiple threads. While simultaneous reading is OK, you should always use protection when reading and writing at the same time.

In the context of parallel programming and data sharing, this protection is implemented by the introduction of one of the synchronization mechanisms. I spent quite some time introducing critical sections, mutexes, semaphores, `TMonitor`, spinlocks, and readers-writer locks. Using synchronization inevitably brings in new problems, so I also spent some time discussing deadlocks. I also mentioned three data structures that implement locking internally—`TthreadList`, `TthreadList<T>`, and `TthreadedQueue<T>`.

After that, I introduced a less powerful but faster way of locking—interlocked operations. We examined the `TInterlocked` class and all the interlocked operations it implements. As interlocked operations are frequently used in optimistic initialization, we also dealt with the object life cycle in the parallel world.

Toward the end, I introduced a much faster replacement for synchronization—communication. We saw how a parallel program can use data duplication and aggregation to achieve the highest possible speed. I went over four communication techniques, namely Windows messages, `Synchronize`, `Queue`, and polling. We also saw how `TThreadedQueue<T>` can be used in practice.

To wrap things up, I introduced two libraries that contain helpful parallel programming helpers—Spring4D and OmniThreadLibrary.

In the next chapter, we'll finally start writing parallel code. I'll talk about threads, tasks, and writing parallel code with `TThread` and with the Parallel Programming Library, which has been part of Delphi since release XE7.

8

Working with Parallel Tools

After using one whole chapter to warn you about the traps of parallel programming, it is now finally time to write some code! Although I always prefer to use modern multithreading approaches – and we'll spend all of the next two chapters learning about them – it is also good to know the basics. Because of that, I have dedicated this chapter to the good old `TThread` class.

In this chapter, we will ask the following questions:

- How can you use `TThread` to write multithreading code?

- What different approaches to thread management does `TThread` support?

- How are exceptions in threads handled?

- What additional functionality does `TThread` implement?

- How can we implement a communication channel to send messages to a thread?

- How can we centralize thread-message handling in the owner form?

- How can we simplify thread writing fully by implementing a specific framework for one usage pattern?

- How can we write multiplatform code handling multiple worker threads?

Technical requirements

All code in this chapter was written with Delphi 11.3 Alexandria. Most of the examples, however, can also be executed on Delphi XE and newer versions. You can find all the examples on GitHub: `https://github.com/PacktPublishing/Delphi-High-Performance---Second-Edition/tree/main/ch8`.

TThread

Multithreading support has been built into Delphi since its early days.

The very first 32-bit version, Delphi 2, introduced the TThread class. At that time, TThread was a very simple wrapper around the Windows CreateThread function. In later Delphi releases, TThread was extended with multiple functions and support for other operating systems, but it still remained a pretty basic tool.

The biggest problem with TThread is that it doesn't enforce the use of any programming patterns. Because of that, you can use it to create parallel programs that are hard to understand, hard to debug, and that work purely by coincidence. I should know – I shudder every time I have to maintain my old TThread-based code.

Still, the TThread approach can be very effective and completely readable, provided that you use it correctly. Over the following pages, I'll first show the basic TThread usage patterns and then improve the basic approach by introducing *communication* techniques to the old framework.

The TThread class is declared in the System.Classes unit. You cannot use it directly, as it contains an abstract method, Execute, which you have to override in a derived class:

```
TMyThread = class(TThread)
protected
  procedure Execute; override;
end;
```

When you create an instance of this class, Delphi uses operating system functions to create a new thread and then runs the Execute method in that thread. When that method exits, the thread is terminated.

The **TThread** and **Stop** buttons in the Threads program demonstrate one way of using TThread to create a thread. The former creates the thread descendant and starts the thread. It also sets up an event handler, which gets triggered when the thread is terminated:

```
procedure TfrmThreads.btnThreadClick(Sender: TObject);
begin
  FThread := TMyThread.Create;
  FThread.OnTerminate := ReportThreadTerminated;
end;
```

By default, TThread.Create creates and **starts** the thread. It is also possible to create a thread in a **suspended** (paused) state by providing True as an argument to the constructor. If so, you have to call TThread.Start at some point to actually start the thread.

The following code fragment is functionally equivalent to the previous code, except that it initially creates a thread in a suspended state. This approach is safer if we do additional initialization on the thread object (FThread), as it makes sure that the thread did not start up before it was fully initialized:

```
procedure TfrmThreads.btnThreadClick(Sender: TObject);
begin
  FThread := TMyThread.Create(True);
  FThread.OnTerminate := ReportThreadTerminated;
  FThread.Start;
end;
```

In our case, the thread will only stop when we tell it to. It is, therefore, safe to assign the OnTerminate handler to a running (not suspended) thread, as we did in the initial version.

> **Warning**
>
> Although the TThread object implements the Suspend and Resume methods, which pause and start the thread, you should never use them in your code! Unless they are used with great care, you cannot know exactly what state the thread is in when you suspend it. Using these two functions will only bring you great pain and weird, hard-to-remedy problems.

The traditional way to stop a thread is to first call the Terminate method, which sets the Terminated flag. The Execute method should check this flag periodically and exit when it is set (I'll show you an example of that in a moment). Next, we should call the WaitFor method, which waits for a thread to finish execution. After that, we can destroy the thread object. The following code is executed when you click the **Stop** button:

```
procedure TfrmThreads.btnStopThreadClick(Sender: TObject);
begin
  FThread.Terminate;
  FThread.WaitFor;
  FreeAndNil(FThread);
end;
```

A simpler way is to just destroy the thread object, as it will internally call Terminate and WaitFor itself. Sometimes, though, it is good to know that we can do this in three separate steps:

```
procedure TfrmThreads.btnStopThreadClick(Sender: TObject);
begin
  FreeAndNil(FThread);
end;
```

Let's now move away from thread management to actually run code in the thread. This is implemented in the `TMyThread.Execute` method. In the `Threads` demo, this code just checks the `Terminated` flag to see whether it should terminate, posts a message to the main window, and sleeps for a second. As we learned in the previous chapter, we can also use the `Synchronize` or `Queue` methods to do this:

```
procedure TMyThread.Execute;
begin
  while not Terminated do
  begin
    PostMessage(frmThreads.Handle, WM_PING, ThreadID, 0);
    Sleep(1000);
  end;
end;
```

Every thread in a system has a unique integer ID, which we can access through the `ThreadID` property.

Terminating a thread is always a cooperative process. The main program sets a flag that tells the thread to terminate, and the thread must cooperate by checking this flag as frequently as possible. Later in this chapter, we'll also see how to do thread termination without constant checking (without *polling*).

The main form of the demonstration program implements a simple `MsgPing` method, which handles the `WM_PING` message handler. Inside the method, the code logs the thread ID passed in the `wParam` message parameter:

```
procedure TfrmThreads.MsgPing(var msg: TMessage);
begin
  ListBox1.Items.Add('Ping from thread ' +
    msg.WParam.ToString);
end;
```

After the thread stops working (after the code exits from the `Execute` method), the `OnTerminated` event handler is called. The thread object being destroyed is passed in the `Sender` parameter. The code just logs the thread ID of the thread being destroyed:

```
procedure TfrmThreads.ReportThreadTerminated(
  Sender: TObject);
begin
  Assert(Sender is TThread);
  ListBox1.Items.Add('Terminating thread ' +
    TThread(Sender).ThreadID.ToString);
end;
```

This concludes the first example. As you can see, managing a thread is not that hard. The problems appear when we want to do some useful work in the `Execute` method, especially when we want to communicate with the thread. Before I deal with that, however, I'd like to show you some simplifications that TThread supports.

Automatic life cycle management

The thread management that I've shown is most appropriate when we have a long-running thread providing some service. The main program starts the thread when it needs that service and stops it when the service is no longer needed. Lots of times, though, the background thread runs some operations and then stops. At times, it is simpler to allow the thread to terminate itself.

We can do that by making one simple change – by setting the FreeOnTerminate property to True. The demo program does that when you click on the **Free on terminate** button. Although not strictly required in this case, the code creates the thread in a suspended state so that it can be safely initialized:

```
procedure TfrmThreads.btnFreeOnTermClick(Sender: TObject);
begin
  FThreadFoT := TFreeThread.Create(True);
  FThreadFoT.FreeOnTerminate := true;
  FThreadFoT.OnTerminate := ReportThreadTerminated;
  FThreadFoT.Start;
end;
```

The TFreeThread class – just like TmyThread – overrides only the Execute method:

```
procedure TFreeThread.Execute;
var
  i: Integer;
begin
  for i := 1 to 5 do
  begin
    PostMessage(frmThreads.Handle, WM_PING, GetCurrentThreadID, 0);
    Sleep(1000);
    if Terminated then
      Exit;
  end;
end;
```

There's no need to destroy the FThreadFoT object, as Delphi will do that for us. We should, however, still clear the field when the thread is destroyed so that we don't keep a reference to the destroyed object in the code. The simplest way to do that is with the OnTerminated event handler:

```
procedure TfrmThreads.ReportThreadTerminated(Sender: TObject);
begin
  Assert(Sender is TThread);
  ListBox1.Items.Add('Terminating thread ' +
```

```
      TThread(Sender).ThreadID.ToString);
    FThreadFoT := nil;
  end;
```

If you need to run a shorter operation but you still want to be able to abort it, both approaches can be combined. The thread can check the Terminated flag even when it is run in FreeOnTerminate mode. The main program can then just call Terminate to terminate the thread, and the thread will terminate itself.

The demo program uses this approach to destroy both threads when the form is closed:

```
  procedure TfrmThreads.FormDestroy(Sender: TObject);
  begin
    FreeAndNil(FThread);
    if assigned(FThreadFoT) then
      FThreadFoT.Terminate;
  end;
```

Sometimes, we don't care much about thread management because we know that the thread will only run for a short time. In such a case, we can forego the creation of a derived class and simply call the CreateAnonymousThread method, which creates a slightly misnamed *anonymous thread*.

This method takes an anonymous method (TProc) as a parameter. Internally, it creates an instance of the TAnonymousThread class (found in System.Classes) and returns it as a function result. When the Execute method of that class runs, it executes your anonymous method.

An anonymous thread is always created in a suspended state and has the FreeOnTerminate flag set. In the simplest form, you can create and start such a thread and ignore any returned value:

```
  TThread.CreateAnonymousThread(PingMe).Start;
```

The code in the demo assigns the result to a temporary thread variable so that it can set the termination handler:

```
  procedure TfrmThreads.btnAnonymousClick(Sender: TObject);
  var
    thread: TThread;
  begin
    thread := TThread.CreateAnonymousThread(PingMe);
    thread.OnTerminate := ReportThreadTerminated;
    thread.Start;
  end;
```

The background code is now implemented as a normal method of the `TfrmThreads` form object. This makes such an approach inherently dangerous, as it blurs the boundaries between the user interface (accessible through the form methods) and the background thread, which has full access to the form object. When you work with anonymous threads, you should always be careful and respect the separation between objects used in the main and background threads. (Never, *never* access the user interface from a background thread!)

As the `PingMe` method isn't implemented in a class derived from `TThread`, it cannot access the `ThreadID` property directly. It can, however, access the `TThread.Current` property, which returns the `TThread` instance of the current thread:

```
procedure TfrmThreads.PingMe;
var
  i: Integer;
begin
  for i := 1 to 5 do
  begin
    PostMessage(Handle, WM_PING,
      TThread.Current.ThreadID, 0);
    Sleep(1000);
  end;
end;
```

Advanced TThread

The `TThread` class offers a programmer a collection of various properties and methods. In this section, I'll take you through the most useful ones.

The `FatalException` property deals with exceptions. If an exception arises inside your `Execute` method and is not caught, it will be handled inside the system code:

```
if not Thread.Terminated then
try
  Thread.Execute;
except
  Thread.FFatalException := AcquireExceptionObject;
end;
```

The best place to check whether this property is assigned is the `OnTerminate` handler. If `FatalException` is not `nil`, we can log the exception and call `ReleaseExceptionObject` to destroy the exception object, or we can re-raise the exception. The following code from the `Threads` demo shows how to implement such logging:

```
procedure TfrmThreads.ReportThreadTerminated(Sender: TObject);
var
```

```
    thread: TThread;
begin
  thread := Sender as TThread;
  if assigned(thread.FatalException) then
  begin
    ListBox1.Items.Add(Format(
      'Thread raised exception: [%s] %s',
      [thread.FatalException.ClassName,
       Exception(thread.FatalException).Message]));
    ReleaseExceptionObject;
  end;
end;
```

The demo contains a `TExceptThread` class, which just raises an exception so that we can test the exception-handling mechanism. To run a thread of this class, click the **Exception in a thread** button:

```
procedure TExceptThread.Execute;
begin
  Sleep(1000);
  raise Exception.Create('Thread exception');
end;
```

A thread can return an integer value by setting its `ReturnValue` property. This property is, however, protected and hard to access from the main thread. Delphi offers better implementation of such functionality with a `Future<T>` pattern, which we'll explore in the next chapter.

Next comes the communication methods, `Synchronize`, `Queue`, and `ForceQueue`. The first two I've mentioned already in the previous chapter, while `ForceQueue` is a special version of `Queue`. If you call `Queue` from the main thread, the code detects that and executes your anonymous method immediately (without queuing). In contrast to that, `ForceQueue` always queues the anonymous method, even when called from the main thread.

All communication methods accept a `TThread` parameter, which in all the previous examples we passed simply as `nil`. This parameter, when set, connects the anonymous method with a thread. We can use that to remove all queued but unhandled events by calling the `TThread.RemoveQueueEvents` function. For example, we can call this method from a thread's `OnTerminate` handler to prevent any queued events for that thread from being executed, after the thread was destroyed.

Finally, I'd like to mention methods that are useful when you tune the multithreading performance. A class property, `ProcessorCount`, will give you the total number of processing cores in a system. This includes all hyper-threading cores, so in a system with four cores where each of them is hyper-threaded, `ProcessorCount` would be 8.

> **Note**
>
> Running a system with multiple processing cores doesn't necessarily mean that all of them are available for your program. In Windows, it is possible to limit the number of cores that a program can see. To get the real number of available cores on Windows, call the `GetProcessAffinityMask` function declared in the `Winapi.Windows` unit.

The `Priority` property can be used to read and set a thread's *priority*. A priority represents how important that thread is in the system. In Windows, this property can have the following values:

```
type
  TthreadPriority = (tpIdle, tpLowest, tpLower, tpNormal,
                     tpHigher, tpHighest, tpTimeCritical);
```

`tpIdle` represents a very low-priority thread, which only runs when the system is idle. `tpTimeCritical`, on the other hand, represents a thread that must always run, if at all possible. In reality, a high-priority thread cannot completely stop a low-priority thread from running, as Windows dynamically adjusts the thread priorities to allow all of them at least a small chance of executing.

I would strongly recommend that you never set the priority to `tpTimeCritical`. If you run multiple CPU-intensive threads with this priority, it can really cripple your system.

Thread priority should always be left at the default value of `tpNormal`. Windows is extremely good at scheduling threads and tuning thread priorities. Only in very special situations should it be necessary to raise or lower a thread priority, but even then, you should not go above `tpHighest`.

On non-Windows systems, the `Priority` property has type `integer`. A range of allowed values may differ from platform to platform, but at least on POSIX-compliant systems (macOS or Linux), you can assume that the allowed range is from *0* to *31*, with *0* representing the lowest priority.

A thread that runs in a loop may want to give some slice of its CPU time to other threads occasionally. One way to do this is to call `Sleep(0)`, which signals to the operating system that a thread has finished its job at the moment but it would like to continue executing soon, please. The code can also call `TThread.Yield` instead. This also pauses the thread and allows the execution of another thread that is ready to run on the *current processor*. Calling `Yield` instead of `Sleep` generally means that your thread will be woken again a bit sooner.

Setting up a communication channel

The biggest problem of `TThread` is that it only allows communication to flow from a background thread to the owner. As you've seen in the previous chapter, we can use different mechanisms for that – Windows messages, `Synchronize`, `Queue`, and polling. There is, however, no built-in way to send messages in a different direction, so you have to build such a mechanism yourself. This is not entirely trivial.

Another problem with built-in mechanisms is that they make for unreadable code. Synchronize and Queue are both inherently messy because they wrap code that executes in one thread inside code executing in a different thread. Messages and polling have a different problem. They decouple code through many different methods, which sometimes makes it hard to understand a system.

To fix all these problems (and undoubtedly introduce some new ones as, sadly, no code is perfect), I have built a better base class, TCommThread. You can find it in the DHPThreads unit, which is part of this book's code archive. The demonstration program for the new class is called ThreadsComm. You can use this unit without any limitations in your projects.

This thread contains two parts. Firstly, there is a communication channel running from the owner to the thread. Secondly, there's a channel running in a different direction, from the thread to the owner. I'll explain them one by one.

The code needs some container mechanism that will store messages sent from the owner but not yet processed by the thread. Although it is hard to use, I've decided to go with TThreadedQueue, which doesn't need a separate synchronization mechanism. As I have no idea what kind of data you'll be storing in the queue, I've implemented the queue as TThreadedQueue<TValue>. The TValue type from the System.Rtti unit can store any Delphi data structure inside, so it's very appropriate for such a task.

It is true that the flexibility of TValue means that it is a bit slower than using a specific data type directly. On the other hand – as I always say – if your code depends on millions of messages to be processed each second, then the problem lies not in the implementation but in your architecture. For all normal usage patterns, TValue is more than fast enough.

The code also needs a way to tell the thread that something has happened when a message is sent. I've chosen to use TEvent, as it allows the thread to completely sleep while nothing is going on. This removes the need for constant *polling*, which would use a bit of CPU time.

The following code fragment shows how the two are declared in the TCommThread object. The Event field is declared protected, as it was designed to be used directly from the derived class:

```
strict private
  FQueueToThread : TThreadedQueue<TValue>;
protected
  Event: TEvent;
```

Both fields are initialized in the constructor, which, in addition to the standard CreateSuspended parameter, accepts a queue size. It is set to 1,024 elements by default, but you can make this value larger or smaller if that's required. In the actual unit, the constructor is a bit more complicated, as it also initializes the second communication channel. I've removed that part of the code as shown here to make things clearer:

```
constructor TCommThread.Create(
  CreateSuspended: boolean = False;
```

```
    ToThreadQueueSize: integer = 1024);
begin
    inherited Create(CreateSuspended);
    FQueueToThread := TThreadedQueue<TValue>.Create(
                        ToThreadQueueSize, 0, 0);
    Event := TEvent.Create;
end;
```

The constructor creates both fields *after* the inherited constructor was called, even when CreateSuspended is False. This doesn't cause problems because the thread doesn't actually start in the TThread constructor. It only starts in the TThread.AfterConstruction method, which is executed *after* TCommThread.Create finishes its job.

To send a message to the thread, the code has to call the SendToThread public function. This very simple method pushes an item into the queue and sets the event if the push was successful. If the queue is full, the function returns False:

```
function TCommThread.SendToThread(const value: TValue): boolean;
begin
    Result := (FQueueToThread.PushItem(value) = wrSignaled);
    if Result then
        Event.SetEvent;
end;
```

To process these messages, the code in the Execute method must be written in a specific manner. It must periodically check the Event; otherwise, it can even spend all of its time waiting for the Event to get set. TMyThread from the ThreadCommMain unit of the example program serves as a good starting point for your code.

This method calls Event.WaitFor(5000) to wait for a new message. WaitFor will either return wrSignaled if a message has been received, wrTimeout if nothing happened in five seconds, or something else if something unexpected happened. In the latter case, the code exits the thread, as there's nothing more that can be done currently.

If wrSignaled or wrTimeout is received, the code enters the loop. Firstly, it resets the event by calling Event.ResetEvent. This is mandatory and allows the code to function properly. Without that call, the event would stay *signalled* (set), and the next time around, WaitFor would immediately return wrSignaled, even though no new messages would be received.

After that, the code calls GetMessage to fetch messages from the queue. You should always empty the entire queue; otherwise, there's a possibility that messages could get stuck inside for an unknown time (until the next message is sent). The code also exits when the Terminated flag is set.

Finally, the code sends a message back to the owner. This part is omitted in the code shown here, as I haven't discussed that part of the mechanism yet:

```
procedure TMyThread.Execute;
var
  pingMsg: integer;
  value: TValue;
begin
  while Event.WaitFor(5000) in [wrSignaled, wrTimeout] do
  begin
    Event.ResetEvent;
    // message processing
    while (not Terminated) and GetMessage(value) do
      pingMsg := value.AsInteger;
    // termination
    if Terminated then
      break;
    // send message to the owner ...
  end;
end;
```

The careful among you may have started thinking about the thread termination now. Doesn't this implementation mean that when we request the thread to terminate, we will have to wait for up to five seconds for it to do so? After all, WaitFor may just have been called, and we have to wait until it reports a timeout. Even worse, what if the thread is completely message-driven and does no other processing? In that case, we can wait with Event.WaitFor(INFINITE), but wouldn't that prevent the thread from exiting?

Indeed, careful reader, these are all valid worries. Do not fear, though, as there's a simple solution already built into TCommThread! Besides the message-handling code, TCommThread also overrides the TerminatedSet virtual method, which TThread calls when the Terminated flag is set. Inside this overridden method, the code signals the Event event, and this causes WaitFor to immediately exit:

```
procedure TCommThread.TerminatedSet;
begin
  Event.SetEvent;
  inherited;
end;
```

The only missing part of this half of the puzzle is the GetMessage method. That one is, again, trivial and just reads from the queue:

```
function TCommThread.GetMessage(var value: TValue): boolean;
begin
```

```
    Result := (FQueueToThread.PopItem(value) = wrSignaled);
  end;
```

Sending messages from a thread

The second part of the solution deals with sending messages back to the parent. As I've mentioned before, you already have several possibilities to do the same with built-in Delphi functions. They all work well in combination with the previously described methods to send messages to a thread. This second part is, therefore, purely optional, but I believe it is a good solution because it centralizes the message handling for each thread.

Again, the code starts by creating `TThreadedQueue<TValue>` to store sent values. We cannot use an event to signal new messages, though. Delphi forms interact with the operating system and the user by constantly processing messages, and we cannot block this message processing by waiting for an event; otherwise, the whole user interface will freeze. Instead, `TCommThread` uses `Queue` to call a method of your own choosing to process messages. You pass this method to the constructor as a parameter.

The following code fragment shows relevant types and fields, together with the (now complete) constructor. Besides the user-provided `MessageReceiver` method that will process messages, you can also set the queue size:

```
public type
  TMessageReceiver = TProc<TValue>;
strict private
  FQueueToMain : TThreadedQueue<TValue>;
  FMessageReceiver: TMessageReceiver;

constructor TCommThread.Create(
  MessageReceiver: TMessageReceiver;
  CreateSuspended: boolean = False;
  ToThreadQueueSize: integer = 1024;
  ToMainQueueSize: integer = 1024);
begin
  inherited Create(CreateSuspended);
  FQueueToThread := TThreadedQueue<TValue>.Create(
                      ToThreadQueueSize, 0, 0);
  Event := TEvent.Create;
  FMessageReceiver := MessageReceiver;
  if assigned(MessageReceiver) then
    FQueueToMain := TThreadedQueue<TValue>.Create(
                      ToMainQueueSize, 0, 0);
end;
```

The message-receiver parameter is optional. If you set it to `nil`, the channel to send messages back to the owner won't be created, and this part of the functionality will be disabled.

The code also overrides the standard `TThread` constructor to create `TCommThread` with only one communication channel – the one toward the thread:

```
constructor TCommThread.Create(CreateSuspended: Boolean);
begin
  Create(nil, CreateSuspended);
end;
```

The code can use the protected method, `SendToMain`, to send messages back to the owner. It checks whether the necessary queue was created, posts a message to the queue, and, if that was successful, queues the `PushMessagesToReceiver` method. If the queue is full, this method returns `False`:

```
function TCommThread.SendToMain(const value: TValue): boolean;
begin
  if not assigned(FQueueToMain) then
    raise Exception.Create(
      'MessageReceiver method was not set in constructor!');
  Result := (FQueueToMain.PushItem(value) = wrSignaled);
  if Result then
    TThread.Queue(nil, PushMessagesToReceiver);
end;
```

The `PushMessagesToReceiver` method is implemented in the `TCommThread` object. It reads messages from the queue and, for each message, calls the user-provided `FMessageReceiver` method:

```
procedure TCommThread.PushMessagesToReceiver;
var
  value: TValue;
begin
  // This method executes from the main thread!
  while FQueueToMain.PopItem(value) = wrSignaled do
    FMessageReceiver(value);
end;
```

The value of this approach is twofold. Firstly, all message processing for a thread concentrates on one method. Secondly, you can pass different types of data through this channel – from basic types to objects, interfaces, and even records.

A word of caution is necessary. Never destroy the thread object from inside the message receiver. Instead, use the `ForceQueue` method to queue code to destroy the thread object. That will cause the code to execute *after* the message receiver does its job.

Using TCommThread

In the `ThreadCommMain` unit of the `ThreadComm` program, all of this comes together. The **Start thread** button creates a new thread and immediately sends a number 42 to it. It only sets the message receiver method (`ProcessThreadMessages`) and leaves other parameters at the defaults:

```
procedure TfrmThreadComm.btnStartClick(Sender: TObject);
begin
  FThread := TMyThread.Create(ProcessThreadMessages);
  FThread.SendToThread(42);
end;
```

The **Change** button sends the current value of a spin edit to the thread via the communication mechanism:

```
procedure TfrmThreadComm.btnChangePingClick(
  Sender: TObject);
begin
  FThread.SendToThread(inpPing.Value);
end;
```

The **Stop thread** button just destroys the thread object:

```
procedure TfrmThreadComm.btnStopClick(Sender: TObject);
begin
  FreeAndNil(FThread);
end;
```

The `TMyThread` object overrides the `Execute` method. We've seen a part of that method before. The full implementation is shown here.

The code immediately sends a message of the `PingMsg` type back to the owner. This message carries the current value of the local variable, `pingMsg`, and the current thread ID. The message is sent over the built-in communication mechanism.

Next, the code starts the `while` loop, which I already described. The only new part here appears under the `// workload` comment, when the code again sends the current `pingMsg` and thread ID to the owner:

```
type
  TPingMsg = TPair<integer,TThreadID>;

procedure TMyThread.Execute;
var
  pingMsg: integer;
  value: TValue;
begin
```

```
      pingMsg := 0;
      if not SendToMain(TValue.From<TPingMsg>(
            TPingMsg.Create(pingMsg, ThreadID)))
      then
        raise Exception.Create('Queue full!');
      while Event.WaitFor(5000) in [wrSignaled, wrTimeout] do
      begin
        Event.ResetEvent;
        // message processing
        while (not Terminated) and GetMessage(value) do
          pingMsg := value.AsInteger;
        // termination
        if Terminated then
          break;
        // workload
        if not SendToMain(TValue.From<TPingMsg>(
              TPingMsg.Create(pingMsg, ThreadID)))
        then
          raise Exception.Create('Queue full!');
      end;
    end;
```

The behavior of this method can be described in a few words. It sends pingMsg back to the owner every five seconds. When it receives a message, it assumes that it contains an integer and stores that in the pingMsg variable. The code terminates when the Terminated flag is set.

The only part I haven't shown yet is the ProcessThreadMessages method, which simply shows the data stored inside TPingMsg:

```
procedure TfrmThreadComm.ProcessThreadMessages(value: TValue);
var
  pingMsg: TPingMsg;
begin
  pingMsg := value.AsType<TPingMsg>;
  ListBox1.Items.Add(Format('%s: %d from thread %d',
    [FormatDateTime('hh:nn:ss.zzz', Now), pingMsg.Key,
     pingMsg.Value]));
end;
```

And there you have it – a thread with two communication channels!

Implementing a timer

If you play with the demo application, you'll soon find that it works fine, just not exactly as I described before. I stated that "(the code) sends `pingMsg` back to the owner every five seconds." As the following screenshot shows, that isn't exactly the case:

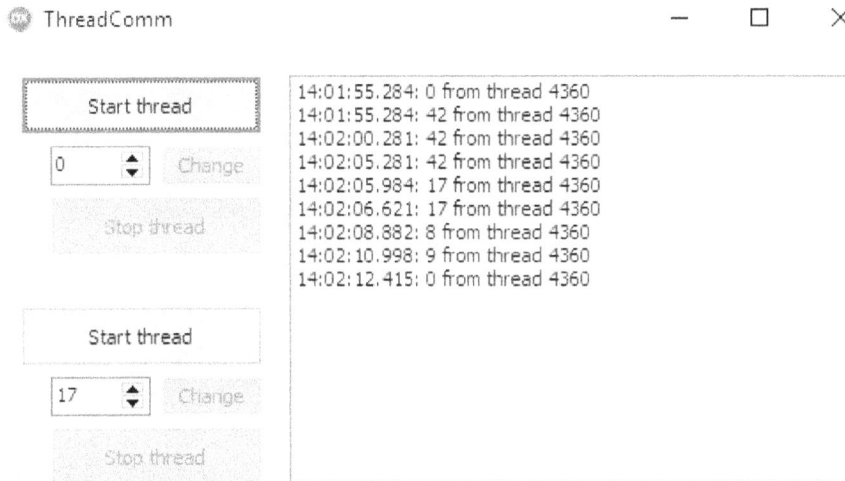

Figure 8.1 – The output of the ThreadComm program

This figure shows one short testing session. During the testing, I started the thread. That caused two messages to be logged. The initial value of *0* was sent from the thread immediately when it started. After that, the main thread sent a new value, *42*, to the worker thread, which resulted in an immediate *ping* response.

After that, I didn't click anything for 11 seconds, which generated two *ping* messages, each sent five seconds after the previous message. Next, I clicked **Change** twice, which resulted in two *ping* messages. After that, I changed the value to *8*, *9*, and *0*, and each time, a *ping* was immediately received.

A proper description of the worker thread would therefore be this – it sends a message every five seconds, unless a new message is received, in which case it sends it back immediately.

If that's OK with you, fine! Feel free to use this framework. However, sometimes, you need a better timer. I could fix that in `TMyThread` with a better calculation of the timeout parameter passed to `WaitFor`. I would also have to fix the code, which now works the same regardless of the `WaitFor` result (`wrSignaled` or `wrTimeout`). I could do that, but then you would have to repeat my steps each time you needed a thread with a built-in timer. Instead of that, I did something better and implemented a generic starting point for all such tasks, `TCommTimerThread`.

This time, my implementation is closer to the approach used in all modern parallel libraries. It is not a multipurpose tool, such as TThread or TCommThread, but a specific solution that satisfies one *usage pattern*. Because of that, you don't have to implement any nasty multithreading plumbing to use it, such as when we had to implement a very specific Execute method in the previous example. Using TCommTimerThread is more akin to using the VCL. You just plug in a few methods, and the framework does the rest. We'll see this approach a lot in the next chapter.

This approach has numerous advantages over the classical TThread. Firstly, it is easier to write code, as all the hard stuff has already been taken care of. Secondly, the code will have fewer bugs, as you don't have to write error-prone multithreading plumbing. As many people use the same framework, any potential bugs in it can be found quickly.

The usage pattern my implementation covers is "a background task that can send and receive messages and process timer-based events." While this looks simple, and even primitive, it can be used to solve many different kinds of problems, especially as you can ignore the message-processing part or the timer event part if you don't need them.

Let us look at the implementation now. TCommTimerThread is derived from the TCommThread class so that it can reuse its communication mechanism. It exposes a public property, Interval, which you can use to set the timer interval, in **milliseconds (ms)**. Setting Interval to a value equal to or smaller than zero disables the timer. In addition to that, the code uses an FTimer: TStopwatch record to determine the time to the next timer event.

Instead of overriding the Execute function, you override four methods (or just some of them). Initialize is called immediately after the thread is created, and Cleanup is called just before it is destroyed. You can do your own initialization and clean up inside them. ProcessTimer is called each time a timer event occurs, and ProcessMessage is called for each received message. This code fragment shows all the important parts of TCommTimerThread:

```
TCommTimerThread = class(TCommThread)
strict private
  FInterval: integer;
  FTimer: TStopwatch;
protected
  procedure Cleanup; virtual;
  procedure Execute; override;
  procedure Initialize; virtual;
  procedure ProcessMessage(const msg: TValue); virtual;
  procedure ProcessTimer; virtual;
public
  property Interval: integer read GetInterval write SetInterval;
end;
```

All important parts of the functionality are implemented in the overridden Execute method. It is very similar to TMyThread.Execute, except that the WaitFor timeout is calculated dynamically and that processing depends on the value returned from WaitFor. If it returns wrSignaled, the code checks the Terminated flag and fetches waiting messages. If it returns wrTimeout, the timer is restarted, and the timer-handling function is called:

```
procedure TCommTimerThread.Execute;
var
  awaited: TWaitResult;
  timeout: Cardinal;
  value: TValue;
begin
  Initialize;
  try
    repeat
      awaited := Event.WaitFor(CalculateTimeout);
      event.ResetEvent;
      case awaited of
        wrSignaled:
          begin
            while (not Terminated) and GetMessage(value) do
              ProcessMessage(value);
            if Terminated then
              break;
          end;
        wrTimeout:
          begin
            FTimer.Reset;
            FTimer.Start;
            ProcessTimer;
          end
        else
          break; //Terminate thread
      end;
    until false;
  finally
    Cleanup;
  end;
end;
```

The demonstration program implements TMyTimerThread, which uses this approach to implement the same functionality as TMyThread. Instead of using a monolithic Execute, the code implements three very small overridden methods. Another change is that the local variable, pingMsg, is now a field, as it must be shared between different methods:

```
TMyTimerThread = class(TCommTimerThread)
strict private
  FPingMsg: integer;
protected
  procedure Initialize; override;
  procedure ProcessMessage(const msg: TValue); override;
  procedure ProcessTimer; override;
end;
```

The code to create this thread looks almost the same as the code that creates TMyThread, except that it also sets the interval between messages:

```
procedure TfrmThreadComm.btnStartTimerClick(
  Sender: TObject);
begin
  FTimerThread := TMyTimerThread.Create(ProcessThreadMessages);
  FTimerThread.Interval := 5000;
  FTimerThread.SendToThread(42);
end;
```

Sending a message to the thread and stopping the thread looks completely the same as before. This thread also uses the same *message receiver* function in the main form, ProcessThreadMessage.

When a thread is initialized, the code sends a *ping* message to the owner. There's no need to initialize FPingMsg, as all of the object fields are initialized to zero when an object is created. The timer event handler, ProcessTimer, just calls the same helper function, SendPing, to send a message to the owner. And the message processing is trivial – it just sets the shared field:

```
procedure TMyTimerThread.Initialize;
begin
  SendPing;
end;

procedure TMyTimerThread.ProcessMessage(const msg: TValue);
begin
  FPingMsg := msg.AsInteger;
end;

procedure TMyTimerThread.ProcessTimer;
begin
```

```
  SendPing;
end;

procedure TMyTimerThread.SendPing;
begin
  if not SendToMain(TValue.From<TPingMsg>(
          TPingMsg.Create(FPingMsg, ThreadID)))
  then
    raise Exception.Create('Queue full!');
end;
```

When the code is written in such a way, it becomes clear why I prefer using patterns and communication over standard `TThread` code. There is no data sharing, no weird communication calls, and best of all, no mess.

Synchronizing with multiple workers

In the previous section, we implemented a very simple synchronization mechanism that was used by a thread to get notifications about new messages – waiting on an event. This is simple and effective but not very flexible.

When we write complex systems where one thread starts multiple worker threads, this simple approach usually fails. We need something better, a solution that can wait on multiple events and do something when (any) one of them triggers. Windows offers a great tool for that – the `WaitForMultipleObjects` API function (which comes in several variations) – but that locks us into one platform. What can we do if we need to run our programs on other operating systems?

A good answer to that question is *condition variables*. They are, however, somehow weirdly implemented in the Delphi RTL (more on that later), and they remain a total mystery to most Windows Delphi programmers. To help anyone that wants to dive into these deep waters, I have prepared two solutions to a simple multithreaded problem one written specifically for Windows and another that can be used on any supported operating system. But first, let's take a look at the test program that binds them all together.

The `MultipleWorkers` demo comes with two buttons – the **WaitForMultiple** button starts the Windows-specific worker, while the **CondVar** button starts the platform-independent solution. The Windows version, which is shown in the following code snippet, disables the button, creates the test object, and starts the background thread. The test object constructor accepts two anonymous functions, which are called from the worker thread context (and, therefore, must not modify the UI). The first function (AsyncLogger) is used to log data on the screen, while the second (an inline anonymous function) is called when the test stops running and simply re-enables the **UI** button:

```
procedure TfrmMultipleWorkers.btnWaitForMultipleClick(
  Sender: TObject);
var
  test: TWaitForMultipleTest;
begin
  btnWaitForMultiple.Enabled := false;
  lblOutput.Text := 'Output: ';
  test := TWaitForMultipleTest.Create(10, AsyncLogger,
    procedure
    begin
      TThread.Queue(nil,
        procedure
        begin
          btnWaitForMultiple.Enabled := true;
        end);
    end);
  test.Run;
end;
procedure TfrmMultipleWorkers.AsyncLogger(msg: string);
begin
  TThread.Queue(nil,
    procedure
    begin
      lblOutput.Text := lblOutput.Text + msg;
    end);
end;
```

The event handler for the second button is identical, except that it creates a TCondVarTest test object. The code uses a few {$IFDEF WINDOWS} compiler directives to enable Windows-specific code for the Windows platform only. Therefore, you'll be able to compare both approaches on Windows or just the platform-independent code on any other platform.

Both solutions implement the same algorithm:

1. The test object creates a background thread.
2. This background thread creates two worker threads and, for the next 10 seconds, waits on signals from both threads. After that, it stops the worker threads and destroys the test object.
3. The first thread wakes up every 700 ms, increments a counter, and notifies the background thread.
4. The second thread does the same, but every 1,200 ms.

5. Whenever the background thread receives a signal from any of the worker threads, it "collects" the counter associated with the thread (more on that later) and calls the logging function (AsyncLogger) to update the screen status.

Let's now examine both approaches, starting with the Windows-specific implementation.

WaitForMultipleObjects

The Run method, which is called from the main thread, simply creates an anonymous thread, tells it to run the TestProc method, and starts that thread. The thread is configured to auto-destroy when TestProc exits:

```
procedure TWaitForMultipleTest.Run;
var
  thread: TThread;
begin
  thread := TThread.CreateAnonymousThread(TestProc);
  thread.FreeOnTerminate := true;
  thread.Start;
end;
```

The TestProc method (which I'll show later) starts by calling the InitWorkers helper. This method creates everything required for synchronization between the background thread and worker threads:

- Two locks, FLocks, used to synchronize data access between the threads, one for each worker thread

- Two counters, FCounters, one for each worker thread

- Two flags, FStopped, used by the worker threads to signal their running/stopped status

- Two events, FEvents, used by the worker threads to wake up the background thread

- FTotal, the total number of triggered events

- FStop, a flag to signal worker threads to stop

- Finally, two anonymous worker threads – one running method, Worker1, and Worker2

All of this is shown in the next code fragment:

```
procedure TWaitForMultipleTest.InitWorkers;
begin
  FLocks[1] := TCriticalSection.Create;
  FLocks[2] := TCriticalSection.Create;
  FCounters[1] := 0; FCounters[2] := 0;
  FStopped[1] := false; FStopped[2] := false;
  FEvents[1] := TEvent.Create(nil, false, false, '');
```

```
FEvents[2] := TEvent.Create(nil, false, false, '');
FTotal := 0;
FStop := false;
TThread.CreateAnonymousThread(Worker1).Start;
TThread.CreateAnonymousThread(Worker2).Start;
end;
```

Both worker threads run the same worker method, RunWorker, only with different parameters:

```
procedure TWaitForMultipleTest.Worker1;
begin
  RunWorker(1, 700);
end;
procedure TWaitForMultipleTest.Worker2;
begin
  RunWorker(2, 1200);
end;
```

The RunWorker method simulates a typical worker thread in a larger program. It loops until instructed to stop (until FStop is set) and, in each loop, does the following:

- Sleeps a little (simulating the work that is done)

- Updates the shared counter, protected with a critical FLocks[idx] section

- Sets the FEvents[idx] event to notify the background thread that the work was done:

```
procedure TWaitForMultipleTest.RunWorker(
  idx, delay_ms: Integer);
begin
  TThread.NameThreadForDebugging('Worker ' +
                                    IntToStr(idx));
  while not FStop do begin
    Sleep(delay_ms);
    FLocks[idx].Acquire;
    FCounters[idx] := FCounters[idx] + 1;
    FLocks[idx].Release;
    FEvents[idx].SetEvent;
  end;
  FStopped[idx] := true;
end;
```

The background thread is the most complicated of them all. It initializes a test by calling the InitWorkers helper method and copies *handles* of both events into an internal array. Then, it runs a loop for a specified duration (10 seconds).

Inside the loop, the code calls the `WaitForMultipleObjects` function to wait for signals from any of the two events and specifies a maximum wait time of 500 ms. The function will return with one of the following values:

- `WAIT_OBJECT_0` (a constant from the `Winapi.Windows` unit) if the first event is set.
- `WAIT_OBJECT_0 + 1` if the second event is set. `WaitForMultipleObjects`, which can wait on up to 64 events or other types of handles that can be *signaled*. They will result in return values from `WAIT_OBJECT_0` to `WAIT_OBJECT_0 + 63`.
- `WAIT_TIMEOUT` if no event was set in 500 ms.

Other values (for details, see the Windows API help at `https://learn.microsoft.com/en-us/windows/win32/api/synchapi/nf-synchapi-waitforsingleobject`) should not be returned in this code. `TestProc` still logs a question mark to output if something weird is returned to help with debugging.

This wait also automatically resets the event, so it will not trigger the wait again unless it is set again in the worker thread.

> **Note**
>
> Events can work in two modes – auto-reset (as in this case) and manual reset. The mode is dictated by the second parameter passed to the event constructor in the `InitWorkers` method.

If the first event was set, the code "collects" the counter for thread 1 (more on that later) and logs `'x'` to the screen. A second event "collects" the counter for thread 2 and logs `'o'`. A timeout results in `'.'` being logged.

Finally, the code stops worker threads by calling `StopWorkers` (which we will show later) and logs the value inside `FTotal` (received and processed events) and the `FCounters` array (generated but unprocessed events):

```
procedure TWaitForMultipleTest.TestProc;
var
  awaited: cardinal;
  handles: array [1..2] of THandle;
  sw: TStopwatch;
begin
  TThread.NameThreadForDebugging('Test runner');
  InitWorkers;
  handles[1] := FEvents[1].Handle;
  handles[2] := FEvents[2].Handle;
  sw := TStopwatch.StartNew;
  while sw.Elapsed.Seconds < FDuration do begin
```

```
      awaited := WaitForMultipleObjects(2, @handles, false,
                    500);
      if awaited = WAIT_OBJECT_0 then begin
        GrabCounter(1);
        FLogger('x');
      end
      else if awaited = (WAIT_OBJECT_0 + 1) then begin
        GrabCounter(2);
        FLogger('o');
      end
      else if awaited = WAIT_TIMEOUT then FLogger('.')
      else FLogger('?');
    end;
    StopWorkers;
    FLogger(' ' + IntToStr(FTotal) +
            '/' + IntToStr(FCounters[1]) +
            '/' + IntToStr(FCounters[2]));
    FOnStop();
  end;
```

The method that "collects" a counter is called GrabCounter. It increments the FTotal counter
according to the current state in FCounters[idx] and resets FCounters[idx]. This modifies
the same data used by the worker threads and, therefore, must be protected by the critical section:

```
procedure TWaitForMultipleTest.GrabCounter(idx: Integer);
begin
  FLocks[idx].Acquire;
  Inc(FTotal, FCounters[idx]);
  FCounters[idx] := 0;
  FLocks[idx].Release;
end;
```

Finally, the StopWorkers method signals both threads to stop, waits on both of them to do so, and
then destroys the shared data:

```
procedure TWaitForMultipleTest.StopWorkers;
begin
  FStop := true;
  while not (FStopped[1] and FStopped[2]) do
    Sleep(1);
  FEvents[1].Free; FEvents[2].Free;
  FLocks[1].Free; FLocks[2].Free;
end;
```

If you run this program, you should get something very similar to the following screenshot:

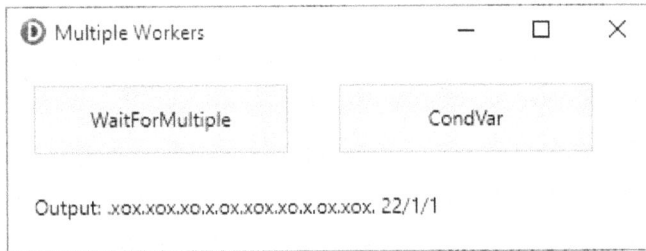

Figure 8.2 – The MultipleWorkers demo in action

The logged numbers show that the 22 FCounters elements were "collected" and moved into FTotal, while finally, both FCounters elements were 1. This is due to the RunWorker implementation. The code sleeps for some time and then increments the FCounters, and only then does it check the FStop flag. The thread shutdown, therefore, follows the following path:

- The background thread stops waiting on events and calls StopThreads

- Both worker threads stop sleeping after some time; they both increment the FCounters array, set FEvents, and exit

- FEvents are ignored because the background thread no longer waits on them

This solution shows a common approach to this type of synchronization problem on Windows. Other platforms, however, have no equivalent to WaitForMultipleObjects, and we have to approach the problem in a different manner.

Condition variables

Condition variables are synchronization primitives that enable a thread to wait on a particular condition, similar to events. They are supported on all operating systems and by Delphi, although the APIs for Windows and other systems are completely different.

Delphi supports condition variables in a somehow weird fashion. In the System.SyncObjs unit, we can find two classes, TConditionVariableCS and TConditionVariableMutex. Their implementation, however, is not what you would expect.

TConditionVariableCS uses operating system support for condition variables on Windows but not on other operating systems; it uses custom implementation based on TMonitor instead. TConditionVariableMutex is similar, but in reverse – it uses operating system support on POSIX systems and a complex internal implementation based on a critical section, a semaphore, and an event on Windows.

Instead of writing some code that wraps both classes and uses TConditionVariableCS on Windows and TConditionVariableMutex on all other systems, I have written a simple lightweight wrapper that uses native operating system support for condition variables on all operating systems.

> **Note**
>
> What about using just TConditionVariableCS or TConditionVariableMutex on all systems? I just don't trust the custom implementation, especially as both implementations are completely different. One uses TMonitor.Pulse function, which was a known source of problems for so long that I still don't trust it, and another has such a complex implementation that I'll never be sure that it doesn't break in some edge case.

My code follows the approach used in Delphi's TLightweightMREW and can be found on GitHub at https://github.com/gabr42/GpDelphiUnits/blob/master/src/GpSync.CondVar.pas. The code is released under the BSD license and can be used without any problems in commercial applications.

> **Note**
>
> If you want to dive deeper into condition variables, you can look up the Windows documentation at https://learn.microsoft.com/en-us/windows/win32/sync/condition-variables or the documentation for Posix systems at https://pubs.opengroup.org/onlinepubs/7908799/xsh/pthread_cond_init.html.

Unlike other synchronization primitives, condition variables are not useful by themselves. If we want to wait on them, we have to pair them with a locking mechanism. On Windows, we can choose between a critical section and a reader-writer lock, while on other systems we have to use mutexes. That means we have to manage two objects in the code – a condition variable and a lock – which can become messy.

The GpSync.CondVar unit wraps operating system support for condition variables in a TLightweightCondVar record, but in most cases, you will want to use a TLockConditionVariable record, which pairs a condition variable with a lock (TLightweightMREW on Windows and pthread_mutex_t on other systems).

> **Note**
>
> TLightweightCondVar and TLockConditionVariable are nice examples of custom record initialization and finalization, which we covered in *Chapter 5*, *Fine-Tuning the Code*. For details, see the code.

The following code fragment shows TLockConditionVariable in full. The interface combines synchronization functionality (Acquire and Release) with signaling functions (Signal and Broadcast) and waiting functions (Wait and TryWait):

```
type
  TLockConditionVariable = record
  private
    FCondVar: TLightweightCondVar;
{$IFDEF MSWINDOWS}
    FSRW    : TLightweightMREW;
{$ENDIF MSWINDOWS}
{$IFDEF POSIX}
    FMutex  : pthread_mutex_t;
{$ENDIF POSIX}
  public
{$IFDEF POSIX}
    class operator Initialize(
      out Dest: TLockConditionVariable);
    class operator Finalize(
      var Dest: TLockConditionVariable);
{$ENDIF POSIX}
    procedure Acquire; inline;
    procedure Release; inline;
    procedure Broadcast; inline;
    procedure Signal; inline;
    procedure Wait; inline;
    function TryWait(Timeout: Cardinal): boolean; inline;
  end;
```

The difference between the Signal and Broadcast functions shows only when more than one thread waits on a condition variable. In such a case, Signal will wake one thread only (we don't know in advance which one), and Broadcast will wake all waiting threads.

Let's see how a condition variable (or, rather, TLockConditionVariable) is used in practice. The core of TCondVarTest closely follows the TWaitForMultipleTest implementation, so I'll mostly focus on the differences between the two.

The InitWorkers method is shorter because we don't have to initialize locks and events. TLockConditionVariable (and TLightweightCondVar) are automatically initialized when they are included in a larger object (TCondVarTest in our case). However, we need a new array, FChanged, which will tell the background thread which worker thread did some work:

```
procedure TCondVarTest.InitWorkers;
begin
  FCounters[1] := 0; FCounters[2] := 0;
```

```
    FChanged[1] := false; FChanged[2] := false;
    FStopped[1] := false; FStopped[2] := false;
    FTotal := 0;
    FStop := false;
    TThread.CreateAnonymousThread(Worker1).Start;
    TThread.CreateAnonymousThread(Worker2).Start;
  end;
```

The RunWorker worker is almost identical to the Windows-only version. We lock TLockConditionVariable, make the changes, unlock, and signal the background thread:

```
  procedure TCondVarTest.RunWorker(idx, delay_ms: Integer);
  begin
    TThread.NameThreadForDebugging('Worker ' +
      IntToStr(idx));
    while not FStop do begin
      Sleep(delay_ms);
      FCondVar.Acquire;
      FCounters[idx] := FCounters[idx] + 1;
      FChanged[idx] := true;
      FCondVar.Release;
      FCondVar.Signal;
    end;
    FStopped[idx] := true;
  end;
```

The biggest changes between the two approaches can be found in the TestProc method. When using condition variables, we have to Acquire access at the beginning and Release it at the end. Then, the code between the two calls is then completely protected.

To wait on a condition variable, we call the Wait or TryWait function (the latter accepts a timeout) – similar to calling WaitForMultipleObjects. The difference is that [Try]Wait on a condition variable implicitly releases the shared lock (thus allowing worker threads to process data), waits, and then acquires the shared lock just before returning:

```
  procedure TCondVarTest.TestProc;
  var
    sw: TStopwatch;
  begin
    TThread.NameThreadForDebugging('Test runner');
    InitWorkers;
    sw := TStopwatch.StartNew;
    FCondVar.Acquire;
```

```
    while sw.Elapsed.Seconds < FDuration do begin
      if not FCondVar.TryWait(500) then FLogger('.')
      else begin
        if FChanged[1] then begin
          GrabCounter(1);
          FChanged[1] := false;
          FLogger('x');
        end;
        if FChanged[2] then begin
          GrabCounter(2);
          FChanged[2] := false;
          FLogger('o');
        end;
      end;
    end;
    FCondVar.Release;
    StopWorkers;
    FLogger(' ' + IntToStr(FTotal) +
            '/' + IntToStr(FCounters[1]) +
            '/' + IntToStr(FCounters[2]));
    FOnStop();
  end;
```

Because we now have only one signal (condition variable) and not two (events) as before, the code has to check the FChanged array to see which thread did the work. It is possible that both the FChanged elements are set when TryWait exits. The code also has to clear the corresponding element in the FChanged array. In the Windows version, the event was reset automatically.

As access to the shared data is always granted in the main TestProc loop, we don't have to access the shared lock anymore in the GrabCounter method:

```
procedure TCondVarTest.GrabCounter(idx: Integer);
begin
  Inc(FTotal, FCounters[idx]);
  FCounters[idx] := 0;
end;
```

Finally, the StopWorkers method works just as in the previous example, except that it doesn't have to clean up any resources:

```
procedure TCondVarTest.StopWorkers;
begin
  FStop := true;
```

```
   while not (FStopped[1] and FStopped[2]) do
     Sleep(1);
end;
```

If you run the code, you'll see that the output looks just the same as the previous Windows-specific example.

Comparing both approaches

We can see that condition variable-based code is similar to `WaitForMultipleObjects`-based implementation, but there are also some differences that we have to be aware of.

Let's compare both approaches – first, the Windows-only solution:

- It uses shared locks and events that must be managed in code
- The worker thread follows the `acquire/modify/release/signal` pattern
- The background thread follows the `wait/acquire/process/release` pattern

Compare this with the condition variables-based implementation:

- It uses a shared `TLockConditionVariable` that doesn't have to be created or destroyed
- The worker thread follows the `acquire/modify/release/signal` pattern
- The background thread follows the `acquire/wait/process/release` pattern

We can see that the biggest changes lie in code that waits on an event or a condition variable. When using events, we call wait from unprotected code (not inside an acquired lock) and then acquire/release when we access the shared data. With condition variables, we have to acquire the lock before we start waiting (which automatically releases the lock for the duration of the wait), and we have to release the lock after the wait.

Summary

While this chapter focused on a single topic, it was still quite diverse. You've learned everything about the `TThread` class, which is a basis for all multithreading code in Delphi. Even the task-based approach that we'll explore in the next chapter uses `TThread` as a basic building block.

I have shown three different ways to create a `TThread`-based multithreading solution. A program can take complete ownership of a thread so that it is created and destroyed by the owner. This approach is best used when a thread implements a *service*, as in cases where the main program knows best when the service is needed.

Another method, which is more appropriate for *background calculations*, is `FreeOnTerminate` mode. With this approach, a thread object is immediately destroyed when a thread's `Execute` function exits. The thread owner can set up an `OnTerminate` event to catch this condition and process the result of the calculation.

Instead of writing a separate class for each background operation, we can also create a thread that executes an *anonymous method*. This approach is most appropriate for short-lived tasks that don't require any interaction with the thread's owner.

After that, we looked into exception handling inside a thread. I discussed a few other important `TThread` methods and properties.

The second half of the chapter focused on writing better threading tools. Firstly, I extended `TThread` with two communication channels. The result, `TCommThread`, can be used as a base class for your own threads, which can use these channels to communicate with the thread owner.

Next, I implemented a specific solution for one usage pattern – a background task that can send and receive messages and react to timer events. I then showed how you can produce very clean and understandable code, by using such threading implementation.

Finally, we looked at a multi-platform, multi-worker synchronization tool – condition variables. As they are not directly supported by Delphi, I introduced custom code that wraps the native operating system implementation of that synchronization primitive.

In the next chapter, we'll focus even more on a usage-pattern approach. We'll dig deep into Delphi's **Parallel Programming Library** (**PPL**), which is all about usage patterns. For good measure, I'll also implement a few patterns that PPL doesn't cover.

9

Exploring Parallel Practices

You made it! I admit that the previous two chapters were not much fun and that they mostly tried to scare you away from multithreaded code. Now it's time for something completely different.

In this chapter, I'll cover the fun parts of multithreaded programming by moving to high-level concepts—tasks and patterns. We will not create any threads in the code; no, we won't. Instead, we'll just tell the code what we need, and it will handle all the nitty-gritty details.

When you have read this chapter to the end, you'll know all about the following topics:

- What are tasks and how are they different from threads?
- How do we create and destroy tasks?
- What are our options for managing tasks?
- How should we handle exceptions in tasks?
- What is thread pooling and why is it useful?
- What are patterns and why are they better than tasks?
- How can we use the Async/Await pattern to execute operations in a thread?
- How can we use the Join pattern to execute multiple parts of code in parallel?
- What is a Future and how can it help when calculating functions?
- How can we convert a normal loop into a parallel loop with Parallel For?
- How can a pipeline help us parallelize many different problems?

Technical requirements

All code in this chapter was written with Delphi 11.3 Alexandria. Most of the examples, however, could also be executed on Delphi XE and newer versions. You can find all the examples on GitHub: `https://github.com/PacktPublishing/Delphi-High-Performance---Second-Edition/tree/main/ch9`.

Tasks and patterns

Traditionally, parallel programming was always implemented with a focus on threads and data sharing. The only support we programmers got from the operating system and the language runtime libraries were thread- and synchronization-related functions. We were able to create a thread, maybe set some thread parameters (such as thread priority), and kill a thread. We were also able to create some synchronization mechanisms—a critical section, mutex, or a semaphore. But that was all.

As you are not skipping ahead and you read the previous two chapters, you already know that being able to start a new thread and do the locking is not nearly enough. Writing parallel code that way is a slow, error-prone process. That's why in the last decade, the focus in parallel code has shifted from threads to *tasks* and *patterns*. Everyone is doing it—Microsoft with the .NET Task Parallel Library, Intel with Thread Building Blocks, Embarcadero with the Parallel Programming Library and third-party extensions such as OmniThreadLibrary, and so on.

The difference between threads and tasks is simple. A **thread** is an operating system concept that allows us to execute multiple parts of a process simultaneously. A **task** is merely a part of the code that we want to execute in that way. Or, to put it in different words—when you focus on threads, you tell the system *how* to do the work. When you work with tasks, you tell the library *what* you want to be done. You don't use time on *plumbing*, setting up threads, communication mechanisms, and so on. Everything is already ready and waiting for you.

As it turns out, working with tasks is usually still too low-level. For example, if you want the program to iterate over a loop with multiple tasks in parallel, you would not want to do that by using basic tasks. I'll create such an example in this chapter and you'll see that it still takes some work. A much simpler approach is to use a specific usage *pattern* —a *parallel for* in this example.

Many such patterns exist in modern parallel libraries. Delphi's Parallel Programming Library offers *join*, *future*, and *parallel for*. .NET has *async* and *await*. Thread Building Blocks offers an incredibly powerful *flow graph*. OmniThreadLibrary gives you *background workers*, *pipelines*, *timed tasks*, and more.

In this chapter, I'll cover all the patterns implemented by the Parallel Programming Library, and then some. I'll extend the set with patterns that you can freely use in your code. Before that, however, we have to cover the basics and deal with *tasks*.

Variable capturing

Before I start writing parallel code, I have to cover something completely different. As you'll see in this chapter, a big part of the incredible usefulness of tasks and patterns is the ability to use them in combination with anonymous methods and variable capturing. As powerful as that combination is, however, it also brings in some problems.

The code in this chapter will frequently run into the problem of capturing a *loop variable*. Instead of trying to explain this problem at the same time as dealing with the already hard concepts of parallel code, I decided to write a simple program that demonstrates the problem and does nothing more.

The code in the AnonMethod project tries to write out the numbers from *1* to *20* in a convoluted way. For each value of i, the code calls TThread.ForceQueue, and passes in an anonymous method that calls Log(i) to write the value into ListBox.

The problem with the following code lies in the programmer's assumptions. When you write code such as this, you should not forget that by writing an anonymous method, you are creating a completely separate part of the code that will be executed at some other time. In this example, Log will be called *after* btnAnonProblemClick finishes the execution:

```
procedure TfrmAnonMthod.btnAnonProblemClick(Sender: tObject);
var
  i: Integer;
begin
  ListBox1.Items.Add'''');
  for i := 1 to 20 do
    TThread.ForceQueue(nil,
      procedure
      begin
        Log(i);
      end);
end;

procedure TfrmAnonMthod.Log(i: integer);
begin
  ListBox1.Items[ListBox1.Items.Count-- 1]  :=
  ListBox1.Items[ListBox1.Items.Count-- 1]
    + i.ToString +'''';
end;
```

To support this, the compiler must somehow make the value of i available to our anonymous method when it is executed. This is done through the magic of *variable capturing*. I'm not going to go into specifics here, as this is a book about high-performance computing and not about Delphi's implementation details. The only important part you have to know is that the compiler doesn't capture the current *value* of i (as the author of the preceding code fragment assumed) but a *reference* to the value (a pointer to i).

Because of that, all Log calls will receive the same parameter—a value that some pointer points to. It will contain whatever was left inside i after the for loop was completed, namely 21. That is just how Delphi works. After a for loop, a loop variable will contain the upper bound plus *1*. (Or *minus 1*, if the loop is counting downward.)

A trick that solves the problem of capturing a loop variable is to write an intermediate function that creates an anonymous method. If we pass our loop variable (i) to such a function, the compiler captures the parameter to the function and not the loop variable, and everything works as expected.

Don't worry if you did not understand any of the mumbo-jumbo from the previous paragraph. Writing about anonymous methods is hard, but luckily, they are easily explained by the code. The following example shows how we can fix the code in AnonMethod. The MakeLog function creates an anonymous method that logs the value of parameter i. In the loop, we then call MakeLog for each value of i and pass the output of this function to the ForceQueue call:

```
function TfrmAnonMthod.MakeLog(i: integer):TThreadProcedure;
begin
  Result :=
    procedure
    begin
      Log(i);
    end;
end;


procedure TfrmAnonMthod.btnAnonFixClick(Sender: tObject);
var
  i: Integer;
begin
  ListBox1.Items.Add('');
  for i := 1 to 20 do
    TThread.ForceQueue(nil, MakeLog(i));
end;
```

The following screenshot shows both the unexpected output from the original code on the first line and the correct output from the fixed code on the second line:

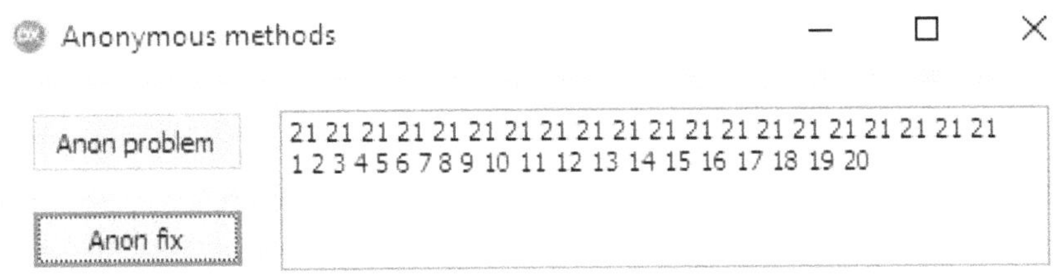

Figure 9.1 – The wrong and correct way to capture a loop variable

Having dealt with this side topic, let us now move to the main point of this chapter – working with Delphi's tasks framework.

Tasks

Let's get back to business, and in this chapter, business means tasks. I don't want to spend the entire thing talking about the theory and classes and interfaces, so I will start by introducing some code. Actually, it is not my code—I got it from our good friend, Mr. Smith.

For the past few chapters, he was quite busy, working on things botanical, so I let him be. (He gave me a comprehensive overview of his activities, but I'm just a programmer, and I didn't really understand him.) Now he has more spare time, and he returned to his great love—prime numbers. He is no longer trying to improve SlowCode. Now he studies how the probability of finding a prime number changes when numbers become larger. To do so, he wrote two simple functions (shown as follows) that check a range of numbers and count how many prime numbers are in this range:

```
function IsPrime(value: integer): boolean;
var
  i: Integer;
begin
  Result := (value > 1);
  if Result then
    for i := 2 to Round(Sqrt(value)) do
      if (value mod i) = 0 then
        Exit(False);
end;

function FindPrimes(lowBound, highBound: integer): integer;
var
  i: Integer;
begin
  Result := 0;
  for i := lowBound to highBound do
```

```
    if IsPrime(i) then
      Inc(Result);
end;
```

He soon found out that his code uses only one CPU on his 64-core machine. Since he found out that I'm writing about parallel programming, he asked me to change his code so that all the CPUs will be used. I didn't have the heart to tell him that his approach was flawed from the beginning and that he should use mathematical theory to solve his problem. Rather, I took his code and used it as the starting point for this chapter. There are different ways to parallelize the code. It probably won't surprise you that I'll start with the most basic one, *tasks*.

In Delphi's Parallel Programming Library, tasks are objects that implement the ITask interface. They are created through methods of the TTask class. Both the class and the interface are implemented in the System.Threading unit.

The code in the ParallelTasks demo shows basic operations on the tasks. The following method is executed if you click on the **Run tasks** button.

The first task is created by calling the TTask.Run method. This method creates a new task and immediately starts it. The task will be started in a thread, run the SleepProc code, and terminate. Run returns an ITask interface, which we can use to query and manipulate the task.

The second task is created by calling the TTask.Create method. This method creates a new task but does not start it. It returns an ITask interface, just as Run does. In this second example, the task payload is a simple anonymous method. As we want it to start running immediately, the code then calls the ITask.Start method, which starts the task. Start returns the task's ITask interface, which the code saves in the tasks array.

The code then waits for both tasks to be completed by calling the TTask.WaitForAll method. It will wait until both the anonymous method and SleepProc exit. The program will be blocked for about 2.5 seconds in WaitForAll, which is not optimal. Later, we'll look into different techniques to work around the problem:

```
procedure SleepProc;
begin
  Sleep(2500);
end;

procedure TfrmParallelTasks.btnRunTasksClick(Sender: TObject);
var
  tasks: array [1..2] of ITask;
begin
  tasks[1] := TTask.Run(SleepProc);
  tasks[2] := TTask.Create(procedure begin Sleep(2000) end)
                .Start;
```

```
    TTask.WaitForAll(tasks);
  end;
```

If we don't want to block the main program, we have to use a different technique to find out when the task has finished its work. We can apply any of the notification methods from the previous chapter— Windows messages, Synchronize, Queue, or polling. In this example, I used Queue.

When you click the **Async TTask** button in the ParallelTasks demo, the btnAsyncTaskClick method (shown as follows) is executed. This method creates a task by calling TTask.Run and stores the returned task interface in the form field FTask. The code also disables the button to give a visual indication that the task is running.

The task then executes the LongTask method in its own thread. After the hard work (Sleep) is done, it queues the notification method, LongTaskCompleted, to the main thread. This method cleans up after the finished task by setting the FTask field to nil, and reactivates the btnAsyncTask button:

```
procedure TfrmParallelTasks.LongTask;
begin
  Sleep(2000);
  TThread.Queue(nil, LongTaskCompleted);
end;

procedure TfrmParallelTasks.LongTaskCompleted;
begin
  FTask := nil;
  btnAsyncTask.Enabled := True;
end;

procedure TfrmParallelTasks.btnAsyncTaskClick(Sender: TObject);
begin
  FTask := TTask.Run(LongTask);
  btnAsyncTask.Enabled := False;
end;
```

When you have to wait for multiple tasks to complete, the code becomes a bit more complicated. You have to keep count of running tasks and only trigger the notification when the last task completes. You'll have to wait a bit for the practical example. Later in this chapter, I'll implement a custom join pattern, and I'll return to the notification problem then.

Exceptions in tasks

Handling exceptions in tasks is a tricky business. The best approach is to handle exceptions explicitly by wrapping the task code in a try..except block.

The following code shows how exceptions are handled in the `ExplicitExceptionTask` task. This task is executed when the user clicks on the **Exception 1** button in the `ParallelTasks` demo.

The code always raises an exception, so `LongTaskCompleted` is never called. Instead of that, the code always executes the `except` block, which queues `LongTaskError` to be executed in the main thread.

As the queued code is not executed immediately, we have to *grab* the exception object by calling the `AcquireExceptionObject` method. Variable capturing will take care of the rest. If we did not call the `AcquireExceptionObject` method, the exception object would be destroyed before the queued anonymous method was called:

```
procedure TfrmParallelTasks.ExplicitExceptionTask;
var
  exc: TObject;
begin
  try
    raise Exception.Create('Task exception');
    TThread.Queue(nil, LongTaskCompleted);
  except
    exc := AcquireExceptionObject;
    TThread.Queue(nil,
      procedure
      begin
        LongTaskError(exc);
      end);
  end;
end;
```

`LongTaskError` just logs the details about the exception and calls `ReleaseExceptionObject` to release the exception acquired with `AcquireExceptionObject`:

```
procedure TfrmParallelTasks.LongTaskError(exc: Exception);
begin
  ListBox1.Items.Add(Format('Task raised exception %s %s',
    [exc.ClassName, exc.Message]));
  ReleaseExceptionObject;
end;
```

Another option is to not use the `try..except` block in the task but to periodically check the `ITask.Status` property. If a task raises an unhandled exception, the `TTask` implementation will catch it, store the exception object away, and set the status to `TTaskStatus.Exception`.

Clicking on the **Exception 2** button in the demo program calls the btnException2Click method, which starts such a task. At the same time, the code enables a timer, which periodically checks the status of the FTask task:

```
procedure TfrmParallelTasks.btnException2Click(
  Sender: TObject);
begin
  FTask := TTask.Run(ExceptionTask);
  TimerCheckTask.Enabled := true;
end;

procedure TfrmParallelTasks.ExceptionTask;
begin
  Sleep(1000);
  raise Exception.Create('Task exception');
end;
```

The code in the timer event handler jumps through lots of hoops to get the exception information from the task. It looks like the designer of ITask decided that you ought to always explicitly handle task exceptions, and they made this second approach needlessly complicated.

The full code for the timer method is shown as follows. Firstly, it checks the task's status. If it is one of Completed, Canceled, or Exception, then the task has finished its work. Completed is used to indicate normal task completion. Canceled means that somebody called the ITask.Cancel method of that task. Expection tells us that an unhandled exception was raised by the task.

In all three cases, the code clears the FTask field and disables the timer. If the status was Exception, it goes through more steps to log that exception.

Unfortunately, there's no clean way to access the exception object raised by the task. We can only see its contents by calling the ITask.Wait method. This method waits for a specified time on a task to complete (in our code, the time is set to *0* milliseconds). It returns True if the task has completed, False if the timeout was reached and the task is still running, or it raises an exception if the task was canceled or an exception was caught.

> **Note**
>
> The Deadlock demo from *Chapter 7, Getting Started with the Parallel World,* shows how to cancel a task.

We must wrap the `Wait` call in a `try..except` block to catch that exception. It will not be a simple exception object but an instance of the `EAggregateException` class, which can contain multiple *inner* exceptions. The code then uses a loop to log all caught exceptions:

```
procedure TfrmParallelTasks.TimerCheckTaskTimer(
  Sender: TObject);
var
  i: integer;
begin
  if not assigned(FTask) then
    Exit;
  if not (FTask.Status in [TTaskStatus.Completed,
             TTaskStatus.Canceled, TTaskStatus.Exception])
  then
    Exit;

  if FTask.Status = TTaskStatus.Exception then
  try
    FTask.Wait(0);
  except
    on E: EAggregateException do
      for i := 0 to E.Count - 1 do
        ListBox1.Items.Add(Format(
          'Task raised exception %s %s',
          [E[i].ClassName, E[i].Message]));
  end;
  FTask := nil;
  TimerCheckTask.Enabled := true;
end;
```

All in all, this is terribly complicated, which is why I strongly support explicit exception handling in task code.

Parallelizing a loop

Let us now return to Mr. Smith's problem. He is calling the `FindPrimes` method, which checks the primality of natural numbers in a `for` loop. I showed that code at the beginning of the chapter, but for convenience, I'll reprint it here:

```
function FindPrimes(lowBound, highBound: integer): integer;
var
  i: Integer;
begin
  Result := 0;
```

```
  for i := lowBound to highBound do
    if IsPrime(i) then
      Inc(Result);
end;
```

When he calls, for example, `FindPrimes(1,10000000)`, the computer spends a long time doing the calculation, during which time only one CPU is busy. That is not surprising given that in any Delphi program, only one thread is running by default. If we want to make other CPUs busy too, we have to run multiple threads.

In his case, it is quite simple to split the job between multiple tasks. Let's say we want to check numbers from 1 to 100,000 with four tasks. We'll let the first task scan the numbers from 1 to 25,000, the second from 25,001 to 50,000, and so on. The tricky part is collating the partial results. Each thread will count the number of primes in its subrange, and we have to collect and aggregate (add together) all the partial results. I'll show two different ways to do that.

The first approach is implemented in the `btnCheckPrimes1Click` method, which is activated by clicking the **Check primes 1** button. The number of tasks created is dictated by the `inpNumTasks` spin edit, so the code allocates a `tasks: TArray<ITask>` array of that length. It also allocates a `results: TArray<Integer>` array, which will be used to collect partial results.

After that, the code creates and starts all the tasks. Some simple mathematics makes sure that each task gets a subrange of approximately the same size. The actual anonymous method that is run in the task is returned by the `PrepareTask` function to circumvent the problem of capturing a `for` variable. Each task receives three parameters—the lower and upper bound for the search range (`lowBound` and `highBound`) and a pointer to the slot, which will receive the partial result (`@results[i-1]`).

When all the tasks are created and started, the code waits for all of them to complete. After that, a simple `for` loop aggregates the partial results:

```
function TfrmParallelTasks.PrepareTask(lowBound,
  highBound: integer; taskResult: PInteger): TProc;
begin
  Result :=
    procedure
    begin
      taskResult^ := FindPrimes(lowBound, highBound);
    end;
end;

procedure TfrmParallelTasks.btnCheckPrimes1Click(Sender: TObject);
var
  aggregate: Integer;
  i: Integer;
  highBound: Integer;
```

```
    lowBound: Integer;
    numTasks: Integer;
    results: TArray<Integer>;
    sw: TStopwatch;
    tasks: TArray<ITask>;
begin
  sw := TStopwatch.StartNew;
  numTasks := inpNumTasks.Value;
  SetLength(tasks, numTasks);
  SetLength(results, numTasks);

  lowBound := 0;
  for i := 1 to numTasks do
  begin
    highBound := Round(CHighestNumber / numTasks * i);
    tasks[i-1] := TTask.Run(PrepareTask(lowBound,
                      highBound, @results[i-1]));
    lowBound := highBound + 1;
  end;
  TTask.WaitForAll(tasks);

  aggregate := 0;
  for i in results do
    Inc(aggregate, i);

  sw.Stop;
  ListBox1.Items.Add(Format(
    '%d prime numbers from 1 to %d found in %d ms ' +
    'with %d tasks',
    [aggregate, CHighestNumber, sw.ElapsedMilliseconds,
     numTasks]));
end;
```

The second approach uses a variable, shared between tasks (`aggregate`), to store the number of found prime numbers. To ensure proper operation, it is updated with a call to `TInterlocked`. Add. The code in `btnCheckPrimes2Click` implements that. As it looks mostly the same as the preceding code, I'll just show the worker function that creates the task method here:

```
function TfrmParallelTasks.PrepareTask2(lowBound,
  highBound: integer; taskResult: PInteger): TProc;
begin
  Result :=
    procedure
    var
```

```
      running: int64;
      max: int64;
    begin
      running := TInterlocked.Increment(FNumRunning);
      max := TInterlocked.Read(FMaxRunning);
      if running > max then
        TInterlocked.CompareExchange(FMaxRunning,
          running, max);
      TInterlocked.Add(taskResult^,
        FindPrimes(lowBound, highBound));
      TInterlocked.Decrement(FNumRunning);
    end;
end;
```

> **Note**
>
> All the code in `PrepareTask2` that is not highlighted in the previous example is used to detect how many tasks are actually active (running) at the same time.

If you start the program and test the code with different numbers of tasks, you will see that the execution time also changes. For testing, I have run the code with 4 tasks, 2 tasks, and 1 task. In my case, calculations run for 2.5, 4.2, and 6.6 seconds, respectively, which is exactly what we would expect. The log window also shows that 4, 2, and 1 tasks ran in parallel during the tests, which again is what we wanted.

To verify that the expected number of tasks actually started, you can use the **Check primes 2** button, which calculates the actual number of concurrent tasks. Its output is shown in the following figure:

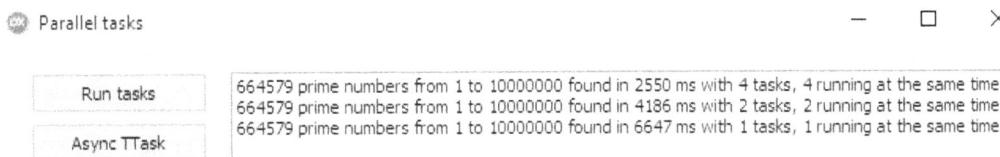

Figure 9.2 – Running the test code with 4 tasks, 2 tasks, and 1 task in parallel

Thread pooling

While starting a thread is much faster than creating a new process, it is still a relatively slow operation. It may easily take a few milliseconds to create a new thread. Because in high-performance applications, every millisecond counts, most task- and pattern-based libraries use the concept of a *thread pool*.

A thread pool is just a form of storage that manages unused threads. For example, let's say that your application—which was working in a single thread until now—has just created (and started) its first task. The Parallel Programming Library will then ask a thread pool to run that task. Unless you pass a custom thread pool parameter to `TTask.Create` or `TTask.Run`, a default thread pool will be used.

The thread pool will then create a new thread to run your task in. After the task completes its job, the thread will not die but will be stored in a list of inactive threads.

If the program now starts another task, the thread pool will not create a new thread but will reuse the previous one. If a new task is started while the previous one is still active, a new thread will be created.

This creation and storage of inactive threads can be controlled by changing the properties of the `TThreadPool` object, which implements a thread pool. The following code fragment shows the public interface exposed by a thread pool:

```
function SetMaxWorkerThreads(Value: Integer): Boolean;
function SetMinWorkerThreads(Value: Integer): Boolean;

property MaxWorkerThreads: Integer read GetMaxWorkerThreads;
property MinWorkerThreads: Integer read GetMinWorkerThreads;
```

Calling `SetMinWorkerThreads` limits the minimum number of worker threads—except that it does not. The comment inside the code states that "*The actual number of pool threads could be less than this value depending on actual demand.*" So, this call is actually just a hint to the library. The `MinWorkerThreads` property gives you read access to that value. By default, this value is initialized to the number of CPU cores in the system.

Calling `SetMaxWorkerThreads` limits the maximum number of worker threads. If that number of threads is active and you start a new task, it will have to wait for another task to complete its work. The `MaxWorkerThreads` property gives you read access to that value. By default, this value is initialized to two times the number of CPU cores in the system.

> **Warning**
>
> In older Delphi versions (pre-11), `MaxWorkerThreads` was initialized to the number of CPU cores **times 25**, which was much too high for normal operation. If you are using Delphi 10.4 or older, set `MaxWorkerThreads` to `TThread.ProcessorCount * 2`, as is the default in Delphi 11.

Async/Await

Tasks are powerful but clumsy. The real power of modern multithreading libraries comes not from them but from specialized code designed to solve specific usage patterns. Although I intend to cover all of the patterns implemented in the Parallel Programming Library, I'll start with a pattern you won't find there.

The *Async/Await* pattern comes from .NET. It allows you to start a background task (*async*) and execute some code in the main thread after the task completes its work (*await*). In Delphi, we cannot repeat the incredible usefulness of the .NET syntax, as it requires support from a compiler, but we can approximate it with something that is useful enough. Such an implementation can be found in my **OmniThreadLibrary**. For the purpose of this book, I have re-implemented it in the DHPThreading unit, which you can freely use in your code.

The AsyncAwait demo shows when this pattern can be useful and how to use it. The **Long task** button calls a method that simulates typical old-school Delphi code.

Firstly, the user interface (the button) is disabled. Then, a time-consuming operation is executed. For example, a program could fetch data from a database, generate a report, export a long XML file, and so on. In this demo program, the LongTask method simply executes Sleep(5000). At the end, the user interface is enabled again:

```
procedure TfrmAsyncAwait.btnLongTaskClick(Sender: TObject);
begin
  btnLongTask.Enabled := false;
  btnLongTask.Update;
  LongTask;
  btnLongTask.Enabled := true;
end;
```

This typical pattern has one big problem. It blocks the user interface. While the code is running, the user cannot click within the program, move the window around, and so on. Sometimes, the long operation can easily be adapted to run in a background thread, and in such cases, a pattern such as *Async/Await* (or, as we'll later see, *Future*) can be really helpful.

Async/Await allows you to rewrite the code, shown as follows. The code still disables the user interface but does not call btnLongTask.Update. This method will exit almost immediately, and the normal Delphi message loop will take over, redrawing the controls as required.

The code then calls Async(LongTask) to start the long task in a background thread. The second part of the command, .Await, specifies a *termination event*, code that is called after the LongTask method exits. This event will execute in a main thread, so it is a good place to update the user interface:

```
TfrmAsyncAwait.btnLongTaskAsyncClick(Sender: TObject);
begin
  Log(Format('Button click in thread %d',
```

```
      [TThread.Current.ThreadID]));

  (Sender as TButton).Enabled := false;

  Async(LongTask)
  .Await(
    procedure
    begin
      Log(Format('Await in thread %d',
           [TThread.Current.ThreadID]));
      (Sender as TButton).Enabled := true;
    end
  );
end;
```

If you click the **Long task async** button in the demo program, it will call this method. The time-consuming task will start, but the program will still be responsive. After 5 seconds, the button will be re-enabled, and you'll be able to start the task again.

Actually, the program has three buttons, which all start the same code. If you click all three, three background threads will be created, and LongTask will be started in each of them. The following screenshot shows what happens inside the program if you do that. The screenshot proves that the long tasks are started in three threads (**Long task started [...]**) and that *await* handlers are executed in the main thread (**Await in thread [...]**):

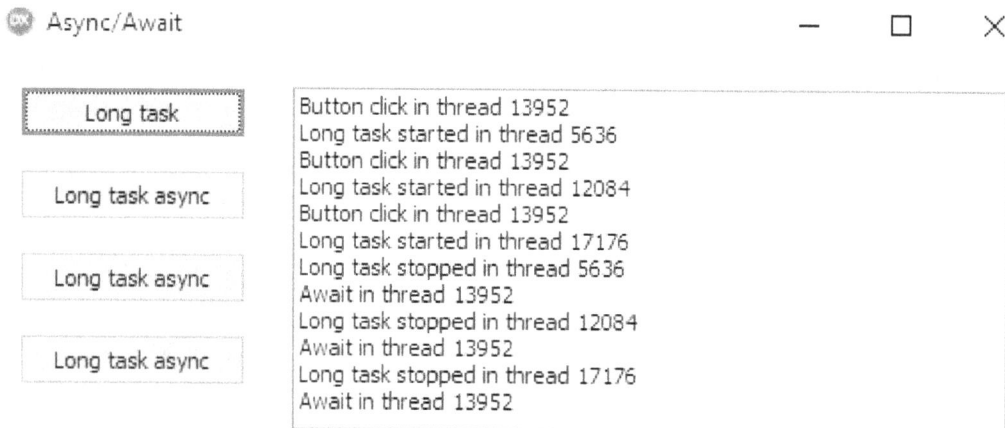

Figure 9.3 – Running background tasks with the Async/Await pattern

The DHPThreading unit implements Async as a method returning an IAsync interface, which implements the Await method:

```
type
  IAsync = interface
             ['{190C1975-FFCF-47AD-B075-79BC8F4157DA}']
    procedure Await(const awaitProc: TProc);
  end;

function Async(const asyncProc: TProc): IAsync;
```

The implementation of Async creates an instance of the TAsync class and returns it. The TAsync constructor is also trivial; it just stores the asyncProc parameter in an internal field:

```
function Async(const asyncProc: TProc): IAsync;
begin
  Result := TAsync.Create(asyncProc);
end;

constructor TAsync.Create(const asyncProc: TProc);
begin
  inherited Create;
  FAsyncProc := asyncProc;
end;
```

Everything else happens inside the Await call. It stores the awaitProc parameter in an internal field and starts the Run method in a task.

The tricky part is hidden in the assignment of FSelf := Self. In the main program, Async creates a new IAsync interface but doesn't store it away in a form field. Because of that, this interface would be destroyed on exiting the btnLongTaskAsyncClick method. To prevent that, Await stores a reference to itself in an FSelf field. This increases the reference count and prevents the interface from being destroyed:

```
procedure TAsync.Await(const awaitProc: TProc);
begin
  FSelf := Self;
  FAwaitProc := awaitProc;
  TTask.Run(Run);
end;
```

The Run method simply runs FAsyncProc and then queues an anonymous method to the main thread. This method first calls FAwaitProc in the main thread and then clears the FSelf field. This decrements the reference count and destroys the ITask interface:

```
procedure TAsync.Run;
begin
  FAsyncProc();

  TThread.Queue(nil,
    procedure
    begin
      FAwaitProc();
      FSelf := nil;
    end
  );
end;
```

The *Async/Await* pattern is intentionally simple. It offers no capability for processing exceptions, no communication channel to return a value, no interface to stop the running task, and so on. Later, we'll see more complicated patterns that offer better control, but sometimes simplicity is exactly what you need.

Join

The next pattern I want to present is **Join**. This is a very simple pattern that starts multiple tasks in parallel. In the Parallel Programming Library, *Join* is implemented as a class method of the TParallel class. To execute three methods, Task1, Task2, and Task3, in parallel, you simply call TParallel. Join with the parameters collected in an array:

```
TParallel.Join([Task1, Task2, Task3]);
```

This is equivalent to the following implementation, which uses tasks:

```
var
  tasks: array [1..3] of ITask;

tasks[1] := TTask.Run(Task1);
tasks[2] := TTask.Run(Task2);
tasks[3] := TTask.Run(Task3);
```

Although the approaches work the same way, that doesn't mean that Join is implemented in this way. Rather than that, it uses a pattern that I haven't yet covered, a *parallel for* to run tasks in parallel.

Join starts tasks but doesn't wait for them to complete. It returns an ITask interface representing a new, composite task, which only exits when all of its subtasks have been executed. You can do with this task anything you can with a *normal* instance of TTask. For example, if you want to wait for the tasks to finish, you can simply call Wait on the resulting interface.

The following line of code starts two tasks in parallel and waits on both of them to finish. An overload of the Join function allows you to pass in two TProc parameters without the array notation:

```
TParallel.Join(Task1, Task2).Wait;
```

The ParallelJoin program demonstrates TParallel.Join in action. The **Join 2 tasks** button executes the following code:

```
procedure TfrmParallelJoin.btnJoin2Click(Sender: TObject);
begin
  ListBox1.Items.Add('Starting tasks');
  TParallel.Join(Task1, Task2);
end;
```

Both tasks look the same. They log a message, wait a little, log another message, and exit:

```
procedure TfrmParallelJoin.QueueLog(const msg: string);
begin
  TThread.Queue(nil,
    procedure
    begin
      ListBox1.Items.Add(msg);
    end);
end;


procedure TfrmParallelJoin.Task1;
begin
  QueueLog('Task1 started in thread ' +
    TThread.Current.ThreadID.ToString);
  Sleep(1000);
  QueueLog('Task1 stopped in thread ' +
    TThread.Current.ThreadID.ToString);
end;


procedure TfrmParallelJoin.Task2;
begin
  QueueLog('Task2 started in thread ' +
    TThread.Current.ThreadID.ToString);
  Sleep(1000);
  QueueLog('Task2 stopped in thread ' +
```

```
        TThread.Current.ThreadID.ToString);
    end;
```

The **Join 3 tasks** button starts similar code that executes three tasks all working in the same fashion as the preceding two tasks.

If you start the program and click **Join 2 tasks**, you'll see that each task is executed in its own thread. If you click the **Join 3 tasks** button, three tasks will be started immediately, again, each in their own thread. It is quite possible that some of the threads will be the same as when running two tasks, as they come from a thread pool.

The following screenshot shows the behavior of the `Join` pattern:

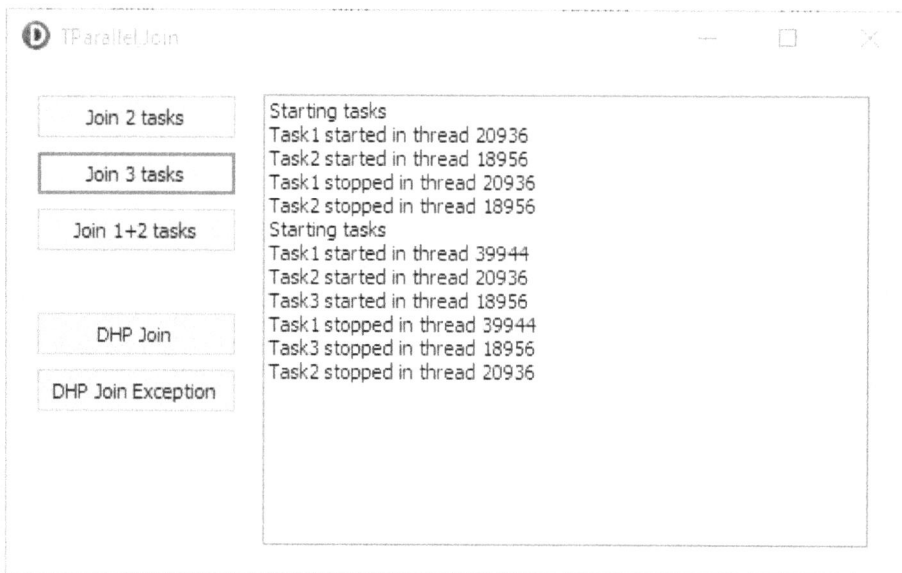

Figure 9.4 – Running multiple tasks with TParallel.Join

Join/Await

The *Join* pattern as implemented in the PPL does not implement any way to notify the main thread when the work is completed. To fix this, I have created a new Join implementation that supports executing a notification handler in the main thread, as with the *Async/Await* pattern.

As with the `TParallel` version, this `Join` accepts multiple `TProc` parameters. It, however, doesn't return an `ITask` interface but `IJoin`. This interface is similar to the `IAsync` interface used for the

Async/Await pattern. It implements two versions of the `Await` method. One is the same as in the `IAsync` version, while the other can be used to catch exceptions raised in tasks:

```
type
  IJoin = interface
            ['{ED4B4531-B233-4A02-A09A-13EE488FCCA3}']
    procedure Await(const awaitProc: TProc); overload;
    procedure Await(const awaitProc: TProc<Exception>); overload;
  end;


function Join(const async: array of TProc): IJoin;
```

A full implementation of the *Join/Await* pattern can be found in the `DHPThreading` unit. In this book, I'll only show the interesting parts.

A trivial function `Join` creates and returns an instance of a `TJoin` class. The `TJoin` constructor makes a copy of Join's `async` parameter and nothing more.

Interesting things start happening in `Await`. It stores `FAwaitProc` away, initializes a counter of running tasks, `FNumRunning`, makes sure that it cannot be destroyed by using the `FSelf := nil` trick, and starts all tasks. The helper function, `GetAsyncProc`, is used to work around the problem of capturing a `for` loop variable:

```
function TJoin.GetAsyncProc(i: integer): TProc;
begin
  Result :=
    procedure
    begin
      Run(FAsyncProc[i]);
    end;
end;

procedure TJoin.Await(const awaitProc: TProc<Exception>);
var
  i: integer;
begin
  FAwaitProc := awaitProc;
  FNumRunning := Length(FAsyncProc);
  FSelf := Self;
  for i := Low(FAsyncProc) to High(FAsyncProc) do
    TTask.Run(GetAsyncProc(i));
end;
```

The main part of the *Join/Await* implementation is the Run method, which is started once for each task passed to the `Join` call. First, it calls the `asyncProc` parameter (executing your task in the process) and catches all exceptions. Any exception will be passed to the `AppendException` function, which I'll cover in a moment.

After that, the code decrements the number of running tasks in a thread-safe manner. If the counter has dropped to *0* (as `TInterlocked` is used, this can only happen in one task), the code queues an anonymous method that calls the *Await* handler and drops the reference to itself so that the `TJoin` object can be destroyed. The `CreateAggregateException` function will be shown in a moment:

```
procedure TJoin.Run(const asyncProc: TProc);
begin
  try
    asyncProc();
  except
    on E: Exception do
      AppendException(AcquireExceptionObject as Exception);
  end;

  if TInterlocked.Decrement(FNumRunning) = 0 then
    TThread.Queue(nil,
      procedure
      begin
        FAwaitProc(CreateAggregateException);
        FSelf := nil;
      end);
end;
```

The `AppendException` function adds the caught exception object to an `FExceptions` array. As it may be called from multiple threads at once (if more than one task fails simultaneously), locking is used to synchronize access to `FExceptions`:

```
procedure TJoin.AppendException(E: Exception);
begin
  TMonitor.Enter(Self);
  try
    SetLength(FExceptions, Length(FExceptions) + 1);
    FExceptions[High(FExceptions)] := e;
  finally
    TMonitor.Exit(Self);
  end;
end;
```

The `CreateAggregateException` method is only called when all tasks have been executed and so doesn't need any locking. If the `FExceptions` array is empty, the function returns `nil`. Otherwise, it creates an `EAggregateException` object containing all caught exceptions:

```
function TJoin.CreateAggregateException: Exception;
begin
  if Length(FExceptions) = 0 then
    Result := nil
  else
    Result := EAggregateException.Create(FExceptions);
end;
```

What about the second `Await` overload? It simply calls `Await` that we've already seen. If any of the tasks raises an exception, it will be reraised inside the `Await` code. Otherwise, your `awaitProc` will be called:

```
procedure TJoin.Await(const awaitProc: TProc);
begin
  Await(
    procedure (E: Exception)
    begin
      if assigned(E) then
        raise E;
      awaitProc();
    end);
end;
```

The **DHP Join Exception** button on the `ParallelJoin` demo shows how to use the exception-handling mechanism. The code starts three tasks. One works as in the previous `Join` examples, while the other two tasks raise exceptions (only one of those tasks is shown as follows):

```
procedure TfrmParallelJoin.Task2E;
begin
  QueueLog('Task2E started in thread ' +
    TThread.Current.ThreadID.ToString);
  Sleep(1000);
  QueueLog('Task2E raising exception in thread ' +
    TThread.Current.ThreadID.ToString);
  raise Exception.Create('Task2 exception');
end;

procedure TfrmParallelJoin.btnDHPJoinExcClick(Sender: TObject);
begin
  ListBox1.Items.Add('Starting tasks');
```

```
    Join([Task1, Task2E, Task3E]).Await(TasksStopped);
  end;
```

The termination handler, `TasksStopped`, is called in the main thread after all tasks have finished their job. If the E parameter is `nil`, there was no exception—but we already know that in the demo program, this is not so.

When at least one exception is raised (and not handled) in a task, the E parameter will contain an `EAggregateException` object. The code in `TaskStopped` iterates over all inner exceptions stored in that object and logs them on the screen:

```
procedure TfrmParallelJoin.TasksStopped(E: Exception);
var
  i: Integer;
begin
  ListBox1.Items.Add('Tasks stopped');
  if assigned(E) then
    for i := 0 to EAggregateException(E).Count - 1 do
      ListBox1.Items.Add('Task raised exception: ' +
        EAggregateException(E)[i].Message);
end;
```

The demo program also contains a **DHP Join** button, which starts the same three tasks that we used to test the `TParallel.Join` implementation. In addition to the `TParallel` demo, the code logs the moment at which all tasks are completed:

```
procedure TfrmParallelJoin.btnDHPJoinClick(
  Sender: TObject);
begin
  ListBox1.Items.Add('Starting tasks');
  Join([Task1, Task2, Task3]).Await(
    procedure
    begin
      ListBox1.Items.Add('Tasks stopped');
    end);
end;
```

In the following screenshot, we can see how with this implementation, the notification handler is executed after all tasks have completed work. The screenshot also shows an output of the exception-handling code:

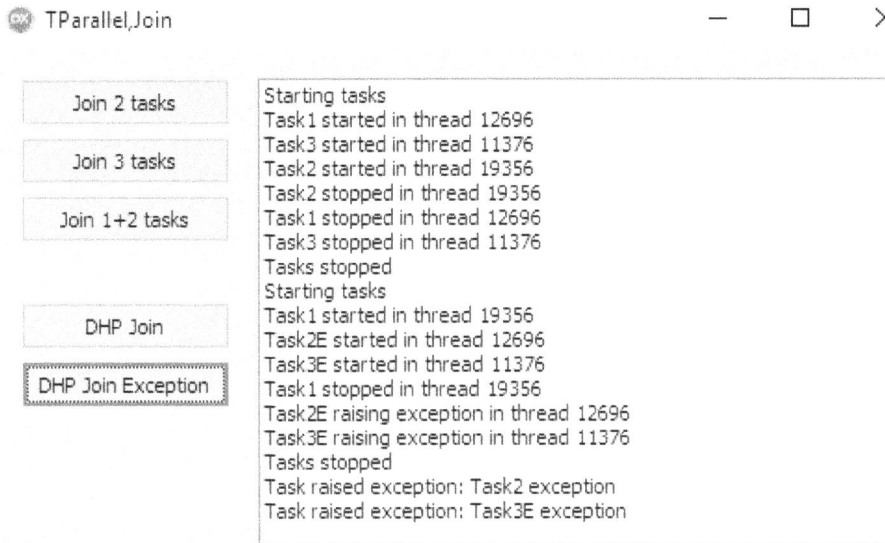

Figure 9.5 – Running tasks with the Join/Await pattern

Future

A significant category of operations that we would like to execute in a background thread can be summarized with the words "do something and return a result." These operations are represented by a pattern called *Future*.

A *Future* always wraps some function. In Delphi's Parallel Programming Library, a Future is represented by an IFuture<T> interface, where T is the data type returned by the function. This interface is created by calling the TTask.Future<T> function. IFuture<T> is derived from ITask, so it supports all ITask methods and properties, such as Status, Cancel, Wait, and so on.

The ParallelFuture demo shows two ways of using IFuture<T>. In the first scenario, the code creates a Future, does some other work, and then asks the Future about the result of the calculation. If the background calculation is not yet finished, the code waits until the result is ready. That could block the user interface, so this approach should only be used if you know that the Future will be calculated relatively quickly.

In the demo, the **Create Future** button starts the calculation, while the **Get value** button reads the result of the calculation. The calculation is performed by the CountPrimes function, which counts prime numbers from *2* to *5,000,000*:

```
function CountPrimes: integer;
var
  i: Integer;
```

```
begin
  Result := 0;
  for i := 2 to 5000000 do
    if IsPrime(i) then
      Inc(Result);
end;
```

A Future is created simply by calling `TTask.Future<integer>`, passing in `CountPrimes` as a parameter:

```
procedure TfrmFuture.btnFutureClick(Sender: TObject);
begin
  FFuture := TTask.Future<integer>(CountPrimes);
end;
```

Getting a result of a Future—if you don't mind waiting until the result is calculated—is as simple as accessing its `Value` property. After that, the code destroys the Future interface by assigning it a `nil` value:

```
procedure TfrmFuture.btnGetValueClick(Sender: TObject);
begin
  ListBox1.Items.Add('Result = ' + FFuture.Value.ToString);
  FFuture := nil;
end;
```

The second approach is more appropriate for general use. First, it calculates the resulting value and then notifies the main thread that the value was calculated. We can use any notification mechanism to do that—Windows messages, `TThread.Queue`, polling, and so on.

Clicking on the **Create Future 2** button in the demo program starts the `btnFuture2Click` method, which uses this approach. As before, the code begins by calling the `TTask.Future<integer>` function, but this time, its argument is an anonymous function returning `Integer`. When a background thread executes this anonymous function, it first calls our real calculation—`CountPrimes`—and then queues a method, `ReportFuture`, to be executed in the main thread:

```
procedure TfrmFuture.btnFuture2Click(Sender: TObject);
begin
  FFuture := TTask.Future<integer>(
    function: Integer
    begin
      Result := CountPrimes;
      TThread.Queue(nil, ReportFuture);
    end);
end;
```

ReportFuture reads the Value property to access the result of the calculation:

```
procedure TfrmFuture.ReportFuture;
begin
  ListBox1.Items.Add('Result = ' + FFuture.Value.ToString);
  FFuture := nil;
end;
```

The big difference between the two approaches is that, in the first case, reading from FFuture. Value may block until the result is calculated. In the second case, the result is guaranteed to be ready at that point, and the value is returned immediately.

If a Future calculation raises an exception, this exception is caught by TTask and will be reraised when the main thread accesses the Value property. You can use the technique described in the *Exceptions in tasks* section or—as I would wholeheartedly recommend—catch and handle all exceptions in the background task itself.

Parallel for

The Parallel Programming Library implements only one pattern that I haven't talked about yet—*Parallel for*, a multithreaded version of a for loop. This pattern allows for very simple parallelization of loops, but this simplicity can also get you into trouble.

When you use *Parallel for*, you should always be very careful that you don't run into a data sharing-trap.

For comparison purposes, the ParallelFor demo implements a normal for loop (shown as follows), which goes from 2 to 10 million, counts all prime numbers in that range, and logs the result:

```
const
  CHighestNumber = 10000000;

procedure TbtnParallelFor.btnForClick(Sender: TObject);
var
  count: Integer;
  i: Integer;
  sw: TStopwatch;
begin
  sw := TStopwatch.StartNew;
  count := 0;

  for i := 2 to CHighestNumber do
    if IsPrime(i) then
      Inc(count);

  sw.Stop;
```

```
  ListBox1.Items.Add('For: ' + count.ToString + ' primes. '
    + 'Total time: ' + sw.ElapsedMilliseconds.ToString);
end;
```

To change this `for` loop into a Parallel for, we just have to call `TParallel.For` instead of using `for`, and pass in both range boundaries and workloads as parameters. A **Parallel for - bad** button in the demo program activates the `btnParallelForBadClick` method, which implements this approach—with a small (intentional) mistake.

The three parameters passed to `TParallel.For` in the following code are the lower bound for the iteration (2), the higher bound (`CHighestNumber`), and a worker method (an anonymous method). A worker method must accept one `integer` parameter. This parameter represents the loop variable (`i`) of a standard `for` statement:

```
procedure TbtnParallelFor.btnParalleForBadClick(
  Sender: TObject);
var
  count: Integer;
  i: Integer;
  sw: TStopwatch;
begin
  sw := TStopwatch.StartNew;
  count := 0;

  TParallel.For(2, CHighestNumber,
    procedure (i: integer)
    begin
      if IsPrime(i) then
        Inc(count); // intentional mistake!
    end);

  sw.Stop;
  ListBox1.Items.Add('Parallel for - bad: ' +
    count.ToString + ' primes. ' + 'Total time: ' +
    sw.ElapsedMilliseconds.ToString);
end;
```

This code demonstrates the classical data-sharing problem. It is so easy to replace `for` with `TParallel.For` that sometimes you forget that the body of this iteration executes in multiple parallel copies and that all access to shared variables (`count` in this example) must be protected.

We could use locking to solve that problem, but we've already seen that interlocked operations are faster. The code activated by the **Parallel for - good** button therefore uses `TInterlocked.Increment` instead of `Inc` to fix the problem.

The following code fragment shows the Parallel for loop from the `btnParallelForClick` method. All other parts of the method are the same as in the preceding code fragment:

```
TParallel.For(2, CHighestNumber,
  procedure (i: integer)
  begin
    if IsPrime(i) then
      TInterlocked.Increment(count);
  end);
```

The following screenshot compares all three versions. We can clearly see that the parallel version is about four times faster than the single-threaded code. The test machine had four physical cores, so this is not surprising. We can also see that the "bad" code produces a different value each time. The code runs without any other problems, and that's why these errors are usually hard to detect:

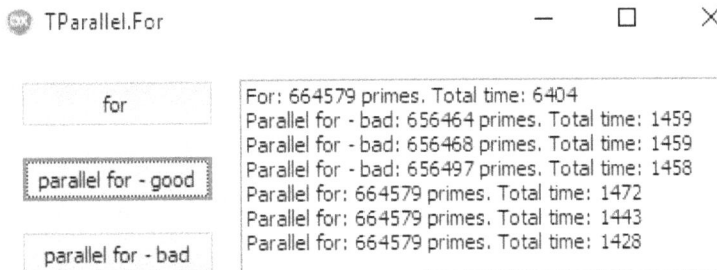

Figure 9.6 – Calculating the number of primes with a parallel for loop

What the screenshot doesn't show is that `TParallel.For` stops the main thread until all the work is done. This is completely understandable. A *Parallel for* is meant as an in-place replacement for a `for` loop, and, with this approach, you can sometimes speed up a program with very little work.

If this behavior blocks the user interface for too long, you can still wrap a *Parallel for* pattern inside a *Future* pattern. A *Future* will take care of a non-blocking execution, while *Parallel for* will speed up the program.

> **Note**
> Alternatively, you can wrap *Parallel for* in an *Async/Await* pattern.

The **Async Parallel for** button in the demo program does exactly that. When you click it, the code starts a computation function, `ParallelCountPrime` in a Future. This function will return the number of primes in the tested range:

```
procedure TbtnParallelFor.btnAsyncParallelForClick(
  Sender: TObject);
```

```
begin
  FStopWatch := TStopwatch.StartNew;
  FParallelFuture := TTask.Future<Integer>(ParallelCountPrimes);
end;
```

The `ParallelCountPrimes` function is executed in a background thread by the Future pattern. It uses `TParallel.For` and `TInterlocked.Increment` to quickly and correctly count prime numbers. After that, it assigns the `count` variable to the Result of the function so that the main thread will be able to read it via the `Value` property. At the end, the code queues an anonymous function that will log the result in the main thread and destroy the Future interface:

```
function TbtnParallelFor.ParallelCountPrimes: integer;
var
  count: Integer;
begin
  count := 0;
  TParallel.For(2, CHighestNumber,
    procedure (i: integer)
    begin
      if IsPrime(i) then
        TInterlocked.Increment(count);
    end);

  Result := count;

  TThread.Queue(nil,
    procedure
    begin
      FStopwatch.Stop;
      ListBox1.Items.Add('Async parallel for: ' +
        FParallelFuture.Value.ToString +
        ' primes. Total time: ' +
        FStopwatch.ElapsedMilliseconds.ToString);
      FParallelFuture := nil;
    end);
end;
```

This code may look suspicious to careful readers. Is there a race condition between the assignment to `FParallelFuture` in the `btnAsyncParallelForClick` event handler and at the end of `ParallelCountPrimes`? What if `FParallelFuture := nil` executes first and only then the event handler assigns the value to that global field?

I'll admit that this code is "smelly" and that you should use such tricks with care (and maybe document them), but in this case, the code cannot fail. Even if the future is created and executed before the `FParallelFuture` field is initialized, it has to queue the final, cleanup part to the main thread for execution via `TThread.Queue`. As such queued pieces of code are only executed in very specific parts of the VCL library, `FParallelFuture` cannot be cleared happen before `btnAsyncParallelForClick` completes its work and returns control to the VCL.

Exceptions in parallel for

If your parallel iteration code raises an exception, it will be caught and reraised in the main thread. In theory, you should be able to wrap `TParallel.For` with a `try .. except` block and process the exception there. In practice, however, the implementation of exception catching was buggy in older versions of Delphi, and we were likely to run into trouble that way.

A click on the **Parallel for exception** button runs code, shown as follows, that demonstrates the exception-catching technique. It starts multiple background tasks to process a small range of integers (from 1 to 10) and raises an exception in each task. When Parallel for catches an exception, it tries to shut down the background operation, but during that time, a different thread may have already raised the exception too. As multithreading is unpredictable by nature, a different number of exceptions will be raised each time.

The Parallel for pattern collects all exceptions and joins them in `EAggregateException`. The following code iterates over all exceptions in this object and logs them on screen:

```
procedure TbtnParallelFor.btnParallelForExceptionClick(
  Sender: TObject);
var
  i: Integer;
begin
  ListBox1.Items.Add('---');
  try
    TParallel.For(1, 10,
      procedure (i: integer)
      begin
        Sleep(100);
        raise Exception.Create('Exception in thread ' +
          TThread.Current.ThreadID.ToString);
      end);
  except
    on E: EAggregateException do
      for i := 0 to E.Count - 1 do
        if not assigned(E) then
          ListBox1.Items.Add(i.ToString + ': nil')
        else
```

```
            ListBox1.Items.Add(i.ToString + ': ' +
              E[i].ClassName + ': ' + E[i].Message);
      on E: Exception do
        ListBox1.Items.Add(E.ClassName + ': ' + E.Message);
    end;
  end;
```

There is a problem with the aggregate exception collection in parallel for that sometimes causes `nil` entries to be stored in the `E[]` array. The preceding code has a special `if` clause that checks whether an inner exception, `E[i]`, is assigned at all. As you can see from the following screenshot, this is necessary because sometimes an inner exception is not correctly captured. In such cases, `E[i]` contains `nil`:

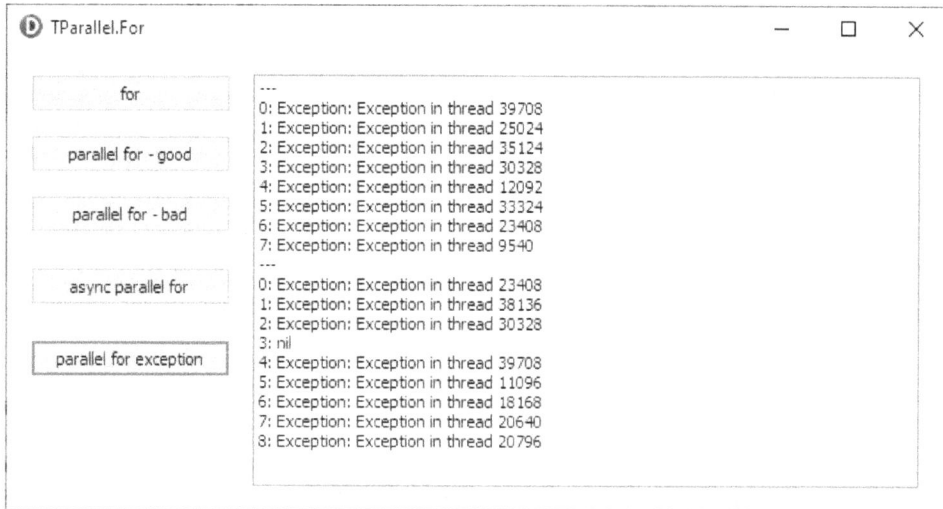

Figure 9.7 – Logging exceptions raised in parallel for tasks

Pipelines

So far, I have discussed two different categories of problems that can be solved using patterns. In the first category, there are long operations that we just want to push into the background. We are happy with the execution time, but we would like to keep the user interface responsive, so we need to execute them in a background thread. *Async/Await* and *Future* are the tools for this occasion.

In the second category, there are problems that can be split into parts that are independent or mostly independent of each other. We can use *Join*, *Join/Await*, or *Parallel for* to parallelize them and implement the "mostly independent" part with locking or interlocked operations.

This, however, doesn't even nearly cover all use cases. There's at least one big category left that is not trivial to parallelize with the Parallel Programming Library tools. I'm talking about problems that are hard to split into parts but where operations executed on a given data item (an atomic part of the input data) can be split into stages.

To work on these problems, we need a pattern that can transfer data from one stage to another. In Thread Building Blocks, such a pattern is called the *flow graph*, while OmniThreadLibrary implements the less powerful *pipeline*.

In this *data flow* approach to parallelism, the task is split into independent processing phases (called **stages**), which work on partial problems. Each stage does some operation on a part of the input data and transfers the result to the next stage via a *pipe* (which is just a thread-safe implementation of a queue). If each stage depends only on local data (no data sharing) and data received through a thread-safe pipe, we have a system that is automatically thread-safe and requires no additional locking and data protection.

As each stage is independent of another, each can run in its own task. The whole system therefore runs on as many parallel threads are there are stages – or even more! We can implement some of the stages as multiple parallel tasks and that way increase the parallelism of the whole solution!

> **Real-world example**
> When thinking about pipelines, you can imagine a big factory with a robotized assembly line. Each robot represents a *stage* that takes a part and does some small operation on it. Then, the assembly line (the *pipe*) takes the part and carries it to the next robot (into the next stage).

In this chapter, we'll see how to write a web traveling "spider" with a pipeline pattern. In the next chapter, we'll return to the topic with a pipeline implementation of a file processing utility.

Web spider

To show a more complex example that you can take and adapt to your own needs, the `ParallelPipeline` demo implements a simple web spider. This code accepts a URL and retrieves all of the pages on that website.

> **Note**
> This project is not meant to be a fully functional web spider application, but a relatively simple demo. It may not work correctly on all sites. It may also cause you to be temporarily locked out of accessing the site that you are trying to crawl as it can generate a large number of HTTP requests, which may trigger security measures on a website.

This example is more complicated than the pipeline concept I have discussed so far. It extracts data (URLs) from the retrieved pages and feeds them back into the pipeline so that new URLs can be processed. As we'll see later, this makes it hard for the pipeline to determine when the work has finished. Still, a self-feeding pipeline is a useful architecture and you should know how to write one.

> **Note**
>
> If you know my book *Hands-On Design Patterns with Delphi*, you'll notice that this example is very similar to the pipeline example in that book. This example solves the same problem but is based on new pipeline wrappers that allow you to solve the same problem while writing less code than before. This version also fixes some problems with the old web spider and makes it more useful in a practical application.

The web spider pipeline is fully implemented in the `PipelineWebSpider` unit, which provides a very simple interface, as shown here in shortened form:

```
type
  TWebSpider = class
  public
    procedure Start(const baseUrl: string);
    procedure Stop;
    property OnPageProcessed: TProc<string>
      read FOnPageProcessed write FOnPageProcessed;
    property OnFinished: TProc
      read FOnFinished write FOnFinished;
  end;
```

We can only start and stop the pipeline, nothing more. The pipeline triggers the `OnPageProcessed` event every time one web page has been processed. It also calls the `OnFinished` event when the work is done. (We can, of course, always stop the process by calling the `Stop` method, even if the website hasn't been fully crawled yet.)

The main form uses this class in a fairly straightforward manner, as we can see in the following code:

```
procedure TfrmPipeline.btnStartClick(Sender: TObject);
begin
  if not assigned(FWebSpider) then
  begin
    FWebSpider := TWebSpider.Create;
    FWebSpider.OnPageProcessed :=
      procedure (url: string)
      begin
        lbLog.Items.Add(url);
      end;
```

```
    FWebSpider.OnFinished := StopSpider;
    FWebSpider.Start(inpUrl.Text);
    inpUrl.Enabled := false;
    btnStart.Caption := 'Stop';
  end
  else
    StopSpider;
end;

procedure TfrmPipeline.StopSpider;
begin
  FWebSpider.Stop;
  FreeAndNil(FWebSpider);
  inpUrl.Enabled := true;
  btnStart.Caption := 'Start';
end;
```

To start the process, the code creates a TWebSpider object, sets up event handlers, and calls the Start method, passing in the initial URL. If the user later clicks the same button, the web spider will be stopped and destroyed – as it will be if the OnFinished handler is called.

The process works in three stages:

1.   The first stage receives a URL. It checks whether this URL has already been processed. If not, it passes the URL to the second stage. The first stage functions as a *filter*.

2.   The second stage receives a URL and retrieves its contents. If the operation is successful, both the URL and returned HTML are passed to the third stage. The second stage is a *downloader*. As the download is a slow process, we'll start multiple copies of this stage to speed up the process.

3.   The third stage parses the HTML, extracts all URLs that are referenced by that page, and sends them back to the first stage. (The first stage then makes sure that the process doesn't cycle indefinitely.) The third stage functions as a *parser*.

The internal structure of the TWebSpider pipeline is shown in the next picture:

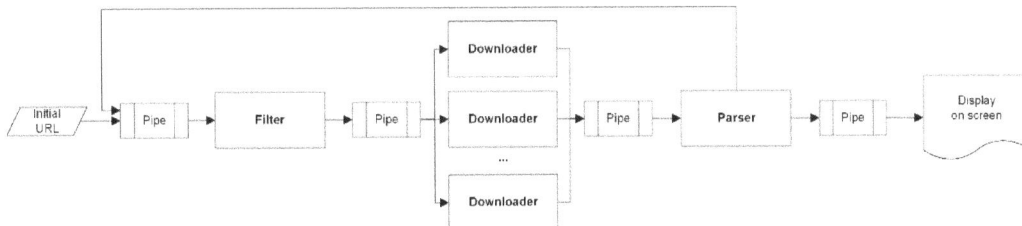

Figure 9.8 – Pipeline setup in the web spider demo

The whole process is set up in the `Start` method, as follows:

```
procedure TWebSpider.Start(const baseUrl: string);
var
  i: integer;
begin
  FPipeline := TPipeline<string, string>.Create(
    10000, 100,
    procedure
    var
      url: string;
    begin
      if assigned(OnPageProcessed) then
        while FPipeline.Output.Read(url) do
          OnPageProcessed(url);
    end);

  FHttpGetInput := FPipeline.MakePipe<string>(100);
  FHtmlParseInput := FPipeline.MakePipe<THttpPage>(10);

  FPipeline.Stage('Unique filter',
    procedure
    begin
      Asy_UniqueFilter(baseUrl, FPipeline.Input,
        FHttpGetInput);
    end);

  for i := 1 to TThread.ProcessorCount do
    FPipeline.Stage<string,THttpPage>(
      'Http get #' + i.ToString,
      FHttpGetInput, FHtmlParseInput, Asy_HttpGet);

  FPipeline.Stage<THttpPage,string>('Html parser',
    FHtmlParseInput, FPipeline.Input, Asy_HtmlParse);

  FPageCount := 1;
  FPipeline.Input.Write('');
end;
```

First, the code creates the main `TPipeline` object and specifies that both the input to the pipeline and the final output from the pipeline (the input and output pipe, represented by the `Input` and `Output` property of the class) will be `string` types. The `TPipeline` class is implemented in the `DHPThreading` unit, and in this book, I'll only show selected parts of its implementation. For the full picture, see the source code.

The two numeric parameters to `TPipeline.Create` specify the size of the input and output pipes. If the pipe becomes full at any point, the task that tries to write to it simply stops until the next stage reads from the pipe. That way, the pipeline becomes self-regulating. If a stage is too fast, it fills up its output pipe and stops, which frees the CPU so that it can catch up.

> **Warning**
>
> The problem with self-feeding pipelines and size-limited pipes (as they are in the PPL implementation) is that the pipeline may put itself in a deadlock, where each stage tries to write to a full output pipe. That's why the input pipe size is set to a high number in this demo. OmniThreadLibrary offers a better solution with `TOmniBlockingCollection`, which we'll explore in the next chapter.

The last parameter is an optional callback that is called in *the main thread* every time any data is written to the pipeline's output pipe. It reads the data from the output pipe (the `Read` call will fail, returning `False` if the pipe is empty) and calls the `OnPageProcessed` event handler to display the data on the screen.

Next, the code creates two pipes; `FHttpGetInput` connects the *Filter* and the *Downloader* stages and stores `string` elements while `FHtmlParseInput` connects the *Downloader* and the *Parser* stages and contains `THttpPage` records. `THttpPage` is defined as an internal type inside the `TWebSpider` class.

> **Note**
>
> Here, I have made the `FHtmlParseInput` pipe smaller than other pipes. The `THttpPage` record contains the full downloaded web page, which can potentially be quite large and we don't want too many big pages to queue up for processing.

The output of a `MakePipe<T>` function is a `TPipe<T>` object, which is just a simple wrapper for `TThreadedQueue<T>`, and is defined in the `DHPThreading` unit. The `TPipeline` object owns all pipes and destroys them automatically so you don't have to worry about that.

Next, the code sets up the processing stages. The *Filter* stage requires an additional initialization parameter – a base URL – and is created with a variant of the `Stage` method, which accepts a thread name and an anonymous function that works as a background thread, as shown here:

```
FPipeline.Stage('Unique filter',
  procedure
  begin
    Asy_UniqueFilter(baseUrl, FPipeline.Input,
      FHttpGetInput);
  end);
```

As the download is a relatively slow operation that requires almost no CPU, the code starts as many copies of the *Downloader* stage as there are CPU cores. This value could probably be even higher, but this tuning should be done with care and with lots of testing.

The multiple copies of the *Downloader* stage and the single *Parser* stage are created with the Stage<TIn, TOut> method, which accepts the name of the thread, the input pipe (storing elements TIn), output thread (storing elements TOut), and the stage worker method, as shown here:

```
FPipeline.Stage<THttpPage,string>('Html parser',
  FHtmlParseInput, FPipeline.Input, Asy_HtmlParse);
```

Each call to a Stage method creates and runs a task that runs a stage worker method. These tasks are managed by the TPipeline object so you don't have to start or stop them explicitly. Just for an example, here is one of the Stage overloads:

```
procedure TPipeline<TIn, TOut>.Stage(const name: string;
  const stage: TProc);
begin
  FTasks.Add(TTask.Run(
    procedure
    begin
      TThread.NameThreadForDebugging(name);
      stage();
    end),
    FThreadPool);
end;
```

Each task is created in an internal thread pool (managed by the TPipeline object). See the end of the *The filter stage* section for why this is necessary.

At the end, the code sets the number of unprocessed work items in the pipeline to 1 and pushes the initial item (an empty string indicating the starting baseUrl) to the input of the first stage. We'll see how FPageCount is used to stop the pipeline when all the work has been done later.

To shut down the pipeline, the Stop method tells the pipeline to shut down and destroys the FPipeline object:

```
procedure TWebSpider.Stop;
begin
  FPipeline.ShutDown;
  FreeAndNil(FPipeline);
end;
```

Now that we've seen how the pipeline is started and stopped, let's take a look at the separate stages.

The filter stage

The filter stage creates a `TStringList` object, which holds the names of all the URLs that have already been processed. For every URL received as input, it checks whether the URL is already in the list. If so, the URL is thrown away. The code also checks whether the URL belongs to the site we are crawling. If not, it is also thrown away. (We certainly don't want to crawl the entire internet!) If all of the tests pass, the URL is added to the list of already processed links and is sent to the output queue.

The `Asy_UniqueFilter` method implements the filter stage, as follows:

```
procedure TWebSpider.Asy_UniqueFilter(baseUrl: string;
  inQueue, outQueue: TPipe<string>);
var
  baseUrl2: string;
  visitedPages: TStringList;
begin
  visitedPages := TStringList.Create;
  try
    visitedPages.Sorted := true;
    if not (baseUrl.StartsWith('https://')
            or baseUrl.StartsWith('http://'))
    then
      baseUrl := 'http://' + baseUrl;
    if baseUrl.StartsWith('http://') then
      baseUrl2 := baseUrl.Replace('http://', 'https://')
    else
      baseUrl2 := baseUrl.Replace('https://', 'http://');

    inQueue.Process<string>(outQueue,
      procedure (url: string)
      begin
        if url.IndexOf(':') < 0 then
          url := baseUrl + url;
        if (url.StartsWith(baseUrl)
            or url.StartsWith(baseUrl2))
          and (visitedPages.IndexOf(url) < 0) then
        begin
          visitedPages.Add(url);
          outQueue.Write(url);
        end
        else if TInterlocked.Decrement(FPageCount) = 0 then
          NotifyFinished;
      end);
  finally
```

```
      FreeAndNil(visitedPages);
    end;
  end;
```

The code first takes care of possible redirects from an `http://` address to an `https://` address (or vice versa). I wanted these URLs to be both treated as part of the website that is crawled. The code sets up two strings, `baseUrl` and `baseUrl2`, which are later used for testing whether a URL belongs to the website we are crawling.

We also keep the `visitedPages` list sorted for faster searching. A sorted list finds data in *O(log n)* time, compared to *O(n)* time in an unsorted list, remember?

Then, the code calls the helper function, `inQueue.Process<TOut>`. This helper accepts two parameters, an output pipe of type `Pipe<TOut>` and a worker function, which will be called for each input item. The `Process` method reads from the input pipe until the queue is shut down and, for each element, calls the worker function. At the end, it shuts down the output pipe (which will, in turn, shut down the next pipe and so on).

The implementation of the `Process` method is shown next:

```
procedure TPipe<T>.Process<TOut>(outQueue: TPipe<TOut>;
  stageLoop: TProc<T>);
var
  item: T;
begin
  while Read(item) do
  begin
    if ShutDown then
      break; //while

    stageLoop(item);
  end;
  outQueue.DoShutDown;
end;
```

The most important part of this stage is the following lines in the internal worker function:

```
if (url.StartsWith(baseUrl) or url.StartsWith(baseUrl2))
    and (visitedPages.IndexOf(url) < 0) then
begin
  visitedPages.Add(url);
  outQueue.PushItem(url);
end
else if TInterlocked.Decrement(FPageCount) = 0 then
  NotifyFinished;
```

If the URL belongs to the website (`StartsWith`) and was not processed before (`IndexOf`), it is passed to the output stage. Otherwise, we are done processing this URL and it can be thrown away. As we do so, however, we must decrement the shared `FPageCount` counter, which holds the number of processing units in the pipeline. If it falls to zero, the pipeline is now empty and can be stopped. The code calls `NotifyFinished` to signal this, as shown here:

```
procedure TWebSpider.NotifyFinished;
begin
  TThread.Queue(nil,
    procedure
    begin
      if assigned(OnFinished) then
        OnFinished();
    end);
end;
```

This code uses the messaging approach we explored previously to execute the `OnFinished` handler in the main thread. This approach of queuing the event handlers to the main thread is also the reason why we must use a custom thread pool in the web spider code.

The RTL makes sure that all queued anonymous methods are processed and executed in the `TThread.WaitFor` call. In the web spider implementation, however, we use tasks and not threads. And with tasks, the threads merely go back to a thread pool when a task is completed. No `WaitFor` is called.

If we used the common thread pool, the worker thread would not be destroyed after a task has completed its job. Although `TWebSpider` would then be destroyed, the worker `TThread` would not be and the queued message would still be delivered, which could cause all kinds of problems.

As we are using a separate thread pool, however, all of the threads are destroyed when the thread pool is destroyed, which makes sure that all the queued anonymous methods are processed before `TWebSpider` is destroyed.

The downloader stage

The downloader stage uses a `THTTPClient` object to download the web page. This task is implemented in the `Asy_HttpGet` method, as follows:

```
procedure TWebSpider.Asy_HttpGet(
  inQueue: TThreadedQueue<string>;
  outQueue: TThreadedQueue<THttpPage>);
var
  httpClient: THTTPClient;
  response: IHTTPResponse;
  url: string;
begin
```

```
      httpClient := THTTPClient.Create;
      try
        inQueue.Process<THttpPage>(outQueue,
          procedure (const url: string)
          begin
            try
              response := httpClient.Get(url);
            except
              if TInterlocked.Decrement(FPageCount) = 0 then
                NotifyFinished;
            end;

            if (response.StatusCode div 100) = 2 then
              outQueue.Write(THttpPage.Create(url, response))
            else if TInterlocked.Decrement(FPageCount) = 0 then
              NotifyFinished;
          end);
      finally
        FreeAndNil(httpClient);
      end;
    end;
```

As the `Start` method creates multiple tasks, multiple `Asy_HttpGet` methods are executed in parallel. This is fine, as each uses its own local variables and there is no conflict between them.

As in the previous example, the `inQueue.Process<THttpPage>` call (`ThttpPage`, as `outQueue` in this method uses this type) drives the internal data processing loop.

If the web download fails (raises an exception or the response code is not 2xx), the code throws the input URL away. As before, the code then decrements the shared `FPageCount` counter in a thread-safe manner and notifies the owner about job completion if the counter drops to zero.

If everything is OK, the code pushes a `THttpPage` record to the output queue. This record is defined as follows:

```
  THttpPage = TPair<string,IHTTPResponse>;
```

The parser stage

The parser stage parses the returned HTML and extracts all hyperlinks (`<a>` tags). Delphi does not contain an HTML parser in a standard distribution, and as I wanted to remove any dependencies on third-party HTML parsers, I cheated a bit and used regular expressions to detect `<a href="..." >` and `<a href='...' >` strings in the page.

The parser doesn't check whether the returned content is actually in HTML format or not. It merely scans the result for a simple regular expression.

> **Warning**
>
> In production code, you should never parse HTML with a regular expression. This doesn't work. Regular expressions are too limited to parse HTML.

The parser stage sets up the regular expression parser and then loops through all the data in the input queue. Each input is parsed and all detected URLs are sent back to the first stage for filtering. This task is implemented in the `Asy_HtmlParse` method, as follows:

```
procedure TWebSpider.Asy_HtmlParse(
  inQueue: TThreadedQueue<THttpPage>;
  outQueue: TThreadedQueue<string>);
var
  hrefMatch: TRegEx;
  match: TMatch;
  page: THttpPage;
begin
  hrefMatch := TRegEx.Create(
                '<a href=["''](.*?)["''].*?>',
                [roIgnoreCase, roMultiLine]);

  inQueue.Process<string>(outQueue,
    procedure (const page: THttpPage)
    begin
      try
        match := hrefMatch.Match(
                  page.Value.ContentAsString);
        while match.Success do
        begin
          if outQueue.ShutDown then
            break; //while;
          TInterlocked.Increment(FPageCount);
          outQueue.Write(match.Groups[1].Value
                        .Split(['#', '?'])[0]);
          match := match.NextMatch;
        end;
      except
      end;

      FPipeline.Output.Write(page.Key);
```

```
        if TInterlocked.Decrement(FPageCount) = 0 then
          NotifyFinished;
      end);
  end;
```

As before, the main loop is driven by the `inQueue.Process<string>` method. Each iteration of the loop parses one HTML page and finds all occurrences of the `<a href=...>` pattern.

Each detected pattern (each `<a href>` hyperlink) represents a new work unit. For each hyperlink, the code increments the number of work units in the pipeline (`FPageCount`) and then pushes the hyperlink into the output queue.

> **Tip**
>
> The `match.Groups[1].Value.Split(['#', '?'])[0]` trick takes the `match.Groups[1].Value` string, splits it into an array of substrings where both `'#'` and `'?'` are used as delimiters, and takes the first element from this array. This effectively trims the string on the first occurrence of a `'#'` or `'?'` character.

After the input is parsed, the processing of this URL is done. The code therefore decrements the number of processing units and calls the `OnFinished` event handler if necessary.

The need to always maintain the correct state in the shared `FPageCount` counter is what makes writing self-feeding pipelines a complicated process. You must absolutely make sure that you increment the counter for each addition to the pipeline and decrement the counter when the processing unit is not needed anymore.

> **Tip**
>
> Always increment the shared counter before pushing data to the output queue.
>
> Always add data to the output queue before dropping the current processing unit.

This code writes all processed URLs to the `TPipeline.Output` pipe, which also calls the notification handler that was specified in the `TPipeline` constructor call. This handler, in turn, calls the `OnPageProcessed` event handler, which then displays the URL on the screen.

The following screenshot shows the web spider in action:

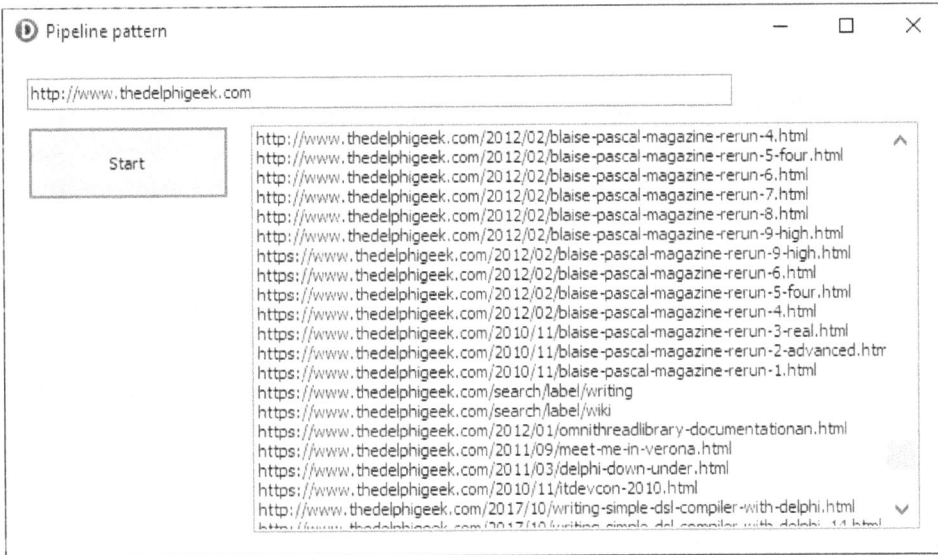

Figure 9.9 – The web spider browsing my blog

Admittedly, the current web spider doesn't do any useful work (except provide a complete example of a complicated pipeline). To convert it into a more useful application, however, you only need to add one stage, which will either save received HTML pages to a disk or store them in a database. This part is left to the reader as an exercise.

Summary

In this chapter, we looked at the high-level multithreaded patterns that are supported (either directly or with a few simple wrappers) in Delphi's Parallel Programming Library framework.

The chapter opened with a discussion of tasks and patterns—what they are and how they can be used to simplify multithreaded programming. As a bonus, I threw in a short treatise about variable capturing, which focused on only one problematic part—capturing a loop variable.

Then, we looked at how we can use tasks to split a loop into multiple parallel loops. We saw that there's quite some work involved, particularly around task creation and setup. On the way, we also learned about the thread-pooling concept.

The last part of the very long section on tasks discussed exception handling. We learned that it is quite hard to correctly capture in the main thread exceptions that were raised inside a task and that it is better to catch all task exceptions explicitly in the code.

After that, we finally moved from the (merely acceptable) tasks to (fun and simple) patterns. For starters, I've shown how to implement a very simple *Async/Await* pattern, something that is missing from the Parallel Programming Library.

Next on the list was the *Join* pattern. This very simple tool sometimes allows for very efficient parallelization of completely independent pieces of code. I also implemented a more powerful and simpler *Join/Await*, which allows you to capture exceptions raised in tasks.

After that, we looked into the *Future* (with a capital F). This interesting and powerful pattern allows you to execute a function in a task, while simplifying accessing the function result in the main thread. The Delphi implementation of this pattern is quite flexible and can be used in different ways.

Next, I covered *Parallel For*. This pattern allows for very simple parallelization of a standard `for` loop. I cautioned you about the problems that can sprout from its use, and I added a short trick that demonstrated how to create a non-blocking Parallel for.

At the end, I showed a powerful parallel programming pattern – a pipeline – and used it to implement a web spider application. Admittedly, a PPL-based pipeline is a bit clumsy, and in the next chapter, when we take a look at the OmniThreadLibrary approach to patterns, we'll also see a simpler-to-use and more flexible pipeline.

10
More Parallel Patterns

After half of a chapter dedicated to parallel patterns, it is now time for another chapter about parallel patterns! In the previous chapter, we focused solely on parallel programming support included in Delphi, which we enhanced a little so that it was easier to use.

In this chapter, I'll look at parallel programming patterns that can be found in an open source parallel programming framework, OmniThreadLibrary. Although the library covers everything from thread-based to task-based multithreading, I will focus almost entirely on patterns and leave other parts for the interested among you to discover.

In this chapter, we will try to answer the following questions:

- How do we install and use OmniThreadLibrary?

- What are blocking collections and how can they be used?

- How does the OmniThreadLibrary implementation of Async, Join, and Future compare to the Delphi implementation?

- How can the Parallel Task pattern be used to process data?

- What is the Background Worker pattern and how is it used?

- How does OmniThreadLibrary's pipeline compare to the implementation from the previous chapter?

- How can we use Map and Timed Task in our applications?

Technical requirements

All the code in this chapter was written with Delphi 11.3 Alexandria. Most of the examples, however, can also be executed on Delphi XE and newer versions. You can find all the examples on GitHub: `https://github.com/PacktPublishing/Delphi-High-Performance---Second-Edition/tree/main/ch10`.

Using OmniThreadLibrary

OmniThreadLibrary is a multithreading library for Delphi, written mostly by the author of this book. It was originally written for Delphi 2007 – and it still supports that version, which is why it duplicates some functionality that was added to Delphi after the 2007 release. For example, there is a TOmniValue type that can store values of any type and which is a clear duplicate of Delphi's TValue, parallel patterns are based on custom task implementation, and so on. There is also quite some similarity between OmniThreadLibrary and the Parallel Programming Library. For example, they both implement similar *Future* and *Join* constructs. There are also some dissimilarities that I'll try to point out in this chapter.

OmniThreadLibrary is released under an OpenBSD license, which allows its free use in commercial applications. It supports VCL, console, and service operations on Windows but does not work on other operating systems.

> **Note**
>
> I will frequently shorten **OmniThreadLibrary** to **OTL** and **Parallel Programming Library** to **PPL**.

To install OTL, you can use the *GetIt* Delphi module (if your version of Delphi is new enough) and click **Install**. This will download the library, unpack it, and add it to the search path.

I would, however, recommend that you download the latest release from `https://github.com/gabr42/OmniThreadLibrary/releases/` or clone the repository from `https://github.com/gabr42/OmniThreadLibrary`. This way, you will know that you are using the latest version. After downloading/cloning, you just have to add the main OTL folder to the search path and you're ready to go.

> **Note**
>
> The code archive for this chapter contains a cut-down copy of OmniThreadLibrary (examples and demos are not included), so you can just open any demo for this chapter, press *F9*, and the code will compile.

OmniThreadLibrary implements two very distinct levels of multithreading support. So-called *low-level multithreading* is a wrapper for TThread that adds features such as thread pooling (before it was implemented in Delphi), communication, lock-free collections (stack and queue), and cancellation mechanisms. The big difference between TThread and OTL's IOmniTask is that with the former, you write a monolithic loop, just like how Windows programming worked in the days before Delphi, while with the latter, you create a system based around event handlers, just like how Windows programming works in Delphi. I will not talk about this anymore in this book, but you can learn all about it in the official documentation.

> **Note**
>
> To learn more about OmniThreadLibrary, read the book *Parallel Programming with OmniThreadLibrary*. It touches all areas of the library and also serves as documentation. There are also many blog articles about OTL on my blog, nicely collected into a list, that can be accessed here: `http://www.omnithreadlibrary.com/documentation/#book`.

The other OTL approach to multithreading is called *high-level multithreading*, which wraps common multithreading problems into *parallel patterns*:

- *Async* and *Async/Await*: Executes code in a background thread and you are (optionally) notified when it completes an execution
- *Future*: Executes a calculation in a background thread and returns a result
- *Join*: Executes multiple tasks in parallel
- *ParallelTask*: Executes the same code in multiple parallel copies
- *For* and *ForEach*: Two ways to execute a `for` loop in parallel – one fast and one powerful
- *Background worker*: Runs a data processing server in the background; processing can run on multiple threads
- *Pipeline*: Runs a multi-stage process on multiple background threads
- *Fork/Join*: A framework to solve *divide and conquer* (recursion) algorithms
- *Map*: Runs a data-conversion function over an array of data on multiple parallel threads
- *Timer*: Implements a background worker that is woken up at specific intervals

High-level multithreading is based on low-level support and uses its functionality quite extensively. If you analyze the `OtlParallel` unit, which hosts all high-level patterns, you'll find many examples of how to work with low-level multithreading support.

High-level patterns use anonymous methods and generic types extensively and are, therefore, available only on Delphi 2009 and newer. In fact, due to multiple bugs in Delphi's 2009 and 2010 versions, it is recommended to use Delphi XE or later.

We will look into more important high-level parallel patterns in this chapter and compare them to the PPL solution, wherever it exists.

Blocking collections

Before I start discussing parallel patterns in OmniThreadLibrary, I'll spend a few pages explaining a data structure that is used by many of these patterns – a *blocking collection*. It is a thread-safe queue of unlimited size that supports multiple simultaneous readers and writers. The `IOmniBlockingCollection` interface, implemented in the OtlCollections unit, is based on the .NET BlockingCollection class and is shown in full here:

```
IOmniBlockingCollection = interface
  procedure Add(const value: TOmniValue);
  procedure CompleteAdding;
  function  GetEnumerator: IOmniValueEnumerator;
  function  IsCompleted: boolean;
  function  IsEmpty: boolean;
  function  IsFinalized: boolean;
  function  Next: TOmniValue;
  procedure ReraiseExceptions(enable: boolean = true);
  procedure SetThrottling(highWatermark, lowWatermark: integer);
  function  Take(var value: TOmniValue): boolean;
  function  TryAdd(const value: TOmniValue): boolean;
  function  TryTake(var value: TOmniValue;
     timeout_ms: cardinal = 0): boolean;
  property ContainerSubject: TOmniContainerSubject
     read GetContainerSubject;
  property Count: integer read GetApproxCount;
end;
```

Elements in IOmniBlockingCollection are of type TOmniValue (implemented in the OtlCommon unit), which can store any Delphi type. Other than that, it does not in any way depend on the OmniThreadLibrary task framework but can be equally well used with Delphi's TThread or with PPL tasks.

The *blocking* part of the name comes from the implementation of the for..in enumerator and other functions that read data from the collection (Take and Next). If the collection is empty, any of these functions will stop (block) until the data is available. Because of that, we can simplify communication between threads, as shown in the following example:

```
var
  bc: IOmniBlockingCollection; // blocking collection
                               // shared between threads

// thread 1:
var data: TOmniValue;
while GenerateData(data) do
  bc.Add(data);
bc.CompleteAdding;

// thread 2:
for var data in bc do
```

```
    ProcessData(data);
```

In this example, two threads share a blocking collection. The first thread generates data by calling the fictitious `GenerateData` function and writes each generated data item into the blocking collection. The second thread reads data items from the blocking collection and sends them to the `ProcessData` function for further processing. Because of the blocking behavior, we don't care whether the first thread takes some time to generate data, and the blocking collection becomes empty. The second thread will simply wait in the `for..in` loop for more data to become available.

However, to break the second thread out of the loop, we must somehow indicate that no more data will be generated. To do that, we put the collection in a *completed* state by calling the `CompleteAdding` function. No more data can be added after that, and any further call to `Add` will raise an exception. The second thread will still read all the data from the blocking collection, and only after that (when the collection is empty *and* in a "completed" state) will the second thread exit from the `for..in` loop.

`IOmniBlockingCollection` supports other nice features, such as throttling, cancelation, and resource exhaustion protection. We will talk more about throttling later in this chapter in the *Pipeline* section. Other features are described in this article: `http://www.omnithreadlibrary.com/book/chap10.html#howto-parallelSearch`.

> **Note**
>
> If you want to know how `IOmniBlockingCollection` is implemented internally, read my blog post *Dynamic lock-free queue – doing it right* (`https://www.thedelphigeek.com/2010/02/dynamic-lock-free-queue-doing-it-right.html`).

Using blocking collections with TThread-based threads

Blocking collection is easier to understand with a practical example, so I put together a `BlockingCollection` demo that uses a blocking collection in combination with `TThread`-based threads. This simple program has only one `TMemo` component to display output and two buttons – one to start background threads and another to stop them. On the global level, the main form uses the `OtlCollections` unit and defines the `FQueue: IOmniBlockingCollection` form field, which is shared between worker threads.

The **Producer/Consumer** button starts worker threads, as shown here:

```
procedure TfrmBlockingCollection.btnStartClick(
  Sender: TObject);
begin
  FQueue := TOmniBlockingCollection.Create;
  TThread.CreateAnonymousThread(
    CreateProducer(17, FQueue)).Start;
  TThread.CreateAnonymousThread(
```

```
    CreateProducer(42, FQueue)).Start;
  TThread.CreateAnonymousThread(
    CreateConsumer(FQueue)).Start;
  btnStart.Enabled := false;
  btnStop.Enabled := true;
end;
```

First, the code creates the shared blocking collection, FQueue. After that, it creates three anonymous background threads – two data producers and one data consumer. The CreateProducer and CreateConsumer helper methods are used to create anonymous methods that will be run in background threads. We'll look at them in a moment, but first, I want to show the Stop button, which stops all background processing:

```
procedure TfrmBlockingCollection.btnStopClick(Sender: TObject);
begin
  FQueue.CompleteAdding;
  FQueue := nil;
  btnStart.Enabled := true;
  btnStop.Enabled := false;
end;
```

The code puts FQueue into a "completed" state, which will, in turn, stop all background threads. After that, it clears the global FQueue interface and re-enables the start button. As FQueue is shared between the main thread and all background threads, the background collection object will only be destroyed when all threads stop using it, but we can be sure that this will happen soon.

Let's now look at the data producer threads. The CreateProducer method accepts two parameters – a numeric interval and a blocking collection, queue – and returns an anonymous method that captures these two parameters:

```
function TfrmBlockingCollection.CreateProducer(
  interval: integer;
  const queue: IOmniBlockingCollection): TProc;
begin
  Result :=
    procedure
    var
      num: integer;
    begin
      num := interval;
      while queue.TryAdd(num) do begin
        Inc(num, interval);
        Sleep(250);
      end;
```

```
      QueueLog('END/' + IntToStr(interval));
    end;
  end;
```

This method (which is executed in a background thread) generates multiples of the `interval` parameter, writes each interval into the queue, and sleeps 250 ms between writes. Because the whole framework is terminated by setting FQueue to a "completed" state, the loop uses `queue.TryAdd` to write into the blocking collection. `TryAdd` will succeed (returning `True`) if the collection is in a normal state and fail (returning `False`) if the collection is in a "completed" state. At the end of the execution, the code logs END/, plus the `interval` parameter.

As the start button creates two producers, one with interval parameter set to 17 and another where it is set to 42, we will get two worker threads, one generating multiples of 17 and another of 42. Both threads will write into the same queue, but this is not a problem, as blocking collections support any number of simultaneous readers and writers.

Now that we have taken care of data producers, let's take a look at the consumer. The `CreateConsumer` function accepts only one parameter, the shared queue, and creates an anonymous method that reads data from this queue:

```
function TfrmBlockingCollection.CreateConsumer(
  const queue: IOmniBlockingCollection): TProc;
begin
  Result :=
    procedure
    var
      value: integer;
    begin
      for value in queue do
        QueueLog(IntToStr(value));
      QueueLog('STOP');
    end;
end;
```

The anonymous method uses a `for..in` loop to read from the blocking collection and logs each value. We already know that this loop will only stop when the blocking collection is "completed," which happens when the **Stop** button is clicked. Finally, the code logs STOP and exits, which terminates the background thread.

If you run the program, start the background threads, let them run for some time, and stop the threads, you'll see output similar to the following screenshot:

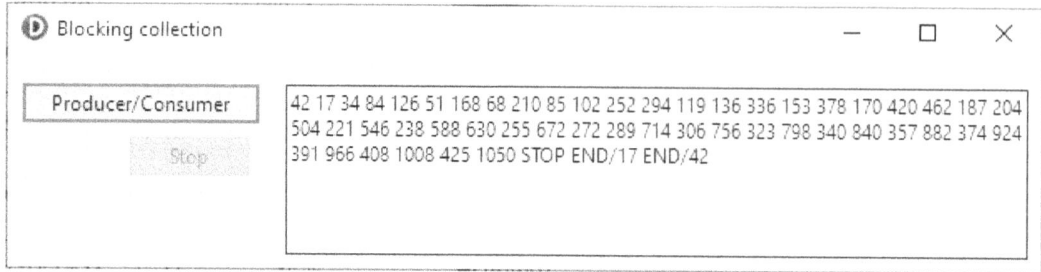

Figure 10.1 – Output of the BlockingCollection demonstration program

In this example, the thread generating multiples of 42 started a bit faster than the thread generating multiples of 17, even though the latter was created first in the code. That is completely normal – when working with multiple threads, we can never know in what order they will execute.

Finally, the code writes STOP END/17 END/42. As when starting, we cannot know which of the producer threads will be stopped first. We can, however, be pretty (but not completely!) sure that the data consumer thread will be stopped first because, by the time the Stop button is clicked, both data producer threads will most likely be found sleeping in the Sleep method.

Before we start discovering more complicated OTL patterns, I'd like to revisit three simple constructs that we covered in the previous chapter. On the next few pages, we'll take a look at the OTL version of the *Async/Await*, *Join*, and *Future* patterns.

Async/Await

The Async/Await program from the code folder for this chapter demonstrates the use of the *Async/Await* pattern in OTL. It is practically identical to the PPL version from the last chapter's code. This is not surprising, given that I had implemented the PPL version of Async/Await, mimicking the OTL version. The biggest change in the code was replacing the System.Threading unit with OtlParallel. The following code fragment shows the method that runs the LongTask method in a background thread by using the OTL Async/Await:

```
procedure TfrmAsyncAwait.btnLongTaskAsyncClick(
  Sender: TObject);
begin
  Log(Format('Button click in thread %d',
    [TThread.Current.ThreadID]));

  (Sender as TButton).Enabled := false;

  Async(LongTask)
  .Await(
    procedure
```

```
    begin
      Log(Format('Await in thread %d',
        [TThread.Current.ThreadID]));
      (Sender as TButton).Enabled := true;
    end
  );
end;
```

As there's really nothing more to say about *Async/Await*, let's move on to the next pattern, *Join*.

Join

The *Join* pattern in OTL serves the same purpose as the PPL version but is implemented in a slightly different manner. The main differences are the way tasks are started and the manner of handling exceptions in background tasks. The OTL version also implements some additional functionality, which I will not explore in this book – namely, setting the number of worker threads and cancellation support.

Let's start exploring the similarities and differences using a simple example, demonstrated in the `ParallelJoin` project. The **Join 2 tasks** button runs the following code that starts two tasks in parallel and waits for them to finish:

```
procedure TfrmParallelJoin.btnJoin2Click(Sender: TObject);
begin
  ListBox1.Items.Add('Starting tasks');
  Parallel.Join(Task1, Task2).Execute;
  QueueLog('Join finished');
end;
```

Just a reminder that, in PPL, we would achieve the same with the following:

```
TParallel.Join(Task1, Task2).Wait;
```

The three big differences between the libraries are as follows:

- In OTL, *Join* is started with `Parallel.Join`; in PPL, it is started with `TParallel.Join`. The main OTL factory class for parallel patterns is `Parallel`, and it is used for all patterns except for *Async/Await*. It is implemented (as all other support for parallel patterns) in the `OtlParallel` unit.

- In OTL, we have to explicitly start tasks by calling the `Execute` method. As we'll see later, the `Parallel.Join` call returns the `IOmniParallelJoin` interface, which we can use to further configure the pattern before starting tasks. Because of that, starting tasks is not an automatic process.

- The third difference is less obvious. In PPL, the default implementation of *Join* is *asynchronous*. The main thread continues its execution while the tasks are running unless we call the `Wait` method on the `ITask` interface returned from the `TParallel.Join` call. In OTL, the default implementation is *blocking*. The main thread stops until the tasks are finished. To change this to an asynchronous version, we have to call the `NoWait` method, as we'll see very soon.

The observant among you will have noticed that the code logs the `'Join finished'` message with a call to the `QueueLog` method. This method uses the `TThread.ForceQueue` mechanism to execute the logging call "just a bit later." It is defined as shown here:

```
procedure TfrmParallelJoin.QueueLog(const msg: string);
begin
  TThread.ForceQueue(nil,
    procedure
    begin
      ListBox1.Items.Add(msg);
    end);
end;
```

This is used so that logging messages appear in the correct order. The background tasks log their messages by calling the same function, but these queued calls are not processed because the code is waiting in the `Parallel.Join` call, and the main form doesn't do any message processing. All logging calls are only displayed after the `btnJoin2Click` method exits.

If the `btnJoin2Click` method logged the final message with the normal `ListBox1.Items.Add` call, this message would be displayed before all background log messages, even though it is actually generated later. By using `QueueLog`, we push the message to the same queue that already contains all background logs so that they are displayed in the correct order.

To start asynchronous tasks and be notified when they are finished, we use the `NoWait` and `OnStopInvoke` methods, as shown here:

```
procedure TfrmParallelJoin.btnJoinNoWaitClick(
  Sender: TObject);
begin
  ListBox1.Items.Add('Starting tasks');
  Parallel.Join([Task1, Task2, Task3]).NoWait.OnStopInvoke(
    procedure
    begin
      ListBox1.Items.Add('Tasks stopped');
    end).Execute;
  ListBox1.Items.Add('Tasks started');
end;
```

The OnStopInvoke method is functionally equivalent to Await in my improved version of the PPL *Join*. In OTL, you can also use the OnStop method, which works the same but is executed in the background thread, and it is triggered slightly faster than OnStopInvoke (which had to be queued for execution in the main thread).

> **Fluent programming**
>
> Most methods in the OtlParallel unit are implemented as functions that return Self. This allows for the simple chaining of methods called on an interface, such as the .NoWait. OnStopInvoke(...).Execute chain in the last example. This concept is also known as a *fluent interface*.

Exception handling

OTL *Join* collects all unhandled exceptions raised in tasks and merges them in an EJoinException exception object. This exception works the same as EAggregateException in Delphi. It is raised in the main thread when we wait on background tasks to complete. As this differs between the blocking and asynchronous versions, we'll look into two examples that show both approaches.

Catching exceptions is simple when using the blocking Join. We just wrap the Execute call in a try..except statement, as shown here:

```
procedure TfrmParallelJoin.btnJoin3EClick(Sender: TObject);
var
  i: integer;
begin
  ListBox1.Items.Add('Starting tasks');
  try
    Parallel.Join([Task1, Task2, Task3E]).Execute;
  except
    on E: EJoinException do begin
      for i := 0 to EJoinException(E).Count - 1 do
        QueueLog('Task raised exception: ' +
          EJoinException(E)[i].FatalException.Message);
      ReleaseExceptionObject;
    end;
  end;
  QueueLog('Join finished');
end;
```

This code fragment, which is triggered with the **Join Exception** button, processes EJoinException exceptions by iterating over its elements, from 0 to Count - 1, and displaying each inner exception message.

When using the asynchronous version, we have to store the `IOmniParallelJoin` interface returned from the `Parallel.Join` call and then, later in the code, call the `WaitFor` method on that interface. This is demonstrated by the following code that is triggered with the **Join.NoWait Exception** button:

```
procedure TfrmParallelJoin.btnJoinNoWaitEClick(
  Sender: TObject);
begin
  ListBox1.Items.Add('Starting tasks');
  FJoin := Parallel.Join([Task1, Task2E, Task3E])
             .NoWait.OnStopInvoke(TasksStopped).Execute;
end;

procedure TfrmParallelJoin.TasksStopped;
var
  i: Integer;
begin
  QueueLog('Tasks stopped');
  try
    try
      FJoin.WaitFor(0);
    except
      on E: EJoinException do begin
        for i := 0 to EJoinException(E).Count - 1 do
          QueueLog('Task raised exception: ' +
            EJoinException(E)[i].FatalException.Message);
        ReleaseExceptionObject;
      end;
    end;
  finally
    FJoin := nil;
  end;
end;
```

When the tasks are started in the `btnJoinNoWaitEClick` event handler, the code stores the Join interface in a form field, FJoin. Later, in the `TasksStopped` handler (which is called after all tasks have completed execution), the code calls `FJoin.WaitFor(0)`, which waits 0 ms for tasks to finish execution (and always succeeds, as we know that at this point, all the work has been done), and then catches `EJoinException` if there were any unhandled exceptions caught in the background tasks. Exception processing then continues, as in the previous example.

The following screenshot shows the output of the exception-logging code in the demo application:

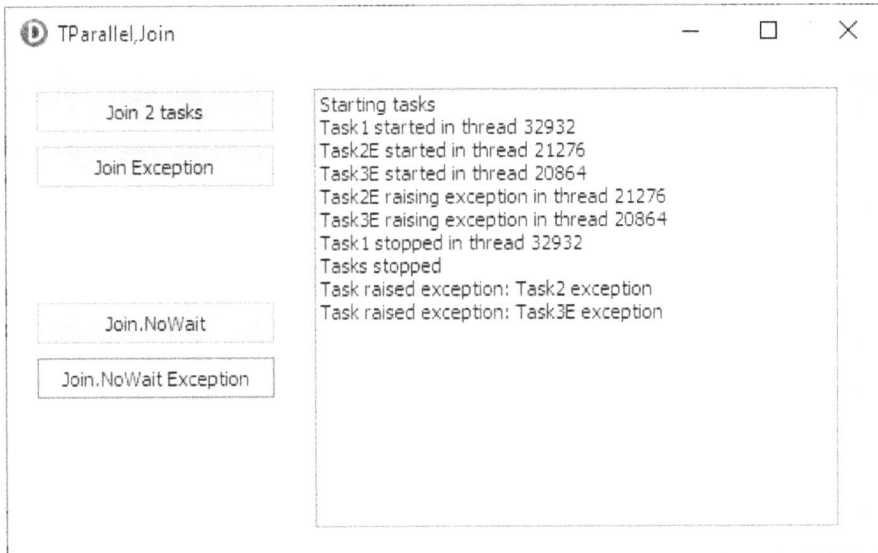

Figure 10.2 – Output from the exception-logging code

The OTL *Join* implements more functionality, as shown in this chapter – for example, a task cancellation mechanism – but this goes outside the scope of this book. Instead, we will turn to the next simple PPL pattern, *Future*.

Future

The OTL version of the *Future* pattern looks very similar to the PPL version but is slightly more powerful, as it implements its own notification mechanism to signal task completion. The basic version, however, is almost the same as the PPL version (the only difference is using `Parallel.Future` instead of `TParallel.Future`):

```
procedure TfrmFuture.btnFutureClick(Sender: TObject);
begin
  FFuture := Parallel.Future<integer>(CountPrimes);
end;
```

The main difference between the OTL and PPL implementations lies in how the code can notify the main thread of the completion of the calculation. In PPL, we used `TThread.Queue` inside the background calculation code. OTL solves this with a *task configuration* block.

This approach, which is used in many OTL parallel patterns, allows the code to pass a special `IOmniTaskConfig` parameter to the future (or other parallel) pattern. This interface can be used to configure a cancellation mechanism, set up a thread pool to be used, define a termination (completion) event handler, and so on. The `ParallelFuture` project uses task configuration to specify a completion handler, as shown here:

```
procedure TfrmFuture.btnFuture2Click(Sender: TObject);
begin
  FFuture := Parallel.Future<integer>(
    function: Integer
    begin
      Result := CountPrimes;
    end,
    Parallel.TaskConfig.OnTerminated(ReportFuture))
end;

procedure TfrmFuture.ReportFuture;
begin
  ListBox1.Items.Add('Result = ' + FFuture.Value.ToString);
  FFuture := nil;
end;
```

The code in the `btnFuture2Click` event handler, associated with the **Create Future 2** button, creates a future as in the previous example, except that it also passes a second parameter – a task configuration block. This block, created with the `Parallel.TaskConfig` factory, configures the `OnTerminated` handler, `ReportFuture`. This handler is called in the main thread after the future calculation finishes its job, similar to `OnStopInvoke` in the *Join* example.

The `ReportFuture` method can then log the calculated value and clear the `FFuture` field, which destroys the future implementing object.

Handling *Future* exceptions is the same as in PPL. Unhandled exceptions from the background method are raised again when the code accesses the `Value` property.

After this short revisit of three patterns from the previous chapter, we'll explore two that are specific to OTL – *Parallel Task* and *Background Worker*.

Parallel Task

The *Parallel Task* pattern is conceptually very simple – it runs the same code on multiple background threads. As with almost all other OTL high-level patterns, a new instance is created by calling the appropriate factory function – in this case, that is `Parallel.ParallelTask`, which returns an `IOmniParallelTask` interface. Similar to most OTL patterns, it implements functionality to catch and report exceptions, features a cancellation mechanism, and most importantly for our example, triggers a notification method when the last background worker stops execution.

A parallel task was designed to speed up a longer operation without having to deal with background threads, notifications, and so on. Because of that, by default it works in *blocking* mode, just like the Join pattern. Similarly, we can call `NoWait` on the interface and turn it into an *asynchronous* pattern.

To demonstrate the pattern – and a few other OTL-specific features – I have created a `ParallelTask` program that uses this pattern to generate a large block of random data as fast as possible.

The logic behind the main worker method, `CreateRandomData`, is as follows:

- Random data is written into an output stream

- As the `TStream` implementation is not thread-safe, worker threads create independent memory blocks filled with random data (represented by `TMemoryStream` objects) and write them into a shared blocking collection

- The main thread reads from that blocking collection and copies data into the output stream

This method, shown next, executes in blocking mode. It only returns when all random data is generated. This greatly simplifies the application:

```
procedure TfrmParallelTask.CreateRandomData(
  fileSize: integer; output: TStream);
const
  CBlockSize = 1*1024*1024 {1 MB};
var
  buffer    : TOmniValue;
  memStr    : TMemoryStream;
  outQueue  : IOmniBlockingCollection;
  unwritten : IOmniCounter;
begin
  outQueue := TOmniBlockingCollection.Create;
  unwritten := CreateCounter(fileSize);
  Parallel.ParallelTask.NoWait
    .NumTasks(SpinEdit1.Value)
    .OnStop(Parallel.CompleteQueue(outQueue))
    .Execute(
      procedure
      var
        buffer      : TMemoryStream;
        bytesToWrite: integer;
        randomGen   : TGpRandom;
      begin
        randomGen := TGpRandom.Create;
```

```
            try
              while unwritten.Take(CBlockSize, bytesToWrite)
              do begin
                buffer := TMemoryStream.Create;
                buffer.Size := bytesToWrite;
                FillBuffer(buffer.Memory, bytesToWrite,
                  randomGen);
                outQueue.Add(buffer);
              end;
            finally FreeAndNil(randomGen); end;
          end);

    for buffer in outQueue do begin
      memStr := buffer.AsObject as TMemoryStream;
      output.CopyFrom(memStr, 0);
      FreeAndNil(memStr);
    end;
  end;
```

Let's examine the method in more detail. Firstly, it creates a blocking collection, outQueue, that will function as a communication channel between the background tasks and the main thread. It also uses an unwritten counter that the threads will use to know how much data to generate and when to stop.

This counter, implemented in the OtlCommon unit, is mainly used as a counter with atomic increment/decrement operations (it uses interlocked operations to do so). This example, however, uses its Take function, which I'll explain in a minute.

Next, the code creates the parallel task pattern by calling the Parallel.ParallelTask function, switching it to asynchronous mode (NoWait), and setting the number of background tasks (NumTasks). We need the pattern to run asynchronously, as we want the main thread not to be blocked because it has to copy generated data into the output stream while data is generated.

The next function, called –OnStop(Parallel.CompleteQueue(outQueue))–, makes sure that the background task actually terminates when the work is done. The CompleteQueue helper function generates an anonymous method that marks the blocking collection (outQueue) completed. The OnStop function then makes sure that this anonymous method is executed immediately after the last background task stops working.

We must execute this method from the background thread (OnStop) because, at that moment, the main thread will be busy reading from outQueue (as we'll see soon) and is not processing messages. If we would have tried to run it from the main thread by calling OnStopInvoke, the whole program would block. OTL would queue the anonymous method for execution in the main thread, but the main thread would wait while reading from the outQueue and would never have a chance to execute this queued anonymous method.

After that, the `Execute` function executes its parameter – another anonymous method – in all background workers. Because we used `NoWait`, `Execute` immediately exits.

The main thread then continues reading data from `outQueue`. As we'll see next, background workers will enqueue `TMemoryStream` objects into that queue. The main thread simply reads them from the blocking collection (and waits if no data is available at the moment) and destroys the memory stream objects. Only when the last worker stops and the anonymous method generated by the `CompleteQueue` helper is executed does the `for buffer in outQueue do` statement exit, and control is returned to the method that called the `CreateRandomData` method.

Now we know everything about how the task is started, but I still have to explain how the background tasks work. Each of the tasks creates a random number generator and then repeatedly calls `unwritten.` `Take` to allocate its block of work. This function works similarly to the following code fragment:

```
function TOmniCounter.Take(count: integer;
  var taken: integer): boolean;
begin
  if Value > count then
    taken := count
  else
    taken := Value;
  Value := Value - count;
  Result := (taken > 0);
end;
```

The function tries to decrease the current counter value by the required `count`. If that decreases the counter below zero, it sets the counter to zero. In all cases, it stores the actual amount of change (how much the counter value was decreased) in the `taken` parameter and returns `True` if the caller has been allocated some work.

The trick behind `Take` is that it is completely atomic. Multiple threads can, therefore, use `Take` to get their share from some work quota and be sure that they will never be given too much work, that all work will be completed, and that they will all ultimately stop.

After getting its allocation of work, the job of a background thread is simple. It creates a `TMemoryStream` buffer, uses the `FillBuffer` helper function (not shown here) to generate the data, and adds this buffer to `outQueue`. The process then repeats until there is nothing more to do (the `unwritten` counter drops to 0 and `Take` returns `False`).

> **Note**
>
> To read more about `Take` and `CompleteQueue`, and to see how the same problem could be solved with a Parallel For pattern, see my blog post at `https://www.thedelphigeek.com/2011/09/life-after-21-parallel-data-production.html`.

The following picture shows a comparison between single-threaded and multi-threaded random data generation. Both examples generate a 100 MB stream in 1 MB chunks. We can see in the following screenshot that speed-up with eight parallel tasks is quite noticeable, while doubling this number doesn't bring us much advantage:

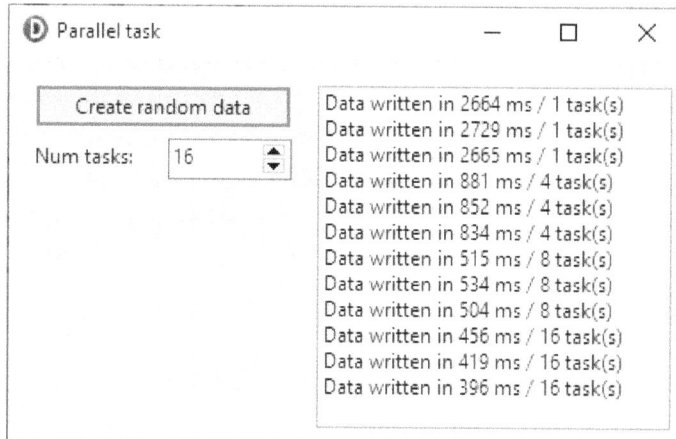

Figure 10.3 – Measuring the speed of data generation with different numbers of parallel tasks

These numbers confirm that data generation can be nicely spread around all eight physical cores in the system, while running it on additional eight hyper-threaded cores doesn't bring many advantages.

The `Parallel Task` pattern is just a thin layer around the `Join` pattern, but it makes one specific operation – running the same code on multiple threads – just a bit simpler and more self-explanatory.

Background Worker

It is now time to introduce the first complex pattern in this chapter – *Background Worker*. It establishes a background server module running on multiple threads, to which we can queue data packets for processing. This module has its own input queue (a blocking collection, of course!) that holds unprocessed data, and an output queue that holds the results of processing until they are returned to the main thread.

Later in this chapter, you'll see that a background worker looks very similar to a single stage of a pipeline. This is true – in fact, a background worker is implemented as a wrapper around the pipeline pattern! It does, however, implement important functionality that is not part of the pipeline pattern – namely, it can cancel work requests while they wait to be processed. We will see in the example how that can be of use.

Another nice feature of the background worker is that it provides a clean way to set up and tear down tasks. With the initialization/finalization mechanism (that we'll explore in the example) you can create objects that will be needed for data processing, prepare database connections, and so on. This makes the background worker pattern a good candidate to write a web service backend.

In this book, I'll create a simpler example – a small utility that executes a query on the popular Delphi forum Delphi-PRAXiS (`https://en.delphipraxis.net/`), downloads the first page of results, and enables a user to quickly browse between them.

The demonstration program is called `ParallelBackgroundWorker` and provides a simple user interface, as shown in the following screenshot:

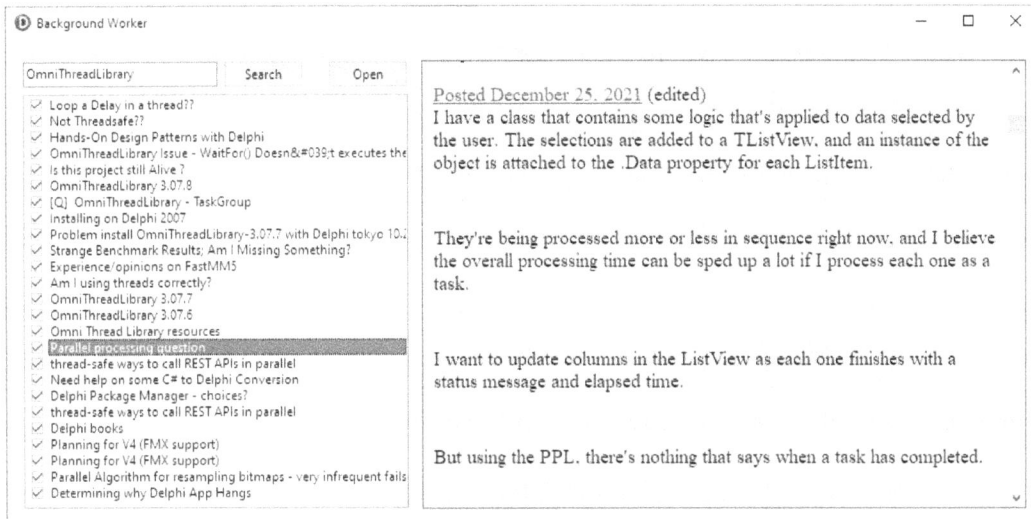

Figure 10.4 – A forum search implemented with a background worker pattern

The program exposes an edit field where you can type a search query. By clicking the **Search** button, the program downloads the search results, parses the output, and queues each page for download to the background worker. A checkbox next to the article title indicates that the page has been downloaded. At any time – even before all articles have been downloaded – you can enter a new search term and click the **Search** button. This will clear the current result list and restart the search.

If you click on an article title, a preview is displayed on the right. A click on the **Open** button opens the selected page in the default web browser.

The Background Worker pattern is created in the `OnCreate` event, as shown here:

```
procedure TfrmBackgroundWorker.FormCreate(Sender: TObject);
begin
  FBackgroundWorker :=
    Parallel.BackgroundWorker
      .NumTasks(4)
      .Initialize(InitializeWorkerTask_asy)
      .Finalize(FinalizeWorkerTask_asy)
      .Execute(DownloadWebPage_asy)
      .OnRequestDone(ArticleDownloaded);
```

```
    FRequests := TDictionary<int64, TRequest>.Create;
    FTitles := TList<int64>.Create;
    CreateHtmlViewer;
  end;
```

Similar to the previous examples, the code uses a fluent programming interface style to set up the pattern. The `Pattern.BackgroundWorker` factory function creates a pattern and returns an `IOmniBackgroundWorker` interface, which is stored in the `FBackgroundWorker` form field.

The code hardcodes four parallel tasks (`NumTasks`) and assigns methods that will be executed in each task at the beginning and at the end (`Initialize` and `Finalize`). Then, it sets the default function to process requests in the worker thread (`Execute`) and the default function to process results in the main thread (`OnRequestDone`). As we'll see later, both handlers (`Execute` and `OnRequestDone`) can be overridden on a per-request basis.

> **Note**
>
> As the methods executed in the background threads are part of the form object, I have decorated their names with the `_asy` suffix to remind any programmer working on the code that this method executes in the background and should not access the user interface.

Finally, the code creates two helper data structures (we'll see later how they are used) and a `THTMLViewer` component that displays the HTML page preview. This third-party component is provided together with the demonstration program and created during program execution, so you don't have to install the component package separately.

> **Note**
>
> The HTML viewer component was written by L. David Baldwin, Bernd Gabriel, et al. and is hosted at `https://github.com/BerndGabriel/HtmlViewer`.

The background worker pattern is destroyed in the form's `OnDestroy` event, as shown here:

```
  procedure TfrmBackgroundWorker.FormDestroy(
    Sender: TObject);
  begin
    FBackgroundWorker.CancelAll;
    FBackgroundWorker.Terminate(5000);
    FBackgroundWorker := nil;
    FreeAndNil(FRequests);
    FreeAndNil(FTitles);
  end;
```

The code firstly cancels all unprocessed requests (CancelAll) so that we don't have to wait for all unprocessed pages to be downloaded. It then asks the pattern to Terminate with a timeout set to 5000 ms. If the pattern cannot safely shut down in that time (for example, if a page download is stuck), the OTL framework kills all background threads. This is, in general, not advisable, but as the program is shutting down, it probably won't cause any problems. Finally, the code clears the FBackgroundWorker interface (technically, this is not necessary, but I like my code to be clean) and destroys the helper structures.

Initial query

When you click the **Search** button, the btnSearchClick function clears the internal structures and user interface, cancels all remaining requests from the previous search (CancelAll), creates a new work request (CreateWorkItem), and sends it to the background worker (Schedule), as shown here:

```
procedure TfrmBackgroundWorker.btnSearchClick(
  Sender: TObject);
begin
  chkSitelist.Clear;
  FRequests.Clear;
  FTitles.Clear;
  FBackgroundWorker.CancelAll;
  FBackgroundWorker.Schedule(
    FBackgroundWorker.CreateWorkItem(
      'https://en.delphipraxis.net/search/?q='
      + inpSearch.Text),
    FBackgroundWorker.Config.OnRequestDone(
      ListDownloaded));
end;
```

CreateWorkItem accepts one parameter of type TOmniValue, which can hold any Delphi data type, including objects, interfaces, records, and arrays, and returns an IOmniWorkItem interface. Besides properties required for cancelation, exception handling, and task scheduling, this interface contains the following four properties that we'll use in the example:

```
IOmniWorkItem = interface
  property Data: TOmniValue read GetData;
  property Result: TOmniValue read GetResult
    write SetResult;
  property TaskState: TOmniValue read GetTaskState;
  property UniqueID: int64 read GetUniqueID;
end;
```

These four properties contain the following:

- `Data` contains data to be processed (a string, in our example).

- `Result` contains the result of data processing (set in the `DownloadWebPage_asy` method).

- `TaskState` contains state (objects and data connections) that are initialized once for each task (initialized in `InitializeWorkerTask_asy` and used in `DownloadWebPage_asy`).

- `UniqueID` contains a unique 64-bit integer. Unique IDs start with 1 and are incremented each time `CreateWorkItem` is called. This allows us to associate results with data request and cancel specific work items by using the `CancelAll` function.

A work request created with `CreateWorkItem` is then sent to the background worker, together with an additional configuration block (`Config.OnRequestDone`), which specifies that we want to override the result processing function just for this one request. In a similar manner, we can also override the background processing function for this request if that was required.

The request then travels to one of the worker tasks that has previously (during its initialization) executed the `InitializeWorkerTasks_asy` method. This method, shown next, creates a `THTTPClient` object, which will be used to download web pages as shown here:

```
procedure TfrmBackgroundWorker.InitializeWorkerTask_asy(
  var taskState: TOmniValue);
begin
  taskState := THTTPClient.Create;
end;
```

This task-global object will be destroyed in `FinalizeWorkerTask_asy` just before the background thread is destroyed:

```
procedure TfrmBackgroundWorker.FinalizeWorkerTask_asy(
  const taskState: TOmniValue);
begin
  taskState.AsObject.Free;
end;
```

The `taskState` variable returned from `InitializeWorkerTask_asy` is stored in the `TaskState` field of each request before it is passed to the data processing function, `DownloadWebPage_asy`. This function casts `TaskState` back into `THTTPClient` and uses its `Get` function to download the page, as shown here:

```
procedure TfrmBackgroundWorker.DownloadWebPage_asy(
  const workItem: IOmniWorkItem);
var
  response: IHTTPResponse;
begin
```

```
response := workItem.TaskState.ToObject<THTTPClient>
              .Get(workItem.Data);
  if (response.StatusCode div 100) = 2 then
    workItem.Result := response.ContentAsString;
end;
```

If the page is downloaded correctly (the status code starts with 2), the code sets the work item's `Result` field to the page content. Otherwise, `Result` is not set. As we'll see next, the main thread uses this fact to detect and report errors.

When `DownloadWebPage_asy` exits, the framework sends the `workItem` interface back to the owner. This request travels back to the main thread, where it triggers the `ListDownloaded` method (the one that was specified in the `Config.OnRequestDone` special configuration call), which we will explore in the next section.

Article download

The `ListDownloaded` method uses a very ugly regular expression filter to extract search results from the returned page. I will not go into how this filter works and instead focus on the background worker-specific parts that are marked in the code:

```
procedure TfrmBackgroundWorker.ListDownloaded(
  const Sender: IOmniBackgroundWorker;
  const workItem: IOmniWorkItem);
var
  filter, url: string;
  hrefMatch: TRegEx;
  match: TMatch;
  index: integer;
  request: IOmniWorkItem;
begin
  if workItem.Result.IsEmpty then
    ShowMessage('Background worker failed to download '
              + workItem.Data)
  else if assigned(FBackgroundWorker) then begin
    filter := '<a href=''(https://en.delphipraxis.net/'
            + 'topic/.*?)''.*?data-linkType="link"'
            + '.*?>(.*)</a>';
    hrefMatch := TRegEx.Create(filter,
                 [roIgnoreCase, roMultiLine]);
    match := hrefMatch.Match(workItem.Result);
    while match.Success do begin
      url := StringReplace(match.Groups[1].Value,
             '&amp;', '&', [rfReplaceAll]);
```

```
        index := chkSitelist.Items.Add(
                match.Groups[2].Value);
        request := FBackgroundWorker.CreateWorkItem(url);
        FRequests.Add(request.UniqueID,
          TRequest.Create(index, url));
        FTitles.Add(request.UniqueID);
        FBackgroundWorker.Schedule(request);
        match := match.NextMatch;
      end;
    end;
  end;
```

The code firstly checks whether the result is set at all (IsEmpty). If not, it reports that the page could not be downloaded. All data passed to the background worker is available in the Data property, which simplifies the logging.

After parsing, the code creates a new download request for each search result (extracted in the url variable). The code adds the title of the result to the check-listbox, chkSitelist, and remembers the list index of the new item. Then, it creates a new request to download the url. After that, it creates a mapping from the unique ID of the new requests to the list index in the FRequests dictionary. Next, it stores this same unique ID into the FTitles array and, finally, schedules the new request.

Later in the code, we will need FRequests to map from the unique ID to the chkSitelist index and the FTitles array to map from the chkSitelist index into the unique ID, which will be used as a key to data stored in the FRequests dictionary.

Each download request (each work item) now travels to the background worker, where it enters the queue of unprocessed requests. When any of the background worker tasks has nothing to do, it takes a request from this queue and sends it to the data processing method, DownloadWebPage_asy. This method downloads the web page and sends the result back to the main thread, where it is passed to the default result processing method, ArticleDownloaded (since we didn't override it in the Schedule call), as shown here:

```
procedure TfrmBackgroundWorker.ArticleDownloaded(
  const Sender: IOmniBackgroundWorker;
  const workItem: IOmniWorkItem);
var
  request: TRequest;
begin
  if not FRequests.TryGetValue(workItem.UniqueID, request)
  then
    Exit;
  request.Page := workItem.Result;
  FRequests.AddOrSetValue(workItem.UniqueID, request);
```

```
    chkSitelist.Checked[request.Index] := true;
  end;
```

This method has quite a simple job. It uses the work item's unique ID to access the data storage record, TRequest (not shown here), in the FRequests dictionary and stores the returned HTML page. Then, it uses previously stored unique ID-to-index mapping (request.Index) to mark the appropriate line in the user interface as checked.

Previewing and opening articles

The few remaining pieces of code have nothing to do with the background worker pattern.

The chkSitelistClick event handler is called when a line in the chkSitelist component is selected. It maps the component index to a unique ID via the FTitles array and tries to access the FRequest data for that request. The code, shown next, shows the HTML page in the HtmlViewer component if that data is available:

```
procedure TfrmBackgroundWorker.chkSitelistClick(
  Sender: TObject);
var
  request: TRequest;
begin
  if FRequests.TryGetValue(FTitles[chkSiteList.ItemIndex], request)
  then
    HtmlViewer.LoadFromString(request.Page)
  else
    HtmlViewer.Clear;
end;
```

The **Open** button opens a URL of the selected article in the default browser, as shown here:

```
procedure TfrmBackgroundWorker.ActionOpenExecute(
  Sender: TObject);
var
  request: TRequest;
begin
  if FRequests.TryGetValue(FTitles[chkSiteList.ItemIndex], request)
  then
    ShellExecute(0, 'open', PChar(request.URL), nil, nil,
      SW_SHOWNORMAL);
end;
```

While this example doesn't show all the *Background Worker* functionality, it demonstrates all the important parts. To discover the rest, you will have to explore web articles or the OTL documentation (links are near the beginning of the chapter). In the meantime, I will continue this discussion with a description of OTL's *Pipeline*.

Pipeline

We have already learned a lot about *pipelines* in the previous chapter. To recap, pipelines are useful when we can split data processing into multiple independent stages and run these stages in parallel, on multiple background threads. We also looked into implementing a pipeline with standard PPL tools (and some additional code), and while the solution is useful, it feels quite clumsy (which is partially my fault, as I tried to keep the PPL pipeline implementation as lightweight as possible).

In this section, we will look into the OTL implementation of pipelines and explore it through the `ParallelPipeline` demo. Although this demo has the same name as the corresponding demo in the previous chapter, it solves a different problem. In the previous chapter, we wrote a web-scanning spider. This time, the demo program implements a very simple file analyzer.

The `ParallelPipeline` demo scans all files with the `.pas` extension in a folder (including subfolders), counts the number of lines, words, and characters in each file, and displays cumulative statistics for all found files.

From the implementation view, the program must do the following:

- Find all files in a folder and subfolders
- Read each file from the disk into the buffer
- Count the number of lines, words, and characters in that buffer
- Update cumulative statistics with information about that file

We can parallelize this operation by first finding all files in the folder, storing their names in a string list, and using *Parallel for* to process that list. If we do so, however, we have to wait until the initial file list is built, and that can take some time, especially if there are lots of files and they are stored on a network server. It would be preferable to start working immediately after the first filename is known, and a pipeline is designed exactly for that.

To create a pipeline, we must think about the problem in a different way – as a series of work stages, connected with communication channels. In a way, we implement my favorite technique of speeding up programs – *changing the algorithm*.

In our problem, the first stage is *enumerating files*. Its input is the name of the folder that is to be scanned, and its output is a list of files with the `.pas` extension found in that folder and its subfolders. Instead of storing the names of found files in some list, this stage involves writing them into a communication channel.

The second stage is a simple *file reader*. It is connected to the first stage by using a shared communication channel – the first stage writes into the channel, and the second stage reads from it. This stage reads the next filename from the channel, opens the file, reads its content into some buffer, and writes that buffer into its own output channel.

This pattern of stage, channel, stage, channel, and so on then continues. The third stage *analyzes* the data. It reads data buffers from its input (which is, simultaneously, the output of the file reading stage), performs analysis on the data, and writes the result of the analysis into its own output.

The fourth and final stage *aggregates* all of the partial analysis results and generates the summary report, which the main program accesses after the job is done.

This way, we can create a sequence of small workers, connected with communication channels. The nice part about this approach is that the workers are completely independent and, because of that, are simple to parallelize. None of them access any shared data at all. The only shared parts are communication channels, but they are implemented and managed by the pattern, so your code doesn't have to care about them. OTL uses `IOmniBlockingCollection` for communication channels in the pipeline pattern.

The pipeline built by the demo program implements only one instance of the first and last stage, two instances of the statistics analysis stage, and multiple instances (depending on the number of cores in the system) of the file-reading stage. Accessing files is a relatively slow operation that doesn't use much CPU, and it makes sense to run it in parallel.

The following diagram shows the pipeline created by the `ParallelPipeline` demo. Note that there are two additional communication channels that I only hinted at. The first stage also has a communication channel on the input, and the last stage has one on the output. They will be typically used by the main program to provide initial data (the starting folder in our example) and read the result (cumulative statistics). In OTL, they are accessed through the pipeline's `Input` and `Output` properties:

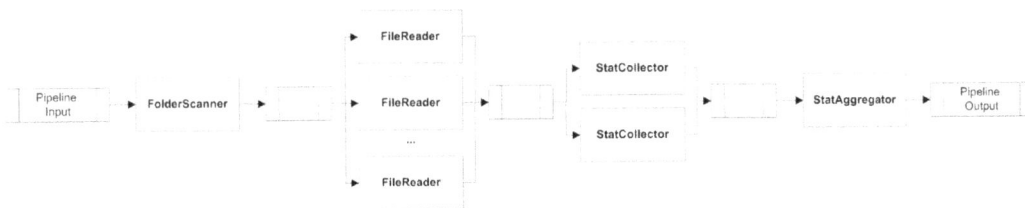

Figure 10.5 – A diagram of stages and connecting channels in the ParallelPipeline demo

Now that we know the stages we need and how they are to be connected, we can proceed. In the next stage, we'll create and initialize the pattern.

Creating the pipeline

Let's look at the code now. The user interface has one button that can both start and stop the parallel operation. Its OnClick event is shown as follows. When starting the process, it calls the CreatePipeline method. To stop the process, it just calls the pipeline's Cancel method. The pipeline object will be destroyed later, through a termination event. I'll return to that part later:

```
procedure TfrmPipeline.btnStartClick(Sender: TObject);
begin
  if btnStart.Tag = 0 then begin
    ListBox1.Items.Add(inpStartFolder.Text);
    CreatePipeline;
    btnStart.Caption := 'Stop';
    btnStart.Tag := 1;
    TimerUpdateProcessing.Enabled := true;
  end
  else
    if assigned(FPipeline) then
      FPipeline.Cancel;
end;
```

The CreatePipeline method (shown as follows) creates, configures, and starts the pipeline:

```
procedure TfrmPipeline.CreatePipeline;
begin
  FPipeline := Parallel.Pipeline;
  FPipeline.Stage(FolderScanner);
  FPipeline.Stage(FileReader,
                  Parallel.TaskConfig.OnMessage(
                    HandleFileReaderMessage))
    .NumTasks(Environment.Process.Affinity.Count)
    .Throttle(1000);
  FPipeline.Stage(StatCollector)
    .NumTasks(2);
  FPipeline.Stage(StatAggregator);

  FPipeline.OnStopInvoke(ShowResult);

  FPipeline.Run;

  FPipeline.Input.Add(inpStartFolder.Text);
  FPipeline.Input.CompleteAdding;
end;
```

First, `FPipeline: IOmniPipeline` (a form field) is created by calling the `Parallel.Pipeline` method.

Next, the code creates all four stages by calling the `Stage` function for each of them. Some stages require additional configuration to set the number of instances, specify message handlers, and limit the size of the output queue. For example, the first stage (`FolderScanner`) and the last stage (`StatAggregator`) are created without any additional configuration, while the code specifies that there should be two instances of the third stage (`StatCollector`) by calling `NumTasks(2)` after specifying the stage method.

This method extensively uses the *fluent* programminginterface style supported by many OTL interfaces. In the previous code, this behavior is used when configuring the second stage (`Stage().NumTasks().Throttle()`) and the third stage (`Stage().NumTasks()`).

The configuration of the second stage (`FileReader`) is the most complex. Besides setting the number of parallel tasks for this stage to the number of CPUs that are available to the process, we also limit the size of the output queue to 1,000 items (`Throttle(1000)`) and specify the function that will process messages sent from the stage to the main thread (`Parallel.TaskConfig.OnMessage`).

The code also sets up an event handler, which will be called (in the main thread) after all the pipeline tasks complete their execution (`OnStopInvoke`). This event handler is also called when the process is canceled by the user.

After all the configuration is done, the code calls `Run` to start the pipeline. Tasks for all stages will be created, and each will try to read from its own communication channel. As there is no data present in the channels yet, they will all simply go to sleep until some data appears. At this point, the pipeline doesn't use any CPU.

To kickstart the process, the code writes the starting folder into the pipeline's input (`FPipeline.Input.Add`). This will wake the `FolderScanner` stage, which will scan the folder, send all found files to the output channel, and so on. After that, it will try to read the next starting folder from the input, and as there will be no data present, the stage will again go to sleep.

As the code knows that there's only one input element to be processed, we want to prevent that. After the first stage is done with its input, it should exit and not try to read the next element. We can enforce that by calling the `CompleteAdding` method on the input channel. This puts the channel into *completed* mode, which indicates that no more data can be written into the channel, thus allowing the first stage to stop.

When the first stage is done processing the data, it will put its output channel into *completed* mode, which will allow the second stage to stop. That way, the indication that the pipeline is finished with its work trickles down the communication channels. A similar mechanism is used by the `Cancel` method, which is used to cancel currently running operations.

A note on communication channel size and throttling is in order at this point. By default, each communication channel can store up to 10,240 elements. After that number is reached, the *producer* (the task writing into the channel) will go to sleep when it tries to write a new data item to the channel. Only when the channel becomes partially empty (by default, that happens when it is under 75% full) will the producer awaken.

This mechanism was implemented so that one very fast task cannot produce so much data that it fills up the computer's memory. The default value sometimes needs manual adjustment, though. That's why the second stage limits its output to 1,000 data buffers. The content of a file can be relatively big, especially compared with other outputs (a filename outputted from stage one and statistics data outputted from stage three).

I must admit that the number of tasks I specified for stages two and three was not determined scientifically but by guesswork. In a real-life application, it would be prudent to test such a program with different configuration parameters (different numbers of tasks and different communication channel sizes) on local and network folders. A true *flow graph* implementation should do such tuning automatically, but IOmniPipeline was never designed to do that; instead, it was designed to be a simple and powerful pattern that can be easily controlled.

Stages

Let us now see how the stages are implemented. The first stage, FolderScanner, scans a folder and writes all found Pascal files to the output. Its implementation is shown here:

```
procedure TfrmPipeline.FolderScanner(
  const input, output: IOmniBlockingCollection;
  const task: IOmniTask);
var
  value: TOmniValue;
begin
  for value in input do begin
    DSiEnumFilesEx(
      IncludeTrailingPathDelimiter(value) + '*.pas',
      0, true,
      procedure (const folder: string; S: TSearchRec;
                 isAFolder: boolean; var stopEnum: boolean)
      begin
        stopEnum := task.CancellationToken.IsSignalled;
        if (not stopEnum) and (not isAFolder) then
          output.TryAdd(
            IncludeTrailingPathDelimiter(folder) + S.Name);
      end);
  end;
end;
```

The stage method receives two parameters of type IOmniBlockingCollection, which represent its input and output communication channel, and an optional parameter that allows access to the task that runs the stage code (task: IOmniTask).

An IOmniBlockingCollection channel contains items of type TOmniValue, which is a fast variation of Delphi's Variant or TValue. This type can store any other Delphi data type inside.

A typical stage would run a for value in input loop to read all the data from the input. This for loop doesn't exit if there is no data on the input but simply enters a *wait* state, which uses no CPU power. Only when input is put into *completed* mode *and* there is no data inside will the for loop exit.

We know that the input will only contain one item, so we could also just call the input.TryTake function to read that value. The approach implemented in the FolderScanner method is more flexible, though.

The code then uses the helper function, DSiEnumFilesEx, from the DSiWin32 library, which is included with the OTL. This function scans folders and their subfolders and calls an anonymous method for each found file.

This anonymous method first checks whether the operation should be aborted by checking task.CancellationToken.IsSignalled. If so, it will set the stopEnum flag, which tells DSIEnumFilesEx to stop enumerating files and exit. If everything is OK, however, it concatenates the folder and filename parts and writes them to the output. A TryAdd method is used instead of Add, as this call may fail when the pipeline is canceled (when the code calls IOmniPipeline.Cancel).

The second stage, FileReader, is implemented with a similar loop. It also checks the cancellation token and exits if it is signaled. Without that check, the stage would always process all data waiting in its input queue before exiting, even if the pipeline was canceled:

```
procedure TfrmPipeline.FileReader(
  const input, output: IOmniBlockingCollection;
  const task: IOmniTask);
var
  sl: TStringList;
  value: TOmniValue;
  outValue: TOmniValue;
begin
  for value in input do begin
    if task.CancellationToken.IsSignalled then
      break;

    sl := TStringList.Create;
    try
      sl.LoadFromFile(value);
      outValue.AsOwnedObject := sl;
```

```
        sl := nil;
        output.TryAdd(outValue);
        task.Comm.Send(MSG_OK, value);
      except
        task.Comm.Send(MSG_ERROR, value);
        FreeAndNil(sl);
      end;
    end;
  end;
```

Inside the loop, the code loads the contents of a file to a string list. An implicit operator converts value into a string parameter, needed in the LoadFromFile method. After that, TStringList is sent to the output queue. To assure that somebody will always destroy this object, even if the list is canceled, it is best to store it as an *owned object*. It will then be reference-counted and destroyed when appropriate.

The preceding code sends a message to the main thread when a file is read or an exception is raised when accessing a file. This message is sent via OTL's built-in message channel (task.Comm) and is dispatched to the HandleFileReaderMessage method, which was specified as a message handler while building the pipeline.

This following method displays all errors on the screen but only stores the currently processed file in a global field. It would be useless to overflow the user interface with information on all processed files, so the code uses a timer (not shown in the book), which every quarter of a second displays the current FLastProcessed on screen:

```
procedure TfrmPipeline.HandleFileReaderMessage(
  const task: IOmniTaskControl;
  const msg: TOmniMessage);
begin
  if msg.MsgID = MSG_ERROR then
    ListBox1.ItemIndex := ListBox1.Items.Add(
                            '*** ' + msg.MsgData)
  else
    FLastProcessed := msg.MsgData;
end;
```

Let's move to the third stage, StatCollector, which analyzes one file. It is implemented as a *simple stage*, meaning that the for .. in loop is implemented in the pipeline pattern itself. The pattern also handles the cancellation.

This approach is ideal when each input value produces zero or one output value. The stage method is called once for each item read from the input channel. If the stage writes anything into its output parameter, that value is written to the stage's output channel. If the stage leaves output untouched, nothing is written out. This makes the stage code, shown next, really simple:

```
type
  TStatistics = record
    Files: Int64;
    Lines: int64;
    Words: int64;
    Chars: int64;
  end;

procedure TfrmPipeline.StatCollector(
  const input: TOmniValue;
  var output: TOmniValue);
var
  stat: TStatistics;
begin
  stat := GenerateStatistics(input.ToObject<TStringList>);
  output := TOmniValue.FromRecord<TStatistics>(stat);
end;
```

The third stage calls the GenerateStatistics function to generate the statistics. I decided not to show it in the book, as the implementation is not relevant to the pipeline pattern. The resulting record is assigned to the output parameter via a special FromRecord<T> function, which can store a record inside TOmniValue.

There's no need to destroy the TStringList that is contained in each TOmniValue input. Because it was assigned to the AsOwnedObject property (see the code for the previous stage), it will be destroyed automatically.

The last stage, StatAggregator, reads all TStatistics records from the input queue. The ToRecord<T> function is used to extract the record from a TOmniValue item. The Merge function, which is not shown in the book, merges current aggregate and new partial statistics data together.

Only after all the data is processed (when the for loop exits) will the resulting aggregate value be written to the pipeline's output:

```
procedure TfrmPipeline.StatAggregator(const input,
  output: IOmniBlockingCollection; const task: IOmniTask);
var
  aggregate: TStatistics;
  stat: TStatistics;
  value: TOmniValue;
```

```
begin
  aggregate := Default(TStatistics);

  for value in input do begin
    if task.CancellationToken.IsSignalled then
      break;

    stat := value.ToRecord<TStatistics>;
    Merge(aggregate, stat);
  end;

  output.TryAdd(
    TOmniValue.FromRecord<TStatistics>(aggregate));
end;
```

If we want to implement the parallelization in some other way (for example, with a *Parallel for* pattern), then this stage would need to implement locking to access the shared aggregate state. As the pipeline pattern gently guides you to use data duplication and communication instead, there is no need for sharing, locking, and other complications.

Displaying the result and shutting down

When all tasks have finished execution, the ShowResult method is called because it was passed to the OnStopInvoke method when the pipeline was constructed. ShowResult tries to read the result of the computation from the pipeline's output channel by calling FPipeline.Output.TryTake. This function returns False if the pipeline was canceled, so no result is displayed in that case:

```
procedure TfrmPipeline.ShowResult;
var
  value: TOmniValue;
  stat: TStatistics;
begin
  if FPipeline.Output.TryTake(value) then begin
    stat := value.ToRecord<TStatistics>;
    ListBox1.Items.Add(Format(
      'Files: %d, Lines: %d, Words: %d, Characters: %d',
      [stat.Files, stat.Lines, stat.Words, stat.Chars]));
  end;

  Cleanup;
end;
```

Map   379

Finally, `ShowResult` calls the `Cleanup` method, which destroys the pipeline interface by setting it to `nil`. It also clears and resets the user interface and prepares it for a new operation:

```
procedure TfrmPipeline.Cleanup;
begin
  FPipeline := nil;
  TimerUpdateProcessing.Enabled := false;
  lblProcessing.Visible := false;
  btnStart.Caption := 'Start';
  btnStart.Tag := 0;
end;
```

Pipelines are a very interesting concept that implement the *data duplication* and *communication* paradigms to the max. It allows for the parallelization of many different problems that don't fit into other patterns but require a different way of thinking about parallel code. The detailed examples in this and the previous chapter will help you get started with them.

Map

Let us move from a very complicated pattern to something really simple. The *Map* pattern was designed for one simple task. It takes an array, runs a mapping function on all of its elements in parallel, and returns a new array containing modified data. It implements the same functionality as Spring's `TEnumerable.Select`, except that it uses multiple worker threads. Besides mapping, it also supports data filtering, so we can also use it as a fast filter or a combined mapper/filter.

The `ParallelMap` demo implements two approaches to solve the same problem. They both create an array of 1,000,000 large-ish integer numbers (from 1,000,000 up), check each element in that array for primality, and generate a new array containing only prime numbers. They both measure the time used to check primality and generate the new array but ignore the time needed to prepare the input data.

The **Serial** button triggers an event handler, shown next, which does this in an idiomatic Delphi way:

```
procedure TfrmParallelMap.btnSerialClick(Sender: TObject);
var
  data  : integer;
  outIdx: integer;
  output: TArray<integer>;
  sw    : TStopwatch;
begin
  PrepareTestData;

  sw := TStopwatch.StartNew;
  SetLength(output, Length(FTestData));
  outIdx := Low(output) - 1;
```

```
  for data in FTestData do
    if IsPrime(data) then begin
      Inc(outIdx);
      output[outIdx] := data;
    end;
  SetLength(output, outIdx + 1);
  sw.Stop;

  LogResult(output, sw.ElapsedMilliseconds);
end;
```

We know that the output array will never be larger than the input array, so the code pre-allocates an output array of the same size. Then, it iterates over all elements in the input array, calls the `IsPrime` function (not shown here) for each element, and writes only prime numbers into the output array. Finally, the code truncates the output array to remove the unused space and logs the results.

The parallel version is even simpler, as the parallel pattern handles the pre-allocation of the output array and truncation of the unused elements. The code, shown next, simply calls `Parallel.Map<integer,integer>` (because we are mapping integer elements into integer elements) and provides the input array and a mapping function. The output array is returned as a result:

```
procedure TfrmParallelMap.btnParallelClick(Sender: TObject);
var
  output: TArray<integer>;
  sw    : TStopwatch;
begin
  PrepareTestData;

  sw := TStopwatch.StartNew;
  output := Parallel.Map<integer,integer>(FTestData,
    function (const source: integer;
              var target: integer): boolean
    begin
      Result := IsPrime(source);
      target := source;
    end);
  sw.Stop;

  LogResult(output, sw.ElapsedMilliseconds);
end;
```

The anonymous mapping function has two parameters – an element to be processed (`source`) and a mapped element (`target`). It has to return `True` if the output element was generated and `False` otherwise.

In our example, where we don't map the data but just filter it, we set the result of this anonymous function to the output of the IsPrime function and then set the output to the input (target := source). In this case, it is actually faster to always do the assignment (even if that is unnecessary when IsPrime returns False) and then add a test statement (if Result then target := source).

To help us verify that the parallel version really does the same job as the serial version, the LogResult method (not shown here) adds all numbers in the output array together and displays the length of the array and the sum of all elements, as shown in the following screenshot:

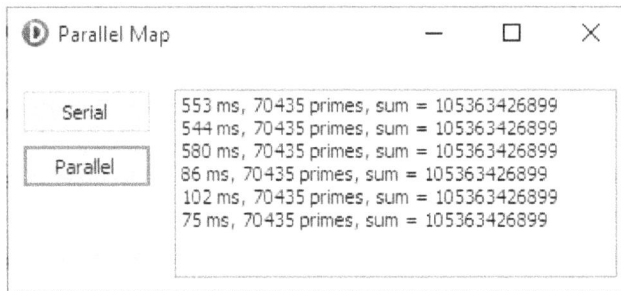

Figure 10.6 – Comparing the serial and parallel mapping of an array

We can see that the parallel version is about 6x faster, which is great considering that the test computer has eight fully parallel cores.

Timed Task

The last parallel pattern I will cover in this book is a simple timer that executes in a background thread. It can be configured very similarly to Delphi's TTimer and is extremely easy to use.

The ParallelTimedTask demo tests the network connection by executing an HTTP GET every 10 seconds and then displaying the results of the operation on the form. To prevent any blocking, it executes the network operation in a background thread. The interface of the program is shown in the following screenshot:

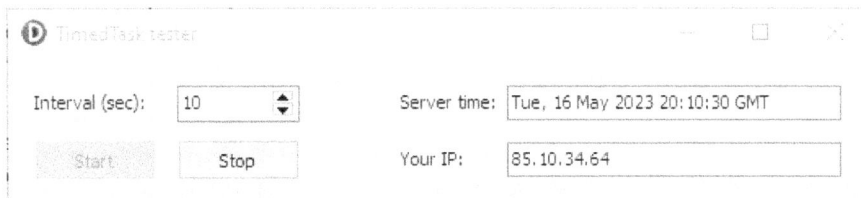

Figure 10.7 – The ParallelTimedTask demo in action

The code uses `Parallel.TimedTask` factory, which returns an `IOmniTimedTask` interface that is stored in the `FTimedTask` form field. To simplify the operation, everything is set up in the form's `OnCreate` event, as shown here:

```
procedure TfrmTimedTask.FormCreate(Sender: TObject);
begin
  FTimedTask := Parallel.TimedTask
                    .Every(inpInterval.Value * 1000)
                    .Execute(SendRequest_asy);
  FTimedTask.Stop;
  FTimedTask.TaskConfig(
    Parallel.TaskConfig.OnMessage(Self));
  btnStop.Enabled := false;
end;
```

The code creates a Timed Task pattern and instructs it to execute the `SendRequest_asy` method in a background thread every 10 seconds (by default, the `inpInterval` control is set to 10). As the `Execute` call automatically starts the timer, the code then calls the `Stop` function to stop it.

The code then instructs `FTimedTask` to send messages directly to the form (`OnMessage(Self)`). We will see later how these messages are handled.

The **Start** button, shown next, then simply activates the timer by setting its `Active` property to `True` (we could also use the `Start` method) and forces the first execution by calling `ExecuteNow`. Without that call, the timer would be triggered for the first time after 10 seconds:

```
procedure TfrmTimedTask.btnStartClick(Sender: TObject);
begin
  FTimedTask.Active := true;
  //equivalent:
  //FTimedTask.Start;
  FTimedTask.ExecuteNow; //force timer event
  btnStart.Enabled := false;
  btnStop.Enabled := true;
end;
```

The **Stop** button similarly sets the `Active` property to `False` (alternatively, you can use the `Stop` method of the `IOmniTimedTask` interface):

```
procedure TfrmTimedTask.btnStopClick(Sender: TObject);
begin
  FTimedTask.Active := false;
  //equivalent:
  //FTimedTask.Stop;
  btnStart.Enabled := true;
```

```
    btnStop.Enabled := false;
  end;
```

The timed task pattern supports changing the timer interval on the fly. If the interval is changed, FTimedTask starts over and waits for the new interval, regardless of how much time had passed before changing it. When the inpInterval component is changed, the following code updates the timer pattern:

```
procedure TfrmTimedTask.inpIntervalChange(Sender: TObject);
begin
  FTimedTask.Interval := inpInterval.Value * 1000;
end;
```

Let us now look at the code that is executed in the background task:

```
procedure TfrmTimedTask.SendRequest_asy(
  const task: IOmniTask);
var
  client    : THttpClient;
  date      : string;
  jsonObject: TJSonObject;
  origin    : string;
  response  : IHTTPResponse;
begin
  //Asynchronous timer method,
  //called from a background thread

  client := THttpClient.Create;
  try
    response := client.Get('https://httpbin.org/ip');
    date := response.GetHeaderValue('Date');

    jsonObject := TJSonObject.Create;
    try
      try
        origin := (jsonObject.ParseJSONValue(
                      response.ContentAsString)
                  as TJSONObject)
                    .Get('origin').JsonValue.Value;
      except //unexpected response
        origin := '';
      end;
    finally FreeAndNil(jsonObject); end;
```

```
      task.Comm.Send(MSG_NEWDATA, TOmniValue.CreateNamed(
        ['Date', date, 'Origin', origin]));
   finally FreeAndNil(client); end;
end;
```

As this method is part of the form itself, I had decorated its name with the _asy suffix, which serves as a visual warning to other programmers (including future me) that this method runs in a background thread and should not access the user interface.

The method creates a THttpClient object, reads the https://httpbin.org/ip page, and parses the output into two variables – date (on the server) and ip (our address). Then, it uses the task parameter, which represents a low-level task interface to send these two values to the form.

The task.Comm.Send call dispatches a MSG_NEWDATA message to the object that was specified in the TaskConfig.OnMessage call. In this example, that was Self, or the form object itself. The message contains TOmniValue, which stores a *named array* – an array of elements that can be accessed by a name.

To process these messages, the code declares MSG_NEWDATA as a Windows user message and declares a message handler method, MSGNewData. As shown next, this method must accept a parameter of type TOmniMessage:

```
const
  MSG_NEWDATA = WM_USER;
type
  TfrmTimedTask = class(TForm)
    //...
    procedure MSGNewData(var msg: TOmniMessage);
      message MSG_NEWDATA;
  end;
```

A TOmniMessage record is defined in the OtlComm unit and contains two fields – a message ID and TOmniValue data. The MSGNewData handler extracts both named values from the data and displays them on the screen:

```
procedure TfrmTimedTask.MSGNewData(var msg: TOmniMessage);
begin
  outServerTime.Text := msg.MsgData['Date'];
  outYourIP.Text := msg.MsgData['Origin'];
end;
```

With that short detour into low-level OTL multithreading, I'll conclude the chapter. OTL implements other interesting parallel patterns, such as *Parallel For* and *Fork/Join*, but they are less useful in everyday programming. You can learn about them by reading the OTL book or my blog articles (the link for which is included near the beginning of this chapter).

Summary

The focus of this chapter was on parallel programming patterns implemented in the open source framework OmniThreadLibrary. We could say that we had once again – after looking at the Spring library – returned to the great advice, *"Don't implement, reuse!"*

Although the chapter was dedicated to parallel patterns, I started by talking about a specialized thread-safe queue that is used again and again inside OTL's parallel patterns – a *blocking collection*. This standalone part of OTL can also be used with Delphi's `TThread` or `ITask`, which was also the focus of the demo.

After that, we focused on the three patterns we already discussed in the previous chapter – *Async*, *Join*, and *Future*. We saw how they are mostly similar to PPL implementation but also how they are different in small details.

Next, we explored the *Parallel Task* pattern. Although it represents just a light wrapper around the Join pattern, it served as a good excuse to explain some other neat features of OTL.

After that, it was time for more complicated topics. First, we looked into the *Background Worker* pattern, which is designed to write internal mini-servers with multiple data processing units. After that, we revisited the *Pipeline* pattern, this time from an OTL perspective.

To end the chapter on a lighter note, we finished with two simple patterns. *Map* is designed to quickly map and filter data in a big array. *Timed Task* is, by design, very similar to Delphi's *TTimer*, except that it executes a timer function in a background thread.

I could spend a lot longer discussing parallel programming and parallel patterns. I could fill up a whole book, without a doubt. However, as this is not a book on parallel programming but a book about high-performance practices, we need to move on to the last topic. In the next chapter, we'll see what we can do if Delphi can't handle it and we have to bring in another compiler.

11
Using External Libraries

We've seen many different ways of improving program speed in this book. We've changed algorithms, reduced the number of memory allocations, fine-tuned the code, made parts of programs parallel, and more. Sometimes, however, all of this is not enough. Maybe the problem is too hard and we give up, or maybe the fault lies in Delphi; more specifically, in the compiler.

In such cases, we will look outside the boundary of a local computer or company, and search the internet for a solution. Maybe somebody has already solved the problem and created a nicely wrapped solution for it. Sometimes, we'll be able to find one. Most of the time, they will *not* be written with Delphi in mind.

Face it, Delphi programmers (even though there are millions of us) are still a minority in a world of C, Java, and Python. It is much simpler to find a C or C++ solution for hard problems than a Delphi one. This is, however, not a reason to give up. It is always possible—with a bit of hard work—to use such a solution in a Delphi program, even if it is pre-packaged for C or C++ users.

In this chapter, we'll answer the following questions:

- What are the types of C and C++ files that you can use in your projects?
- What are the most frequently used object file formats?
- Which object formats can be used in Delphi projects?
- What traps await you when you try to link object files to Delphi?
- How can you use C++ libraries in Delphi projects?

Technical requirements

All code in this chapter was written with **Delphi 11.3 Alexandria**. Most of the examples, however, could also be executed on Delphi XE and newer versions. C++ code was tested with Microsoft Visual Studio 2019 and 2022. You can find all the examples on GitHub: https://github.com/PacktPublishing/Delphi-High-Performance---Second-Edition/tree/main/ch11.

Linking with object files

Before I jump into the complicated world of interfacing with C and C++, I'll introduce a simple example—a library written in Delphi. The motivation for its use comes not from a bad algorithm that Mr. Smith wrote but from a badly performing 64-bit compiler. This is not something that I am claiming without proof. Multiple Delphi programmers have pointed out that the 64-bit Windows compiler (dcc64) generates pretty bad floating-point code that is 2-3 times slower than the floating-point generated from an equivalent source by a C compiler.

When you have already explored all the standard approaches to speeding up the program, and the compiler is the only source of the problem, you cannot do much. You can only rewrite parts of the program in assembler, or use an external library that works faster than the native code. Such a library will either use lots of assembler code or—most of the time—contain a bunch of object files compiled with an optimizing C compiler.

Sometimes, such an external library will only cover your basic needs. For example, the FastMath library, which is written in Delphi (`https://github.com/neslib/FastMath`) contains floating-point functions that operate on vectors and matrices. Others, such as *Intel's Math Kernel Library* (`https://software.intel.com/en-us/mkl/`), also add more complicated mathematical algorithms, such as Fast Fourier Transform.

For a short example, let's look into FastMath and how it implements a Radians function that converts a vector of four floating-point numbers (TVector4) from degrees to radians. For your convenience, the copy of the FastMath repository is included in the code archive for this chapter, in the FastMath-master folder.

As a fallback, the library implements a pure Pascal Radians function in the Neslib.FastMath. Pascal.inc file. The trivial implementation shown in the following code, only calls the Radians function for a single-precision argument four times:

```
function Radians(const ADegrees: Single): Single;
begin
  Result := ADegrees * (Pi / 180);
end;

function Radians(const ADegrees: TVector4): TVector4;
begin
  Result.Init(Radians(ADegrees.X), Radians(ADegrees.Y),
    Radians(ADegrees.Z), Radians(ADegrees.W));
end;
```

A different implementation in the `Neslib.FastMath.Sse2_32.inc` unit (shown in the following code) implements the same method with much faster Intel SSE2 assembler instructions. Similar code in `Neslib.FastMath.Sse2_64.inc` (which is not shown in the book but can be found in the `FastMath` folder in the code archive) does the same for a 64-bit Intel target:

```
function Radians(const ADegrees: TVector4): TVector4;
  assembler;
asm
  movups xmm0, [ADegrees]
  movups xmm1, [SSE_PI_OVER_180]
  mulps xmm0, xmm1
  movups [Result], xmm0
end;
```

For ARM platforms (Android and iOS), the same code is implemented in `trig_32.S` (and `trig_64.S` for 64-bit platforms). A fragment of that file showing 32-bit implementation is reproduced here:

```
_radians_vector4: // (const ADegrees: TVector4;
                  //  out Result: TVector4);
  adr r2, PI_OVER_180
  vld1.32 {q0}, [r0]
  vld1.32 {q1}, [r2]
  vmul.f32 q0, q0, q1
  vst1.32 {q0}, [r1]
  bx lr
```

The batch files in the `fastmath-android` and `fastmath-ios` folders use native development tools, made by Google and Apple, to compile assembler sources into the object libraries, `libfastmath-android.a` and `libfastmath-ios.a`. Functions are then linked to the library in the `Neslib.FastMath.Internal.inc` unit with a mechanism that we'll explore in the continuation of this chapter.

A simple series of `$IF`/`$INCLUDE` statements then links those include files in one unit, `Neslib.FastMath.pas`:

```
{$IF Defined(FM_PASCAL)}
  // Pascal implementations
  {$INCLUDE 'Neslib.FastMath.Pascal.inc'}
{$ELSEIF Defined(FM_X86)}
  // 32-bit SSE2 implementations
  {$INCLUDE 'Neslib.FastMath.Sse2_32.inc'}
{$ELSEIF Defined(FM_X64)}
  // 64-bit SSE2 implementations
  {$IF Defined(MACOS64)}
    {$INCLUDE 'Neslib.FastMath.Sse2_64.MacOS.inc'}
```

```
  {$ELSE}
    {$INCLUDE 'Neslib.FastMath.Sse2_64.inc'}
  {$ENDIF}
{$ELSEIF Defined(FM_ARM)}
  // Arm NEON/Arm64 implementations
  {$INCLUDE 'Neslib.FastMath.Arm.inc'}
{$ENDIF}
```

This design allows the library to present a unified application interface that uses the best implementation for each platform to your Delphi application.

In this case, the implementation is linked against a static library (.a). This will not be the case when trying to include a C-specific solution in a Delphi project. You'll have a bunch of object files (.obj, .o) instead.

Object files are a complicated topic, as there is not a single standard governing them, but a bunch of competing specifications. Before we start using them in Delphi, we should therefore have at least some understanding of what is encountered in the wild.

Object file formats

Object files come in different formats and not all are useful for linking with Delphi. The three most common file formats that you'll encounter are *OMF*, *COFF*, and *ELF*:

- The **Relocatable Object Module Format** (**OMF**) is a format developed by Intel. It was extensively used in DOS times. Support for linking OMF object files was already included in Turbo Pascal DOS compilers and is still present in modern Delphi.

  This format can be linked with Delphi's Win32 and compilers. It is generated by the C++Builder's Win32 compiler (bcc32).

  To analyze an OMF file, you can use the tdump utility (part of the Delphi and C++Builder standard installation) with the -d switch. A command such as tdump -d object_file.obj will output a bunch of information about the OMF file. It is best if you *redirect* this output into a file by using the command-line redirect functionality. For example, you can type tdump -d object_file.obj > object_file.txt to redirect the output to object_file.txt.

  Once you have the file, search for entries with the label PUBDEF or PUBD32 to find the names of exported functions. You can also use the command-line parameter -oiPUBDEF (or -oiPUBD32) to limit output to PUBDEF (PUBD32) records.

The following output shows the results of the `tdump -d -oiPUBDeF LzmaDec.obj` command. The `LzmaDec.obj` file is part of the code archive for this chapter and will be used as an example in the next section. We can see that this object file contains the `LzmaDec_InitDicAndState` and `LzmaDec_Init` functions, and so on. These functions can, with some work, be used from a Delphi program:

```
Turbo Dump Version 6.5.4.0 Copyright (c) 1988-2016 Embarcadero
Technologies, Inc.
Display of File lzmadec.obj
002394 PUBDEF 'LzmaDec_InitDicAndState' Segment: _TEXT:172E
0023B5 PUBDEF 'LzmaDec_Init' Segment: _TEXT:176B
0023CB PUBDEF 'LzmaDec_DecodeToDic' Segment: _TEXT:17D2
0023E8 PUBDEF 'LzmaDec_DecodeToBuf' Segment: _TEXT:1A6E
002405 PUBDEF 'LzmaDec_FreeProbs' Segment: _TEXT:1B40
002420 PUBDEF 'LzmaDec_Free' Segment: _TEXT:1B78
002436 PUBDEF 'LzmaProps_Decode' Segment: _TEXT:1B9B
002450 PUBDEF 'LzmaDec_AllocateProbs' Segment: _TEXT:1C9A
00246F PUBDEF 'LzmaDec_Allocate' Segment: _TEXT:1CE5
002489 PUBDEF 'LzmaDecode' Segment: _TEXT:1D6D
```

- The second file format that you'll frequently encounter is **Common Object File Format** (COFF). It was introduced in *Unix System V* and is generated by Microsoft's Visual C++ compilers. Delphi Win32 and Win64 compilers can link to it.

  To dump this file, use Embarcadero's `tdump` with the `-E` switch, or Microsoft's `dumpbin` utility, which is part of Visual Studio.

- The third most popular object format is **Executable and Linkable Format** (ELF). It is mostly used on Unix platforms. This format is used in the *LLVM* compiler infrastructure and as such, is supported by some Delphi and C++Builder compilers—namely the ones that use the LLVM toolchain. This file format is not supported by the `tdump` utility.

  ELF format is not supported by current Win32 and Win64 Delphi compilers. Usually, that would not present a big problem, except that the Win64 C++Builder compiler (bcc64) uses the LLVM infrastructure and therefore generates object files in ELF format. As far as I can tell, there's currently no way to use C++Builder to generate 64-bit object files that can be linked to Delphi.

Object file linking in practice

To demonstrate some typical problems that you may run into when trying to link object files to Delphi, I picked two simple examples from the *Abbrevia* open source library. Abbrevia was originally a commercial compression library developed by TurboPower. They were selling multiple libraries—AsyncPro, Orpheus, and SysTools, which were later donated to the open source community. Their current home is at `https://github.com/TurboPack`.

For simplicity, I have included the complete Abbrevia repository in the code archive. All demos are already configured so you can just press *F9* in Delphi and enjoy.

The first demo is called `LzmaDecTest` and it links in a single object file—`LzmaDec.obj`. That file is stored in the `Abbrevia\source\Win32` folder (or `Win64` if you are compiling for 64-bit Windows). This demo is a standard Windows console application with one additional line linking the object file:

```
{$L LzmaDec.obj}
```

If you try to compile the test project, you'll notice that compilation fails with the following error message:

```
[dcc32 Error] LzmaDecTest.dpr(22): E2065 Unsatisfied forward or
external declaration: 'memcpy'
```

This happens a lot when linking object files created with C compilers. An object file can contain references to functions that are not implemented inside the object file. The same situation happens with Delphi—a `.dcu` file can (and most probably will) contain references to methods implemented in different units.

An experienced programmer will try to minimize these dependencies when writing a library, simply because that helps with portability, but there are always some basic functions that are necessary in almost all C programs. Think of Delphi's *system unit*—all units in your program implicitly depend on it and most of the time you are not aware of that (and you don't have to be). Most of the object files that you'll have to link to will be compiled with Microsoft's C compiler, and typically they will depend on the `msvcrt` library.

There are two ways out of this problem. Firstly, you can write the missing functions yourself. They are implemented by the `MSVCRT.DLL` dynamic library and you only have to link to them.

For example, in our case, we would need the missing `memcpy` function. We can simply import it from the appropriate DLL by adding the following line to the program. We would also have to add the `Windows` unit to the `uses` list as the `size_t` type is defined in it:

```
function memcpy(dest, src: Pointer; count: size_t):
  Pointer; cdecl; external 'msvcrt.dll';
```

Alternatively, we can just add the `System.Win.Crtl` unit to the `uses` list. It links to all functions from the `MSVCRT.DLL` dynamic library.

Be aware though, `MSVCRT.DLL` is not part of the operating system. If you want to distribute an application that links to this DLL, you should also include the *Microsoft Visual C++ Redistributable* package, which you can download from the Microsoft web server at `https://learn.microsoft.com/en-us/cpp/windows/latest-supported-vc-redist?view=msvc-170`.

Sometimes the missing functions are quite simple—or already implemented in the Delphi RTL—and we can code them ourselves. In this example, we could write the following memcpy implementation:

```
function memcpy(dest, src: Pointer;
                count: size_t): Pointer; cdecl;
begin
  Move(src^, dest^, count);
  Result := dest;
end;
```

For the second demo, I wanted to link to the decompress.obj file, which is also part of Abbrevia and is stored in the same folder as LzmaDec.obj. This demo is named DecompressTest.

Again, I started with an empty console application, to which I added {$LINK decompress.obj}. ({$L} and ${LINK} are synonyms.) The compiler reported four errors:

```
[dcc32 Error] DecompressTest.dpr(45): E2065 Unsatisfied forward or
external declaration: 'BZ2_rNums'
[dcc32 Error] DecompressTest.dpr(45): E2065 Unsatisfied forward or
external declaration: 'BZ2_hbCreateDecodeTables'
[dcc32 Error] DecompressTest.dpr(45): E2065 Unsatisfied forward or
external declaration: 'BZ2_indexIntoF'
[dcc32 Error] DecompressTest.dpr(45): E2065 Unsatisfied forward or
external declaration: 'bz_internal_error'
```

To fix such problems, you'll always have to use the documentation that comes with the object files. There's no way to know what those symbols represent simply by looking at the object file. In this case, I found out that the first is an initialization table (so the symbol is actually the name of a global variable). The second and third functions are implemented in other object files, and the last one is an error-handling function that we have to implement in the code. Adding the following code fragment to the file left me with two errors:

```
var
  BZ2_rNums: array [0..511] of Longint;

procedure bz_internal_error(errcode: Integer); cdecl;
begin
  raise Exception.CreateFmt('Compression Error %d',
                            [errcode]);
end;
```

Take note of the calling convention used here. C object files will almost invariably use the cdecl convention.

To find the object files with missing functions, you can use a full-text search tool that handles binary files and searches for missing names. This will give you false positives as it will also return object files that are *using* those functions. You can then use the `tdump` utility to examine potential candidates and find the true source of those units.

Or, you can look into the documentation. But who does that?

I found `BZ2_hbCreateDecodeTables` in `huffman.obj` and `BZ2_indexIntoF` in `bzlib.obj`. Only two lines are needed to add them to the project:

```
{$LINK huffman.obj}
{$LINK bzlib.obj}
```

That fixes the two errors about missing symbols but introduces three new errors:

```
[dcc32 Error] DecompressTest.dpr(40): E2065 Unsatisfied forward or
external declaration: 'BZ2_crc32Table'
 [dcc32 Error] DecompressTest.dpr(40): E2065 Unsatisfied forward or
external declaration: 'BZ2_compressBlock'
 [dcc32 Error] DecompressTest.dpr(40): E2065 Unsatisfied forward or
external declaration: 'BZ2_decompress'
```

`BZ2_crc32Table` is an initialization table (says the documentation) so we need another variable:

```
var
  BZ2_crc32Table: array[0..255] of Longint;
```

Further detective work found `BZ2_compressBlock` in the `compress.obj` file, so let's add this to the project. We only have to add one line:

```
{$LINK compress.obj}
```

The second error is trickier. Research showed that this function is actually implemented in the `decompress.obj` unit, which is already linked to the project! Why is the error reported then? Simply because that's how the single-pass Delphi compiler works.

As it turns out, the compiler will happily resolve symbols used by object file *A* if they are defined in object file *B*, which is loaded *after* file *A*. It will, however, *not* resolve symbols used by object file *B* if they are defined in object file *A*.

In our case, the `compress.obj` file needs the `BZ2_decompress` symbol, which is defined in `decompress.obj`, but as the latter is linked *before* the former, we get the error. You could try rearranging the units but then some other symbol will not be found.

Luckily, this problem is very simple to fix. We just have to tell the compiler that `BZ2_decompress` exists. The compiler is happy with that and produces workable code. When all units are compiled, the *linker* kicks in to collect them in the EXE file. The linker works in a different way than the compiler and has no problem finding the correct implementation regardless of the include order.

If we add the following line to the code, the error about *unsatisfied* BZ2_decompress *declaration* goes away. I also like putting in a comment that marks the source of the function:

```
procedure BZ2_decompress; external; //decompress.obj
```

Linking compress.obj causes three new errors to appear:

```
[dcc32 Error] DecompressTest.dpr(45): E2065 Unsatisfied forward or
external declaration: 'BZ2_hbMakeCodeLengths'
[dcc32 Error] DecompressTest.dpr(45): E2065 Unsatisfied forward or
external declaration: 'BZ2_hbAssignCodes'
[dcc32 Error] DecompressTest.dpr(45): E2065 Unsatisfied forward or
external declaration: 'BZ2_blockSort'
```

The first two are defined in huffman.obj, which is already linked, so we have to add the following two lines to the code:

```
procedure BZ2_hbMakeCodeLengths; external; //huffman.obj
procedure BZ2_hbAssignCodes; external;     //huffman.obj
```

The last one comes from blocksort.obj, which we have yet to link in. After that, the code finally compiles:

```
{$LINK blocksort.obj}
```

This is, of course, just the first step toward a working program. Now we have to check how functions exported from these object files are actually defined. Then, we have to add appropriate external declarations to the Delphi source.

If we look at the previous example, LzmaDecTest, we can implement functions to initialize and free the decoder by adding the following two lines:

```
procedure LzmaDec_Init(var state); cdecl; external;
procedure LzmaDec_Free(var state; alloc: pointer); cdecl;
  external;
```

This declaration is actually incomplete, as the first parameter to both functions should be a complicated record, which I didn't want to try to decipher just for this simple demo. Rather than do that (and a whole lotta additional work), I would depend on the ready-to-use Abbrevia library.

By now, you've probably noticed that linking in object files is hard work. If at all possible, you should try to find a Delphi (or at least a Free Pascal) wrapper. Even if it covers only a part of the functionality you need, it will spare you hours and hours of work.

Using C++ libraries

Using C object files in Delphi is hard but possible. Linking to C++ object files is, however, nearly impossible. The problem does not lie within the object files themselves, but in C++.

While C is hardly more than an assembler with improved syntax, C++ represents a sophisticated high-level language with runtime support for strings, objects, exceptions, and more. All these features are part of almost any C++ program and are as such compiled into (almost) any object file produced by C++.

The problem here is that Delphi has no idea how to deal with any of that. A C++ object is not equal to a Delphi object. Delphi has no idea how to call functions of a C++ object, how to deal with its inheritance chain, how to create and destroy such objects, and so on. The same holds for strings, exceptions, streams, and other C++ concepts.

If you can compile the C++ source with C++Builder, then you can create a package (.bpl) that can be used from a Delphi program. Most of the time, however, you will not be dealing with a source project. Instead, you'll want to use a commercial library that only gives you a bunch of C++ header files (.h) and one or more static libraries (.lib). Most of the time, the only Windows version of that library will be compiled with Microsoft's Visual Studio.

A more general approach to this problem is to introduce a *proxy* DLL created in C++. You will have to create it in the same development environment as was used to create the library you are trying to link into the project. On Windows, that will in most cases be Visual Studio. That will enable us to include the library without any problems.

To allow Delphi to use this DLL (and as such use the library), the DLL should expose a simple interface in the Windows API style. Instead of exposing C++ objects, the API must expose methods implemented by the objects as normal (non-object) functions and procedures. As the objects cannot cross the API boundary, we must find some other way to represent them on the Delphi side.

Instead of showing how to write a DLL wrapper for an existing (and probably quite complicated) C++ library, I have decided to write a very simple C++ library that exposes a single class, implementing only two methods. As compiling this library requires Microsoft's Visual Studio, which not all of you have installed, I have also included the compiled version (DllLib1.dll) in the code archive.

The Visual Studio solution is stored in the StaticLib1 folder and contains two projects. StaticLib1 is the project used to create the library while the Dll1 project implements the proxy DLL.

The static library implements the CppClass class, which is defined in the header file, CppClass.h. Whenever you are dealing with a C++ library, the distribution will also contain one or more header files. They are needed if you want to use a library in a C++ project—such as in the proxy DLL Dll1.

The header file for the demo library StaticLib1 is shown in the following. We can see that the code implements a single CppClass class, which implements a constructor (CppClass()), destructor (~CppClass()), a method accepting an integer parameter (void setData(int)), and a function returning an integer (int getSquare()). The class also contains one integer private field, data:

```cpp
#pragma once
class CppClass
{
   int data;
public:
   CppClass();
   ~CppClass();
   void setData(int);
   int getSquare();
};
```

The implementation of the CppClass class is stored in the CppClass.cpp file. You don't need this file when implementing the proxy DLL. When we are using a C++ library, we are strictly coding to the interface—and the interface is stored in the header file.

In our case, we have the full source so we can look inside the implementation too. The constructor and destructor don't do anything and so I'm not showing them here. The other two methods are as follows. The setData method stores its parameter in the internal field and the getSquare function returns the squared value of the internal field:

```cpp
void CppClass::setData(int value)
{
   data = value;
}

int CppClass::getSquare()
{
   return data * data;
}
```

This code doesn't contain anything that we couldn't write in 60 seconds in Delphi. It does, however, serve as the perfect simple example for writing a proxy DLL.

Creating such a DLL in Visual Studio is easy. You just have to select **File** | **New** | **Project**, and select the **Dynamic-Link Library (DLL)** project type from the **Visual C++** | **Windows Desktop** branch.

The Dll1 project from the code archive has only two source files. The dllmain.cpp file was created automatically by Visual Studio and contains the standard DllMain method. You can change this file if you have to run project-specific code when a program and/or a thread attaches to or detaches from the DLL. In my example, this file was left just as Visual Studio created it.

The second file, `StaticLibWrapper.cpp`, fully implements the proxy DLL. It starts with two `include` lines (shown in the following), which bring in the required RTL header `stdafx.h` and the header definition for our C++ class, `CppClass.h`:

```
#include "stdafx.h"
#include "CppClass.h"
```

The proxy has to be able to find our header file. There are two ways to do that. We could simply copy it to the folder containing the source files for the DLL project, or we can add it to the project's search path. The second approach can be configured in **Project | Properties | Configuration Properties | C/C++ | General | Additional Include Directories**. This is also the approach used by the demonstration program.

The DLL project must be able to find the static library that implements the `CppClass` object. The path to the library file should be set in project options, in the **Configuration Properties | Linker | General | Additional Library Directories** settings. You should put the name of the library (`StaticLib1. lib`) in the **Linker | Input | Additional Dependencies** settings.

The next line in the source file defines a macro called EXPORT, which will be used later in the program to mark a function as *exported*. We have to do that for every DLL function that we want to use from the Delphi code. Later, we'll see how this macro is used:

```
#define EXPORT comment(linker, "/EXPORT:" __FUNCTION__ "=" __
FUNCDNAME__)
```

The next part of the `StaticLibWrapper.cpp` file implements an `IndexAllocator` class, which is used internally to cache C++ objects. It associates C++ objects with simple integer identifiers, which are then used outside the DLL to represent the object. I will not show this class in the book as the implementation is not that important. You only have to know how to use it.

This class is implemented as a simple static array of pointers and contains, at most, MAXOBJECTS objects. The constant MAXOBJECTS is set to 100 in the current code, which limits the number of C++ objects created by the Delphi code to *100*. Feel free to modify the code if you need to create more objects.

The following code fragment shows three public functions implemented by the `IndexAllocator` class. The `Allocate` function takes an `obj` pointer, stores it in the cache, and returns its index in the `deviceIndex` parameter. The result of the function is FALSE if the cache is full and TRUE otherwise.

The `Release` function accepts an index (which was previously returned from `Allocate`) and marks the cache slot at that index as empty. This function returns FALSE if the index is invalid (does not represent a value returned from `Allocate`) or if the cache slot for that index is already empty.

The last function, Get, also accepts an index and returns the pointer associated with that index. It returns NULL if the index is invalid or if the cache slot for that index is empty:

```
bool Allocate(int& deviceIndex, void* obj)
bool Release(int deviceIndex)
void* Get(int deviceIndex)
```

In the next section, we'll look at how to write exported functions.

Writing exported functions

Let's move on now to functions that are exported from the DLL. The first two—Initialize and Finalize—are used to initialize internal structures, namely GAllocator of type IndexAllocator, and to clean up before the DLL is unloaded. Instead of looking into them, I'd rather show you the more interesting stuff, namely functions that deal with CppClass.

The CreateCppClass function creates an instance of CppClass, stores it in the cache, and returns its index. The important three parts of the declaration are extern "C", WINAPI, and #pragma EXPORT:

- extern "C" is there to guarantee that the CreateCppClass name will not be changed when it is stored in the library. The C++ compiler tends to change (*mangle*) function names to support method overloading (the same thing happens in Delphi) and this declaration prevents that.

- WINAPI changes the *calling convention* from cdecl, which is standard for C programs, to stdcall, which is commonly used in DLLs. Later, we'll see that we also have to specify the correct calling convention on the Delphi side.

- The last important part, #pragma EXPORT, uses the previously defined EXPORT macro to mark this function as *exported*.

The CreateCppClass returns 0 if the operation was successful and -1 if it failed. The same approach is used in all functions exported from the demo DLL:

```
extern "C" int WINAPI CreateCppClass (int& index)
{
#pragma EXPORT
  CppClass* instance = new CppClass;
  if (!GAllocator->Allocate(index, (void*)instance)) {
    delete instance;
    return -1;
  }
  else
    return 0;
}
```

Similarly, the `DestroyCppClass` function (not shown here) accepts an index parameter, fetches the object from the cache, and destroys it.

The DLL also exports two functions that allow the DLL user to operate on an object. The first one, `CppClass_setValue`, accepts an index of the object and a value. It fetches the `CppClass` instance from the cache (given the index) and calls its `setData` method, passing it the value:

```
extern "C" int WINAPI CppClass_setValue(int index,
  int value)
{
#pragma EXPORT
  CppClass* instance = (CppClass*)GAllocator->Get(index);
  if (instance == NULL)
    return -1;
  else {
    instance->setData(value);
    return 0;
  }
}
```

The second function, `CppClass_getSquare`, also accepts an object index and uses it to access the `CppClass` object. After that, it calls the object's `getSquare` function and stores the result in the output parameter, `value`:

```
extern "C" int WINAPI CppClass_getSquare(int index,
  int& value)
{
#pragma EXPORT
  CppClass* instance = (CppClass*)GAllocator->Get(index);
  if (instance == NULL)
    return -1;
  else {
    value = instance->getSquare();
    return 0;
  }
}
```

A proxy DLL that uses a mapping table is a bit complicated and requires some work. We could also approach the problem in a much simpler manner—by treating an address of an object as its external identifier. In other words, the `CreateCppClass` function would create an object and then return its address as an untyped pointer type. `CppClass_getSquare`, for example, would accept this pointer, cast it to a `CppClass` instance, and execute an operation on it. An alternative version of these two methods is shown in the following:

```
extern "C" int WINAPI CreateCppClass2(void*& ptr)
{
  #pragma EXPORT
  ptr = new CppClass;
  return 0;
}

extern "C" int WINAPI CppClass_getSquare2(void* index,
  int& value)
{
  #pragma EXPORT
  value = ((CppClass*)index)->getSquare();
  return 0;
}
```

This approach is simpler but offers far less security in the form of error checking. The table-based approach can check whether the index represents a valid value, while the latter version cannot know whether the pointer parameter is valid or not. If we make a mistake on the Delphi side and pass in an invalid pointer, the code would treat it as an instance of a class, do some operations on it, possibly corrupt some memory, and maybe crash.

Finding the source of such errors is very hard. That's why I prefer to write more verbose code that implements some safety checks on the code that returns pointers.

Using a proxy DLL in Delphi

To use any DLL from a Delphi program, we must first import functions from the DLL. There are different ways to do this—we could use *static* linking, *dynamic* linking, and static linking with *delayed loading*. There's plenty of information on the internet about the art of DLL writing in Delphi so I won't dig into this topic. I'll just stick with the most modern approach—delay loading.

The code archive for this book includes two demo programs, which demonstrate how to use the `DllLib1.dll` library. The simpler one, `CppClassImportDemo`, uses the DLL functions directly, while `CppClassWrapperDemo` wraps them in an easy-to-use class.

Both projects use the `CppClassImport` unit to import the DLL functions into the Delphi program. The following code fragment shows the `interface` part of that unit, which tells the Delphi compiler which functions from the DLL should be imported, and what parameters they have.

Note

To successfully run the demos, the `DllLib1.dll` library has to be found in the EXE folder, in `C:\WINDOWS\SYSWOW64`, or in one of the folders on the `PATH`.

As with the C++ part, there are three important parts to each declaration. Firstly, `stdcall` specifies that the function call should use the `stdcall` (or what is known in C as **WINAPI**) calling convention. Secondly, the name after the `name` specifier should match the exported function name from the C++ source. And thirdly, the `delayed` keyword specifies that the program should not try to find this function in the DLL when it is started but only when the code calls the function. This allows us to check whether the DLL is present at all before we call any of the functions:

```
const
  CPP_CLASS_LIB = 'DllLib1.dll';

function Initialize: integer; stdcall;
  external CPP_CLASS_LIB name 'Initialize' delayed;
function Finalize: integer; stdcall;
  external CPP_CLASS_LIB name 'Finalize' delayed;
function CreateCppClass(var index: integer): integer; stdcall;
  external CPP_CLASS_LIB name 'CreateCppClass' delayed;
function DestroyCppClass(index: integer): integer; stdcall;
  external CPP_CLASS_LIB name 'DestroyCppClass' delayed;
function CppClass_setValue(index: integer;
  value: integer): integer; stdcall;
  external CPP_CLASS_LIB name 'CppClass_setValue' delayed;
function CppClass_getSquare(index: integer;
  var value: integer): integer; stdcall;
  external CPP_CLASS_LIB name 'CppClass_getSquare' delayed;
```

The `implementation` part of this unit (not shown here) shows how to catch errors that occur during *delayed loading*—that is, when the code that calls any of the imported functions tries to find that function in the DLL.

If you get an `External exception C06D007F` exception when you try to call a delay-loaded function, you have probably mistyped a name—either in C++ or in Delphi. You can use the `tdump` utility that comes with Delphi to check which names are exported from the DLL. The syntax is `tdump -d <dll_name.dll>`.

If the code crashes when you call a DLL function, check whether both sides correctly define the calling convention. Also, check if all the parameters have correct types on both sides and if the `var` parameters are marked as such on both sides.

To use the DLL, the code in the `CppClassMain` unit first calls the exported `Initialize` function from the form's `OnCreate` handler to initialize the DLL. The cleanup function, `Finalize`, is called from the `OnDestroy` handler to clean up the DLL. All parts of the code check whether the DLL functions return the OK status (value 0):

```
procedure TfrmCppClassDemo.FormCreate(Sender: TObject);
begin
  if Initialize <> 0 then
    ListBox1.Items.Add('Initialize failed')
end;

procedure TfrmCppClassDemo.FormDestroy(Sender: TObject);
begin
  if Finalize <> 0 then
    ListBox1.Items.Add('Finalize failed');
end;
```

When you click on the **Use import library** button, the following code executes. It uses the DLL to create a `CppClass` object by calling the `CreateCppClass` function. This function puts an integer value into the `idxClass` value. This value is used as an identifier that identifies a `CppClass` object when calling other functions.

The code then calls `CppClass_setValue` to set the internal field of the `CppClass` object and `CppClass_getSquare` to call the `getSquare` method and return the calculated value. At the end, `DestroyCppClass` destroys the `CppClass` object:

```
procedure TfrmCppClassDemo.btnImportLibClick(
  Sender: TObject);
var
  idxClass: Integer;
  value: Integer;
begin
  if CreateCppClass(idxClass) <> 0 then
    ListBox1.Items.Add('CreateCppClass failed')
  else if CppClass_setValue(idxClass, SpinEdit1.Value) <> 0
  then
    ListBox1.Items.Add('CppClass_setValue failed')
  else if CppClass_getSquare(idxClass, value) <> 0 then
    ListBox1.Items.Add('CppClass_getSquare failed')
  else begin
    ListBox1.Items.Add(Format('square(%d) = %d',
      [SpinEdit1.Value, value]));
    if DestroyCppClass(idxClass) <> 0 then
      ListBox1.Items.Add('DestroyCppClass failed')
```

```
    end;
  end;
```

This approach is relatively simple but long-winded and error-prone. A better way is to write a wrapper Delphi class that implements the same public interface as the corresponding C++ class. The second demo, `CppClassWrapperDemo`, contains a `CppClassWrapper` unit that does just that.

This unit implements a `TCppClass` class, which maps to its C++ counterpart. It only has one internal field, which stores the index of the C++ object as returned from the `CreateCppClass` function:

```
type
  TCppClass = class
  strict private
    FIndex: integer;
  public
    class procedure InitializeWrapper;
    class procedure FinalizeWrapper;
    constructor Create;
    destructor Destroy; override;
    procedure SetValue(value: integer);
    function GetSquare: integer;
  end;
```

I won't show all of the functions here as they are all equally simple. One—or maybe two—will suffice.

The constructor just calls the `CreateCppClass` function, checks the result, and stores the resulting index in the internal field:

```
constructor TCppClass.Create;
begin
  inherited Create;
  if CreateCppClass(FIndex) <> 0 then
    raise Exception.Create('CreateCppClass failed');
end;
```

Similarly, `GetSquare` just forwards its job to the `CppClass_getSquare` function:

```
function TCppClass.GetSquare: integer;
begin
  if CppClass_getSquare(FIndex, Result) <> 0 then
    raise Exception.Create('CppClass_getSquare failed');
end;
```

When we have this wrapper, the code in the main unit becomes very simple—and very Delphi-like. Once the initialization in the OnCreate event handler is done, we can just create an instance of TCppClass and work with it:

```
procedure TfrmCppClassDemo.FormCreate(Sender: TObject);
begin
  TCppClass.InitializeWrapper;
end;

procedure TfrmCppClassDemo.FormDestroy(Sender: TObject);
begin
  TCppClass.FinalizeWrapper;
end;

procedure TfrmCppClassDemo.btnWrapClick(Sender: TObject);
var
  cpp: TCppClass;
begin
  cpp := TCppClass.Create;
  try
    cpp.SetValue(SpinEdit1.Value);
    ListBox1.Items.Add(Format('square(%d) = %d',
      [SpinEdit1.Value, cpp.GetSquare]));
  finally
    FreeAndNil(cpp);
  end;
end;
```

As you can see, writing a proxy DLL for a C++ library requires lots of work. Most of the time, however, this is not a complicated job, but merely grunt work that you have to get through to be able to enjoy results. Do it slowly, with lots of testing, and everything will be fine!

Summary

I would be the first to admit that this chapter presents only sketches of ideas rather than fully reusable solutions. That is a direct consequence of the topic, which is too broad to give definite answers. Rather than that, I have tried to give you enough information to do your own research on the topic.

The chapter started with a discussion of the possible reasons for including external libraries in your application. They may cover a topic you are unfamiliar with, or they may implement some specific algorithm faster than you are able to—or faster than the Delphi compiler can do it.

That brought us to the `FastMath` Delphi library, which implements fast functions for working with vectors (series of numbers) and matrices (two-dimensional arrays of numbers). This library uses assembler intermixed with Pascal to give you the best performance in each case. We saw how assembler code is sometimes included internally and sometimes as an external file, and we can write a single frontend that encompasses all platform-specific implementations.

After that, I took a short detour and inspected most of the popular types of object files that you'll encounter in your work. Object files are frequently used when linking C code with Delphi and it is good to know what you can run into.

I followed up with a longer example on how to link C object files with Delphi, and what problems you can encounter. We saw how to replace the standard Microsoft C library, `msvcrt`, with a Delphi equivalent, how to write Delphi functions that plug a hole expected by the object file, and how to include multiple files that use each other in a circular fashion.

Then I moved from C to C++. This is currently the most popular Windows development tool and you'll find it used in most open source and commercial libraries. I explained why we cannot use C++ directly in Delphi applications and how to work around that.

The chapter ended with a practical demonstration that linked a very simple C++ library through a proxy DLL written in C++ to a Delphi application. I looked into two different ways of implementing a proxy DLL and into two ways of using this DLL in a Delphi application.

Linking to C and C++ code is hard and certainly not for everyone, but in critical situations, it can save your skin.

The long and winding path that has led you through the many different topics in this book is coming to an end. There are only two things left: to take a deep breath and get some R&R. The next chapter will take care of that. Instead of introducing new topics, it will return to the beginning. In it, I'll revisit the most important topics of the book to help you *relax* and *remember*.

12

Best Practices

This book has covered lots of different topics. I believe that, right now, it has all become a big mess, from which you can only remember bits and pieces. I know *I* can't handle it all in my head at this moment. There's just too much information packed in here.

To help you remember as much as possible, this last chapter will revisit all the topics of the book, chapter by chapter. I'll repeat the most important ideas from each chapter, while you can just relax and remember all the details. For good measure, I'll throw in some additional tips, tricks, and techniques that didn't find a place in the "regular" chapters.

In this chapter, you won't learn much new. You will, however, be able to review the following topics:

- How do we classify algorithm complexity?
- With what tools can we measure code performance?
- Why is the simplest way to make programs faster to execute less code?
- Where can we find excellent implementations of different data structures and how to use them?
- How to make code even faster by using tweaks specific to the Delphi compiler
- How does the Delphi memory manager work and why is that important for program speed?
- What are the dangers of writing multithreaded code?
- How to start writing multithreaded programs in Delphi
- What kind of parallel patterns are implemented in the Delphi RTL?
- Where can we find more parallel patterns?
- How to import external libraries not written in Delphi

About performance

As I stated in the opening statement of this book—performance matters. But what exactly is performance? What do we mean when we use this word? Sometimes, we just want to say that the program runs so fast that the users don't detect any delays. We can say that the program is not *lagging* and that the user interface does not *block*. In other situations, *performance* means that a program finishes its execution quickly, or that a server can handle a large number of clients.

Knowing what we mean when we use the word will help us determine the best way to fix the problem. Whatever the approach will be, we still have to find out why the application is not performing as we expect. The first suspect should always be the *algorithm*, and to examine it, we need to know more about how algorithms behave.

Algorithms are all different. What we want to know about them when we are designing a program is how they behave when their input set gets larger and larger. Some run in *constant time*, meaning that the size of the data doesn't matter. Accessing array elements is an example where this applies. Some are *linear* and have processing time proportional to the size of the input. A typical example is searching in an unordered list. And some are worse. *Exponential* algorithms get slower so quickly that they are completely useless for large inputs.

The complexity of an algorithm is described by *big-O* notation, which—very simply put—tells you how much slower an algorithm will run when it gets more input to process. We say that (for example) naive sort algorithms, such as Bubble sort, have a time complexity of $O(n^2)$. In other words, if we enlarge the input size by a factor of 1,000, the execution time will get slower by a factor of a million ($1,000^2$).

It is good to know the big-O complexities of popular algorithms. I have already mentioned array access, searching in unordered lists, and naive sorting algorithms. You should also know that searching in a sorted list (bisection) has a complexity of only $O(log\ n)$, while a good sorting algorithm runs in $O(n\ log\ n)$ time. These two approaches are **much** faster than $O(n)$ and $O(n^2)$, and *Chapter 1, About Performance*, contains a table to prove that.

Good performance starts with using data structures that are appropriate for the job. You should know how standard Delphi structures behave when you add or remove elements and when you are trying to locate a specific element. We saw that all *lists* behave more or less the same, and that performance depends on whether you want to keep them sorted (fast lookup) or not (fast element insertion). A special case is a *dictionary*, which executes all operations in $O(1)$ time. It gets that speed increase by forgetting the insertion order and by using lots of memory space—more than is required for just storing the data.

Profiling the code

Even if you know how fast or slow the algorithms that your program uses are, you should never just assume that you know which parts of the program are slowing you down. You should always measure the execution speed.

> **Note**
> Don't guess, measure!

The simplest way to time parts of a program is to use Delphi's TStopwatch. It is useful when you think you know where the problem lies, and just want to confirm your suspicions and (later) verify that the improved solution is indeed faster than the original.

If you don't know exactly which parts of the program are slow, you need to measure with a special tool, a *profiler*. They come in two varieties—*sampling* and *instrumentation*. As always, it is a question of balance. Sampling profilers are fast but will give you approximate results, while instrumenting profilers are very precise and can generate a *call graph* to show you exactly what goes on inside the program, but make the program run slower.

Chapter 2, *Profiling the Code*, described two open source profilers, AsmProfiler and Sampling Profiler, and two commercial tools, AQTime and Nexus Quality Suite, in more detail.

Fixing the algorithm

My preferred approach to improving performance is—always—fixing the algorithm. Look at it this way—if we need *1* time unit to process one data item, and the algorithm is $O(n^2)$, we need *10,000* time units to process an input of size *100*. If you fine-tune the code and speed up the operation by 50% (which is an excellent result), the code will need *5,000* time units to do the job. If you, however, change the algorithm to $O(n \log n)$, it will need in the order of *1,000* time units or less. Even if processing one item takes 100% longer than in the original code, the whole process will run in, say, *2,000* time units.

An algorithm with lower complexity will beat an algorithm with higher complexity, even if the latter executes its steps faster.

As it's impossible to give one piece of advice that will fix all your problems, *Chapter 3*, *Fixing the Algorithm*, looked into different user stories. The first topic was responsive **user interfaces** (**UIs**). A program's UI can lag or block for different reasons, but the two most common are that some non-visual operations inside the program take too much time, or updating the UI itself causes a problem.

In the latter case, the solution is simple, although sometimes not obvious at first glance. If updating the UI takes too long, then do fewer updates. In that chapter, we saw how constant updates to a progress bar can slow down the program, and how we should **always** use BeginUpdate/EndUpdate when updating more than one line of a listbox or memo.

Another option for speeding up the UI is to not do anything until the data is visible on the screen. Instead of a standard listbox, we can use it in *virtual* mode. Alternatively, we can throw the listbox away and use a *virtual tree* control, TVirtualStringTree.

When part of the code takes too long to run, sometimes, it is best not to execute it at all. This brings us to the concept of *caching*. If we find out that some values are frequently calculated, we can just store them away in a list, array, or dictionary and not calculate them the next time they are needed. The chapter introduced a size-limited cache component, TDHPCache<K, V>, which does everything in *O(1)* time.

The whole concept of caching is also helpful when you are doing lots of repainting of a UI. Sometimes, you can precalculate some of the display and store it in a bitmap (or multiple bitmaps). Later, you can just paint it on screen, potentially one over another to create layers.

Don't reinvent, reuse

In previous chapters, I mentioned from time to time how important it is to choose the correct algorithm. I cannot tell you how to do that for every possible application that you want to create, but at least I can give my advice regarding the *data structures* you will use in your applications.

We all know and use the TStringList class, have a good relationship with TList, TObjectList, and TList<T> lists, store data in TArray<T> structures, and sometimes even use a dictionary or two. Unfortunately, that is more or less everything that we can get from Delphi. As your power as a good programmer will only grow if you have a bigger collection of data structures at your disposal, in *Chapter 4, Don't Reinvent, Reuse*, I looked into the **Spring4D** library; more specifically, into its support for *collections*.

Spring4D collections will give you everything that is included in Delphi and then some. There are lists and dictionaries, of course, but also sets and multisets, bidirectional dictionaries, double-ended queues, and more. Many of these are available as unsorted and sorted versions while some offer even more selection. (There are six different multimaps, for example.)

Each of these data structures has its own strengths and weaknesses, so I made sure to list algorithm complexity for accessing data in the collection, searching for a specific item, inserting data into the collection, and deleting data from the collection.

The big difference between Delphi and Spring4D collections is that the latter implements a very rich set of functions that is common to all collection types. We get extensive support for querying data, accessing values, filtering by specific conditions, doing simple statistics, executing tests, and more. There is even a helper class, TEnumerable, which implements additional, rarely used operations.

Another nice feature of Spring4D collections is that they are all implemented as interfaces (we would use IList<T> instead of TList<T>) and therefore we don't have to take care of destroying them.

One of the nicer features of Spring4D collections is the implementation of a red-black tree, which is used in different collection types (sorted dictionaries, sorted sets, and sorted multimaps), but can also be used directly in our code. As this is the only Delphi implementation of a balanced tree with special care toward performance (a special node allocator makes sure that adding a node most of the time doesn't result in a memory manager operation), I also demonstrated how to use the `TRedBlackTree` class directly in your code.

All in all, I consider Spring4D to be one of the best additions to the standard Delphi RTL and can only recommend spending some time learning its functionality.

Fine-tuning the code

When the algorithm is as perfect as it can get, and the program is still not running fast enough, it is time to tweak the code. In *Chapter 5*, *Fine-Tuning the Code*, we looked at different methods to achieve faster performance by making small changes. In a process we call *optimization*, we can change the Delphi code so that it will execute in less time, or we can convert it into assembler code. The latter option should be used sparingly, as assembler code is hard to maintain; it also prevents us from compiling to multiple platforms. The last option, which I discussed later in the book, is to replace the code with an external library.

As a first step in fine-tuning the code, you should always check the compiler settings. The most important setting is probably *optimization*, which can sometimes make a big difference. My recommendation, however, is to disable optimization in the DEBUG version and enable it for RELEASE. Disabling optimization can make a big improvement in the debugging experience.

Another setting that makes a big difference is *range checking*. Disabling range checking can make accessing array elements significantly faster. As arrays are used in all kinds of places, for example, inside list classes (`TList` and its brethren), this can seriously affect the code. On the other hand, range checking catches the worst kinds of programming errors that could lead to corrupt data and hard-to-find errors. I *always* enable range checking for my code, and then only disable it for small and well-tested parts of code.

Changing *overflow checking* doesn't affect the code speed much in most cases, but sometimes you'll still want to disable it. This option is enabled by default, which is a good thing.

This chapter also covered the effects of using *record field alignment*. Surprisingly, it has relatively little effect on the program. Modern CPUs are great when accessing unaligned data, but the biggest difference—on the slow side—is observed when the record field alignment is set to 8. That points to the fact that the most important way to speed up operations with a modern CPU is to put as much data in the processor cache as possible.

If you want to write fast code, you should know about Delphi data types and how they behave. Delphi has a plethora of built-in types—I enumerated some 40 of them, and I only covered basic types. I covered *simple* types—integers, reals, booleans, and so on—just briefly, and focused on string and structured types.

Strings in Delphi are particularly interesting as they can be implemented as statically allocated data (`string[42]`), dynamically allocated data (`WideString`), or dynamically allocated data with reference counting and *copy-on-write* mechanisms (`string` and `AnsiString`). The latter minimizes the amount of operations when assigning a string to a new variable.

Another interesting data type is arrays, especially dynamic ones. The main topic of interest was dynamic arrays, which are reference-counted (just like strings and interfaces) but don't implement the *copy-on-write* mechanism.

Structured types are also interesting, each in its own way. Records are not initialized by default, except if they contain *managed* data (strings, interfaces, and dynamic arrays). We can, however, write our own *initializers* and *finalizers*, which can take care of data initialization and clean-up.

There can be a big speed difference between a record that uses managed fields and one that does not. Classes are less interesting as their content is always initialized to zero. Interfaces bring in an additional slowdown because of the reference counting mechanism, which makes sure that an interface is destroyed when no one is using it.

Another coding mechanism that affects the execution speed is parameter passing. It can make a difference whether the parameter is marked as `var` or `const` or not marked at all. This will greatly affect static arrays and records and—to a lesser extent—strings and interfaces. Interesting things happen when you are passing dynamic arrays to a method, but they are far too weird to explain in a few words. You will have to reread that part of the chapter to experience all the weirdness.

Memory management

Understanding how built-in data types work also means knowing how Delphi manages the memory for them. It helps if you know at least something about Delphi and memory in general. Admittedly, this is a deep topic; something that most programmers—even excellent programmers—only understand in general terms. In other words, don't worry, you can safely forget 90% of the stuff I told you in *Chapter 6, Memory Management*, and you'll still know enough.

When a Delphi program is running, memory is constantly being allocated, reallocated, and freed. This happens when classes are created and destroyed, when the length of a dynamic array is changed, when strings are modified and lists are appended, and so on. If you want your code to perform well, you should be aware of this.

Instead of appending to a string, character by character, you would call `SetLength` to *preallocate* the target length (or some approximation of it) and then store characters in a string. Dynamic arrays can be handled similarly. Various list classes, such as `TList<T>`, offer a different way—the `SetCapacity` method, which preallocates storage.

Memory can also be managed by calling specialized functions, such as `GetMem` to get a memory block, `ReallocMem` to change its size, and `FreeMem` to release (free) it. Or you can use `New` to allocate memory and `Dispose` to release it. These two functions will also correctly initialize/clean up records containing managed fields. Alternatively, you can do it manually by calling `Initialize` and `Finalize`.

Memory management functions allow you to dynamically allocate and release records. By using these functions, you can use records instead of classes when you create a myriad of small objects. Records are created and initialized faster because they don't have to use inheritance mechanisms as objects do. It is also possible to dynamically allocate generic records (`TRecord<T>`).

The internal details of the FastMM memory manager—the default Delphi memory manager—are interesting, but most of the time serve no practical purpose. They do, however, help us understand why multithreaded Delphi programs sometimes exhibit bad performance. The reason, surprisingly, lies not in allocating memory but in releasing it. FastMM implements some performance enhancements that help multiple threads allocate memory (for example, when objects are created) at the same time, but they will have to wait for their turn to have their memory released (when objects are destroyed).

Two recent additions to FastMM, contributed by yours truly, help with that. To use any of them, you'll need the latest version from the official repository (`https://github.com/pleriche/FastMM4`) and not the version that comes with Delphi.

Compiling with the `LogLockContention` symbol defined enables a special detection mode that helps you find parts of a program that are fighting to access the memory manager. You can think of it as a specialized *profiler*, one that is not interested in execution times, but in memory management.

Alternatively, you can compile your program with the `UseReleaseStack` symbol defined. This will enable a special mode that minimizes the problem of releasing memory. Everything comes with a price; this solution does its magic by using more memory than standard FastMM.

You could also replace the built-in memory manager, which is a simple matter of calling two functions—`GetMemoryManager` and `SetMemoryManager`. In the chapter, I covered the successor to FastMM4 – FastMM5 (`https://github.com/pleriche/FastMM5`) – and Intel's TBBMalloc. The latter is written in C but can be linked to Delphi by using the techniques from *Chapter 11, Using External Libraries*. The simplest way to use it is to download a compiled version and Delphi wrappers from `https://sites.google.com/site/aminer68/intel-tbbmalloc-interfaces-for-delphi-and-delphi-xe-versions-and-freepascal`.

Getting started with the parallel world

I mentioned multithreaded programming a lot in the previous chapter, so it is not hard to guess where that path is taking me next. Multithreaded programming, or multithreading, is the art of running multiple parts of your program at the same time. In *Chapter 7, Getting Started with the Parallel World*, we took the first steps into the dangerous world of multithreading.

Multithreading is definitely *not* my first choice for improving existing code. It is hard to write multithreaded programs and very simple to introduce problems, which are then hard to find. Multithreaded programs are also very hard to debug.

To understand multithreading, we should first learn about the processes and threads. In general terms, a process equates to a running program. A process encompasses application code, loaded in memory, and all the resources (memory, files, windows, and so on) used by the program.

A thread, on the other hand, represents the state of the program's execution. A thread is nothing more than the current state of the CPU registers, variables local to the thread (the `threadvar` statement introduces them), and the stack belonging to this thread. (Actually, the stack is part of the process memory, and the thread only owns the *stack pointer* register.)

Every process starts with one thread that makes the process execute. We call it a *main thread*. Additional threads that are created in code are called *background threads*.

I've said it before and I'll repeat it here—adding multithreading to your code can create more problems than it is worth. That's why you should know about the problems before you learn how to do multithreading.

The first rule in Delphi is *never access the UI from a background thread*. This is so important that I'll repeat it.

> **Note**
> Never access the UI from a background thread!

If the code is running in a background thread and definitely needs to access the user interface, you have to use mechanisms that take a part of the code (an anonymous method) and execute it in the main thread. I'm talking about `TThread.Synchronize` and `TThread.Queue`.

The next problem appears when two threads access the same data structure at the same time. It is not really surprising that you should not read from `TList` when another thread is deleting elements from the same list, but it may shock you to know that problems can also appear when one thread is reading from a simple `int64` variable while another thread is writing to it.

Don't think that the first example (simultaneous access to a list) is purely theoretical. I recently fixed a similar example in my code (yes, it happens to the best of us). One thread was reading from `TFileStream` and sending data to a web server. Another thread calculated the progress, and in that calculation, called the `TFileStream.Size` function (they were different threads because I used the asynchronous mode of the WinHTTP API). In my defense, I know very well that it is not a good idea, but, well, mistakes happen.

What's the problem with calling `Size`? Its implementation. When you read from `Size`, the `GetSize` method (shown following) calls `Seek` three times. Imagine what happens when one thread reads from a stream and another changes the current position at the same time:

```
function TStream.GetSize: Int64;
var
```

```
  Pos: Int64;
begin
  Pos := Seek(0, soCurrent);
  Result := Seek(0, soEnd);
  Seek(Pos, soBeginning);
end;
```

To solve such issues, we typically introduce *locking*. This technique creates a sort of barrier that allows only one thread at a time to enter a critical path through the code. *Critical sections* are the standard locking mechanism, but we can also use *mutexes* and *semaphores*, although both are slow and more useful when synchronizing multiple processes.

Delphi also offers two mechanisms that are slightly faster than critical sections—TMonitor and TSpinLock. In a situation when threads are mostly reading from shared data and rarely writing to it, we can use a reader-writer lock, TLightweightMREW.

Locking techniques introduce at least two problems into the program. They lead to slower operations, and they can result in *deadlocks* when improperly used. In a deadlocked program, one thread waits on another thread, which is waiting on the first one, and all operation stops.

An alternative to locking is *interlocked* (atomic) operations. They are faster than locking but only support a small range of operations—incrementing and decrementing a memory location, modifying a memory location, and *conditionally* modifying a memory location.

The latest variation, TInterlocked.CompareExchange, can be used to implement *optimistic initialization*. This mechanism can be used to safely create a shared object or interface. The latter is preferred as it will be safely destroyed at the right time.

Instead of using one of the synchronization techniques described previously, you can introduce *communication* mechanisms into the program. Instead of accessing a shared value, you can use *data duplication* (sending a copy of full or partial input data to each worker thread) and *aggregation* (putting partial results calculated in threads together). Each thread can then work on its own copy of data and doesn't have to use any locking to access it.

> **Note**
> Communication is better than synchronization!

Standard techniques for implementing messaging in a Delphi program are Windows messages, TThread methods, Synchronize and Queue, and polling in connection with a thread-safe message queue, such as TThreadedQueue<T>.

The best way to find problems in a multithreaded application is to do **stress-testing**, repeat multithreaded operations multiple times, and test your program on only one CPU core (the Windows SetProcessAffinityMask API will help you with that). You'll be surprised how many problems can be found this way. If at all possible (the implementation may prevent you from doing this), you should also test the code by creating more worker threads/tasks than there are CPU cores in the computer.

Use automated tests as much as possible and you'll be able to trust your code more. Never trust the multithreaded code fully. It actually never fully works; it's just that the bugs are so deeply hidden that they almost never appear.

Working with parallel tools

There are multiple ways to implement multithreading in an application, and *Chapter 8*, *Working with Parallel Tools*, dealt with the most basic of them all—TThread. This class was introduced in Delphi 2 where it simply wrapped the Windows CreateThread API. Later, it was enhanced with additional methods and added support for other operating systems but, in essence, it stayed the same good old, stupid, clumsy TThread, which we all learned to love and hate.

Threads created with TThread can be used in two modes. In one, the code has full control over a TThread object—it can create it, tell it to terminate (but the object must observe that and willingly terminate), and destroy it. In the other mode, the code just creates a thread that does its job, terminates it, and is automatically destroyed. The former is more appropriate for service-like operations. You start a thread, which then responds to requests and performs some operations in response to those requests. When you don't need it anymore, the code destroys the thread. The latter is more appropriate for background calculations. You start a thread, the thread executes some calculations, notifies the main thread of the result, and exits.

Working with the basic TThread object is quite clumsy, so in this chapter, I presented a way to combine TThread with a communication channel. While it is simple to send messages from a background thread to the main thread (and I described four such mechanisms in the previous chapter), sending messages to a thread is harder.

I introduced a DHPThreads unit, which implements the TCommThread class, representing a thread with a two-way communication channel that can transfer TValue records to and from the thread. To use this thread, you had to write its Execute method in a fairly specific way, which limits its use.

To simplify the code, I introduced a more capable TcommTimerThread, which shows how to work with threads on a more abstract level. You no longer override the Execute method as with the standard TThread mechanism, but merely introduce simple *event handlers*. The base class runs the Execute loop, processes messages, handles timers, and calls event handlers when appropriate. One event handler processes messages while another handles timer events.

If you approach multithreading that way, you move from the standard *big block of code* way to a modular event-driven architecture. This simplifies the programming logic, reduces the quantity of shared data, and makes multithreading a safer and more pleasant experience.

At the end of the chapter, I compared two ways of handling signals from multiple worker threads in another controlling thread. I looked at the Windows-standard `WaitForMultipleObjects` function and compared it with lesser-known but platform-independent *condition variables*.

Exploring parallel practices

The `TCommTimerThread` class from the previous chapter is also a demonstration of the current trend in multithreaded programming. Instead of working directly with threads, we try to put as much of the ugly multithreading *plumbing* into ready-to-use components. The first level of such abstraction is replacing threads with *tasks*.

A thread is just an operating system concept; one that allows executing multiple parts of a process simultaneously. When we program with threads, we have to handle all the cumbersome minutiae related to managing operating system threads.

A task, on the other hand, is the part of code that we want to execute in parallel. When we work with tasks, we don't care how threads are created and destroyed. We just tell the system that we want to run the task and it does the rest.

The task is a useful step forward, but for the majority of users, tasks are still too low-level. That is why parallel programming libraries that support specialized building blocks (which I like to call *patterns*) have appeared for all major programming languages.

In *Chapter 9, Exploring Parallel Practices*, we moved away from thread-based programming and started exploring tasks and patterns.

The first pattern that I looked into was *Async/Await*, a concept that (to my knowledge) first appeared in C#. It enables you to execute an anonymous method in a background thread. When the background thread finishes its work, the pattern executes a second anonymous method in the main thread. This allows us to clean up after the worker and update the UI—something that we must never do from a background thread! This pattern is not a lot different from a task that calls `TThread.Queue` at the end. The main difference is that, in the latter case, we have to write all the necessary code while Async/Await implements as much of the infrastructure as possible, and you only have to provide the appropriate anonymous methods.

Another interesting pattern is *Join*. It creates a task, wrapping multiple anonymous procedures. Join executes each procedure in its own thread.

The next pattern I created from scratch was *Join/Await*. It extended the Join concept with a simple-to-use termination event that executes in the main thread. This pattern is also capable of catching exceptions in worker code and presenting them to the termination event.

Another simple and interesting pattern is *Future*. It starts a calculation in a background thread and then enables you to access the result in the main thread. The Parallel Programming Library's implementation does not implement a notification handler that will notify the main thread when computation is finished, but that is easily fixed by using any of the messaging mechanisms I described before.

Another Parallel Programming Library pattern is *Parallel for*. This very powerful pattern allows you to convert a normal `for` loop into a parallel version. The pattern starts multiple workers and processes just a part of the `for` range in each one. This pattern waits for a parallel loop to finish execution (which blocks the main thread), but that can be easily fixed by wrapping Parallel for in a Future or Async/Await pattern.

Some problems can be split into multiple data processing stages that are independent of each other. In such cases, we can use the *Pipeline* pattern. This pattern runs each stage in its own thread and connects them with communication queues. As each stage works on its own copy of the data, there's no need for locking and all the troubles that come with it. It is not implemented in the Delphi RTL and is quite cumbersome to write from scratch, so I included a simple implementation in the `DHPThreading` unit.

More parallel patterns

If you have to write multithreaded code, patterns are definitely the way to go. By implementing the most complicated parts of the infrastructure internally, they help you focus on the important part— the problem-specific code. The number of patterns in the Delphi RTL is, however, quite limited, so it doesn't hurt if you also know about other parallel programming libraries. In *Chapter 10*, *More Parallel Patterns*, we looked at patterns implemented in the open source parallel programming framework **OmniThreadLibrary** (**OTL**).

OTL implements multithreading support on two levels. *Low-level* multithreading support brings `TThread`-based multithreading to the modern age by adding communication support, timers, and more. It also allows you to write a thread code as a collection of small event handlers instead of as a big monolithic main loop. In a way, this is similar to how Delphi VCL replaced writing a big main loop that processed all kinds of Windows events and replaced it with a core that we can configure by writing event handlers. Low-level multithreading, however, is not the focus of this book, so I spent most of the chapter talking about *high-level* multithreading patterns.

As an introductio, I described a *blocking collection*, which is a thread-safe queue of unlimited size that supports multiple simultaneous readers and writers. This collection is the basis of many OTL patterns, but it can be also used in combination with low-level multithreading or with simple `TThread`-based threads, so I used the latter for a demo.

OTL implements twelve parallel patterns. *Async/Await*, *Join*, and *Future* are very similar to the Delphi PPL implementation from the previous chapter. There are, however, small changes in behavior, especially when it comes to exception handling.

Similar to PPL, OTL implements a *Parallel For* pattern. There is also much more sophisticated (but noticeably slower) *Parallel ForEach*, which can iterate over different data types, supports result sorting, and more.

The *Parallel Task* pattern is very similar to *Join* except that it executes the same code in multiple threads. This may not seem very useful, but it can be used to parallelize code that could also be parallelized with a *Parallel For*. In the chapter, I showed how to use it to initialize a large block of data and used this as an excuse to explain some other neat OTL functionality.

When you are writing a complex background processing facility that will accept requests from the main code, do processing, and return results, you need the *Background Worker* pattern. It can split processing between multiple threads, provide request cancelation, and much more.

OTL's *Pipeline* is similar to the pipeline from the previous chapter but is immensely more powerful. It can run one stage on multiple threads, provides throttling on communication channels, propagates exceptions, and more.

Then, there are two small and specialized patterns – *Map* is designed to quickly map and filter data in a big array. *Timed Task* is by design very similar to Delphi's *TTimer* except that it executes a timer function in a background thread.

The last OTL pattern is *Fork/Join*. This framework for solving divide-and-conquer (recursion) algorithms is quite complex and rarely used, so I ignored it in the book.

Knowing about OTL will help your parallel programming if you are a Windows programmer who doesn't use the FMX framework. The current version of the library sadly doesn't support this graphical framework and other operating systems.

Using external libraries

Sometimes, performance problems are too hard to be solved in code. Maybe the problem lies in a specific domain that requires specialized technical knowledge to be solved efficiently. Or maybe the compiler is generating code that just doesn't cut it.

In both cases, the answer is the same—if you can't solve it internally, find a better solution and use it as an external module. If you can get a DLL, great! Using DLLs from a Delphi application is a breeze. Lots of times, however, you'll have to use either *object files* generated with a C compiler, or C++ *libraries*.

While the former can be used directly in Delphi, there are usually several obstacles to overcome. Firstly, the object file has to be in the correct format. The 32-bit Delphi compiler can link OMF and COFF formats (generated by Embarcadero's and Microsoft's compilers, respectively), while the 64-bit Delphi compiler can only link COFF format files.

Secondly, the object file may refer to external methods that you have to write. If they come from Microsoft's standard *msvcrt* library, then you can resolve the problems by adding `System.Win.Crtl` to the `uses` list. Other times, you'll have to write the necessary code in your application. Just remember, C uses the `cdecl` calling convention, not the standard Delphi `pascal` convention or `stdcall` as is used for Windows DLLs.

Sometimes, you have to link to multiple object files that contain circular references. In the simplest scenario, object file A refers to a function implemented in object file B, while file B refers to a function implemented in file A. In such cases, we have to help the compiler a little by providing a simple `external` declaration for one of the functions.

While linking to C object files is hard but at least possible, linking to C++ object files or libraries can present an impossible task. C++ files usually contain language-specific features such as classes, streams, and strings, and Delphi has no idea how to handle them. A proper approach is to write a piece of software in C++—a proxy DLL that accepts Delphi calls and passes them to the C++ library. All the knowledge of complicated C++ stuff, such as objects, can stay hidden inside this proxy library.

There's one other option to speed up the program that I didn't cover in *Chapter 11, Using External Libraries*, and that is *GPU programming*. Modern graphics cards have powerful processors that can be used for generic programming—provided that you know how to program them. This is not a simple job, and if you want to go that way, you'll need to do some serious studying. If you are interested, you should start with the `OpenCL` framework, which is supported by all important graphics card manufacturers.

Final words

Before I leave you for good, I would like to say a few parting words.

It is obvious that you want to learn; otherwise, you wouldn't have made it to this very last page. Don't stop here! While this book contains lots of interesting information, it is by no means a definitive work. There is so much more to be learned, so much more to find!

Let this be your guiding line—never stop learning. Learn what others have to say, even if they contradict something I have said in this book. Maybe they are right and I am wrong. Follow what other languages are doing and what libraries exist for them. That is the source of many great ideas. Write good code. Write bad code. Find out why it is bad and fix it. Never stop moving.

Explore. Experiment. Evolve!

Index

‹packt›

www.packtpub.com

Subscribe to our online digital library for full access to over 7,000 books and videos, as well as industry leading tools to help you plan your personal development and advance your career. For more information, please visit our website.

Why subscribe?

- Spend less time learning and more time coding with practical eBooks and Videos from over 4,000 industry professionals

- Improve your learning with Skill Plans built especially for you

- Get a free eBook or video every month

- Fully searchable for easy access to vital information

- Copy and paste, print, and bookmark content

Did you know that Packt offers eBook versions of every book published, with PDF and ePub files available? You can upgrade to the eBook version at packtpub.com and as a print book customer, you are entitled to a discount on the eBook copy. Get in touch with us at customercare@packtpub.com for more details.

At www.packtpub.com, you can also read a collection of free technical articles, sign up for a range of free newsletters, and receive exclusive discounts and offers on Packt books and eBooks.

Other Books You May Enjoy

If you enjoyed this book, you may be interested in these other books by Packt:

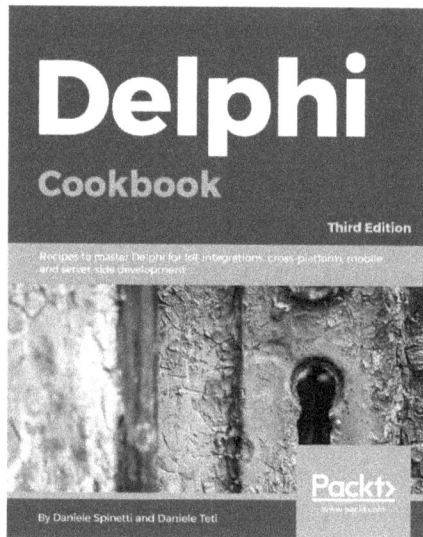

Delphi Cookbook - Third Edition

Daniele Spinetti, Daniele Teti

ISBN: 9781788621304

- Develop visually stunning applications using FireMonkey
- Deploy LiveBinding effectively with the right object-oriented programming (OOP) approach
- Create RESTful web services that run on Linux or Windows
- Build mobile apps that read data from a remote server efficiently
- Call platform native API on Android and iOS for an unpublished API
- Manage software customization by making better use of an extended RTTI
- Integrate your application with IOT

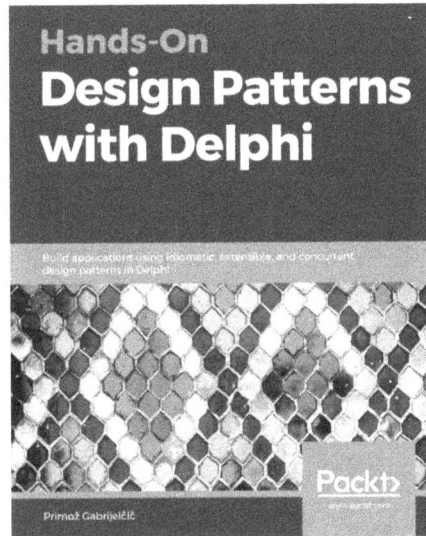

Hands-On Design Patterns with Delphi

Primož Gabrijelčič

ISBN: 9781789343243

- Gain insights into the concept of design patterns
- Study modern programming techniques with Delphi
- Keep up to date with the latest additions and program design techniques in Delphi
- Get to grips with various modern multithreading approaches
- Discover creational, structural, behavioral, and concurrent patterns
- Determine how to break a design problem down into its component parts

Packt is searching for authors like you

If you're interested in becoming an author for Packt, please visit `authors.packtpub.com` and apply today. We have worked with thousands of developers and tech professionals, just like you, to help them share their insight with the global tech community. You can make a general application, apply for a specific hot topic that we are recruiting an author for, or submit your own idea.

Share Your Thoughts

Now you've finished *Delphi High Performance*, we'd love to hear your thoughts! Scan the QR code below to go straight to the Amazon review page for this book and share your feedback or leave a review on the site that you purchased it from.

`https://packt.link/r/1805125877`

Your review is important to us and the tech community and will help us make sure we're delivering excellent quality content.

Download a free PDF copy of this book

Thanks for purchasing this book!

Do you like to read on the go but are unable to carry your print books everywhere?

Is your eBook purchase not compatible with the device of your choice?

Don't worry, now with every Packt book you get a DRM-free PDF version of that book at no cost.

Read anywhere, any place, on any device. Search, copy, and paste code from your favorite technical books directly into your application.

The perks don't stop there, you can get exclusive access to discounts, newsletters, and great free content in your inbox daily

Follow these simple steps to get the benefits:

1. Scan the QR code or visit the link below

https://packt.link/free-ebook/9781805125877

2. Submit your proof of purchase

3. That's it! We'll send your free PDF and other benefits to your email directly

www.ingramcontent.com/pod-product-compliance
Lightning Source LLC
Chambersburg PA
CBHW082124210326
41599CB00031B/5862